移动平台开发书库

Android 开发完全实战宝典

吴善财　等编著

机 械 工 业 出 版 社

本书详细讲解了Android技术在各个领域的具体应用，介绍了各个实例的具体实现过程。全书分为10章，第1、2章是基础知识，讲解了Android前景和搭建开发环境的过程；第3章详细讲解了Android在人机交互界面领域典型实例的设计过程；第4章详细讲解了Android各个组件的使用方法；第5章讲解了Android在交互式应用领域的具体应用；第6章讲解了Android在手机自动服务领域中的应用；第7章讲解了Android在娱乐和多媒体领域的具体应用；第8章讲解了Android在互联网领域各个实例的实现过程；第9章讲解了Android在官方服务绑定领域各个实例的实现过程；第10章讲解了Android在绘图和游戏开发领域的具体应用。

本书适用于Android的初、中、高级用户，既可以作为初学者的自学手册，也可以作为有一定程序开发基础人员的参考书。

图书在版编目（CIP）数据

Android 开发完全实战宝典 / 吴善财等编著. —北京：机械工业出版社，2012.9

（移动平台开发书库）

ISBN 978-7-111-39607-9

Ⅰ. ①A…　Ⅱ. ①吴…　Ⅲ. ①移动终端－应用程序－程序设计　Ⅳ. ①TN929.53

中国版本图书馆CIP数据核字（2012）第210872号

机械工业出版社（北京市百万庄大街22号　邮政编码100037）
责任编辑：丁　诚
责任印制：杨　曦

保定市中画美凯印刷有限公司印刷

2012年10月·第1版第1次印刷
184mm×260mm·36.75印张·916千字
0001—3000册
标准书号：ISBN 978-7-111-39607-9
　　　　　ISBN 978-7-89433-673-6（光盘）
定价：96.00元（含1DVD）

凡购本书，如有缺页、倒页、脱页，由本社发行部调换

电话服务
社服务中心：（010）88361066
销售一部：（010）68326294
销售二部：（010）88379649
读者购书热线：（010）88379203

网络服务
教材网：http://www.cmpedu.com
机工官网：http://www.cmpbook.com
机工官博：http://weibo.com/cmp1952
封面无防伪标均为盗版

前　言

进入 21 世纪，生活和工作的快节奏令我们目不暇接，各种各样的信息充斥着我们的视野、撞击着我们的思维。追忆过去，Windows 操作系统的诞生成就了微软的霸主地位，也造就了 PC 时代的繁荣。而以 Android 和 iPhone 手机为代表的智能移动设备的广泛应用则标志着移动互联网时代（3G 时代）已经来临，谁将会成为这些移动设备上的主宰？目前看来，最有可能的就是 Android——PC 时代的 Windows！

看 3G 时代的璀璨绚丽

随着 3G 时代的到来，无线带宽越来越高，使得更多精彩的应用程序可以在手机上运行，如视频通话、视频点播、移动互联网冲浪、在线看书/听歌及内容分享等。为了承载这些数据应用，手机功能越来越智能，越来越开放。这就需要有一个好的开发平台来支持，而由 Google 公司发起的 OHA 联盟走在了业界的前列，于 2007 年 11 月推出了开放的 Android 平台，任何公司及个人都可以免费获取到源代码及开发包。由于 Android 的开放性和优异的性能，得到了业界广泛的支持，其中包括各大手机厂商和著名的移动运营商。自 2008 年 9 月第一款基于 Android 平台的手机 G1 发布之后，三星、摩托罗拉、索爱、LG、华为等公司都推出了 Android 平台手机，中国移动也联合各手机厂商共同推出基于 Android 平台的 OPhone。

巨大的优势

从技术角度而言，Android 与 iPhone 相似，采用 WebKit 浏览器引擎，具备触摸屏、高级图形显示和上网功能，用户能够在手机上查收电子邮件、搜索网址和观看视频节目等。Android 手机比 iPhone 等其他手机更强调搜索功能，界面更强大，可以说是一种融入了全部 Web 应用的平台。随着版本的更新，从最初的触屏到现在的多点触摸，从普通的联系人到现在的数据同步，从简单的谷歌地图到现在的导航系统，从基本的网页浏览到现在的 HTML 5，这都说明 Android 已经逐渐稳定，而且功能越来越强大。此外，Android 平台不仅支持 Java、C、C++ 等主流的编程语言，还支持 Ruby、Python 等脚本语言。Google 甚至专为 Android 的应用开发推出了 Simple 语言，这使得 Android 有着非常广泛的开发群体。

本书的内容

本书循序渐进地讲解了 Android 技术的基本知识，并通过实例的方式介绍了 Android 在各个领域的具体应用。本书内容新颖、知识全面、讲解详细。全书分为 10 章，其中 1、2 章是基础知识，讲解搭建开发环境的过程和基础知识；第 3 章详细讲解了 Android 在人机交互界面领域典型实例的设计过程；第 4 章讲解了 Android 各个组件的使用过程；第 5 章讲解了 Android 在交互式应用领域的实现过程；第 6 章讲解了 Android 在手机自动服务领域中的实现过程；第 7 章讲解了 Android 在娱乐多媒体领域中的具体实现过程；第 8 章讲解了 Android 在互联网领域各个范例的实现过程；第 9 章讲解了 Android 在官方服务绑定领域各个范例的实现过程；第 10 章是典型手机游戏开发应用。书中每个范例先提出制作思路及包含知识点，在实例最后补充总结知识点，让读者举一反三。

本书特色

本书内容相当丰富，实例覆盖面广，满足 Android 技术人员成长通路上的方方面面。作者的目标是通过一本图书，提供多本图书的价值，读者可以根据自己的需要有选择地阅读，以完善自己的知识和技能结构。在内容的编写上，本书具有以下特色。

（1）结构合理

从用户的实际需要出发，科学安排知识结构，内容由浅入深，叙述清楚，并附有相应的总结和练习，具有很强的实用性，反映了当前 Android 技术的发展和应用水平。同时全书精心筛选的最具代表性、读者最关心的知识点，几乎包括 Android 技术的各个方面。

（2）易学易懂

本书条理清晰、语言简洁，可帮助读者快速掌握每个知识点；每个部分既相互连贯又自成体系，使读者既可以按照本书编排的章节顺序进行学习，也可以根据自己的需求对某一章节有针对性地学习。

（3）实用性强

本书彻底摒弃枯燥的理论和简单的操作，注重实用性和可操作性，讲解了各个实例的具体实现过程，使用户掌握相关操作技能的同时，还能学习到相应的理论知识。

（4）实例全面

书中的开发实例都很典型并具有创意，涵盖了 Android 所涉及的所有领域，每个实例都体现了移动互联网应用所需的创新精神及良好的用户体验理念，这个设计思路很值得大家去思考和学习。

本书由吴善财主编，参与本书编写的还有陈强、刘海洋、曹阳、李强、刁国庆、薛多骞、张子言、李佐彬、李淑芳、陈德春、王梦、唐凯、王石磊、张家春、管西京、张玲玲。

由于作者水平有限，书中难免存在不妥之处，敬请读者批评指正。

作　者

目　录

第 1 章　揭开 Android 的神秘面纱

本章将简单介绍 Android 的发展历程和背景，让读者了解 Android 的发展之路。

1.1　了解智能手机

为了使读者能够更好地学习本书，在本节的内容中，将首先讲解和 Android 关系密切的智能手机的基本知识，为读者学习本章后面的内容打好基础。

1.1.1　智能手机的定义

所谓智能手机（SmartPhone），是指“像个人电脑一样，具有独立的操作系统，可以由用户自行安装软件、游戏等第三方服务商提供的程序，通过此类程序来不断对手机的功能进行扩充，并可以通过移动通信网络来实现无线网络接入的这样一类手机的总称”。简单地说，智能手机就是一部像电脑一样可以通过下载和安装软件来拓展其功能的手机。

智能手机可以是传统的手机增加智能功能，例如 Symbian 操作系统的 S60 系列、Windows Mobile 操作系统的 Windows Mobile Smartphone 系列；也可以是传统 PDA 加上手机通信功能，例如 Windows Mobile 操作系统的 Windows Mobile Pocket PCPhone 系列、Palm 操作系统的 Treo 系列；也可以是其他独立类型，例如 Symbian 操作系统的 S80、UIQ，以及一些 Linux 操作系统的智能手机。然而，就新近的发展来看，这些智能手机的类型有相融合的趋势。

“智能手机（SmartPhone）”这个说法主要是针对“功能手机（Featurephone）”而来的，本身并不意味着这个手机有多“智能（Smart）”；从另一个角度来讲，所谓的“智能手机（SmartPhone）”就是一台可以随意安装和卸载应用软件的手机（就像电脑那样）。“功能手机（Featurephone）”是不能随意安装和卸载软件的，Java 的出现使后来的“功能手机（Featurephone）”具备了安装 Java 应用程序的功能，但是 Java 程序的操作友好性、运行效率以及对系统资源的操作都比“智能手机（SmartPhone）”差很多。

1.1.2　智能手机的特点

智能手机的主要特点如下。

- ❑ 具备普通手机的全部功能，能够进行正常的通话、短信等应用。
- ❑ 具备无线接入互联网的能力，即需要支持 GSM 网络下的 GPRS 或者 CDMA 网络下的 CDMA 1X 或者 3G 网络。
- ❑ 具备 PDA 的功能，如 PIM（个人信息管理）、日程记事、任务安排、多媒体应用以及浏览网页等。
- ❑ 具备一个具有开放性的操作系统，在这个操作系统平台上，可以安装更多的应用程序，从而使智能手机的功能可以得到扩充。

1.1.3 主流智能手机系统

当今世界中比较著名的手机操作系统有如下几种。

1．Symbian：Symbian OS(中文译音“塞班系统”)

Symbian 是由诺基亚、索尼爱立信、摩托罗拉、西门子等几家大型移动通讯设备商共同出资组建的一个合资公司，专门研发手机操作系统，现已被诺基亚全资收购。Symbian 很像是 Windows 和 Linux 的结合体，有着良好的界面，采用内核与界面分离技术，对硬件的要求比较低，支持 C++、VB 和 J2ME，兼容性较差。目前根据人机界面的不同，Symbian 体系的用户界面（User Interface，UI）平台分为 Series 60、Series 80、Series 90、UIQ 等。Series 60 主要是为数字键盘手机而设计，Series 80 是为完整键盘而设计，Series 90 则是为触控笔方式而设计。

2．Windows Phone

Windows Phone 是微软发布的一款手机操作系统，它将微软旗下的 Xbox Live 游戏、Zune 音乐与独特的视频体验整合至手机中。2010 年 10 月 11 日晚上 9 点 30 分，微软公司正式发布了智能手机操作系统 Windows Phone。2011 年 2 月，诺基亚与微软达成全球战略同盟并深度合作共同研发。2012 年 3 月 21 日，Windows Phone 7.5 登陆中国。6 月 21 日，微软正式发布最新手机操作系统 Windows Phone 8，Windows Phone 8 将采用和 Windows 8 相同的内核。

Windows Phone 具有桌面定制、图标拖拽、滑动控制等一系列前卫的操作体验。其主屏幕通过提供类似仪表盘的体验来显示新的电子邮件、短信、未接来电、日历约会等，让人们对重要信息保持时刻更新。它还包括一个增强的触摸屏界面，更方便手指操作；Windows Phone，力图打破人们与信息和应用之间的隔阂，提供给人们最优秀的端到端体验。

3．Linux

Linux 是源于 PC 的移动操作系统，具有上面 2 个操作系统无法比拟的优势：其一，Linux 具有开放的源代码，能够大大降低开发成本；其二，Linux 既满足了手机制造商根据实际情况有针对性地开发自己的 Linux 手机操作系统的要求，又吸引了众多软件开发商对内容应用软件的开发，丰富了第三方应用。 然而 Linux 操作系统有其先天的不足：入门难度大、熟悉其开发环境的工程师少、集成开发环境较差；由于微软 PC 操作系统源代码的不公开，基于 Linux 的产品与 PC 的连接性较差；尽管目前从事 Linux 操作系统开发的公司数量较多，但真正具有很强开发实力的公司却很少，而且这些公司之间是处于相互独立的开发状态，很难实现更大的技术突破。最初摩托罗拉非常推崇 Linux 平台，然而在和诺基亚的较量中不断失败，现在也不再那么热心 Linux 了，转而投向基于 Linux 的 Android 平台，其推出的 Android 手机很受关注。

4．BlackBerry

“黑莓”（BlackBerry）是加拿大 RIM 公司推出的一种移动电子邮件系统终端，其特色是支持推送式电子邮件、手提电话、文字短信、互联网传真、网页浏览及其他无线资讯服务。黑莓最强大也是最有优势的方面在于收发邮件，然而在中国，用手机收发邮件还不是很流行，所以黑莓在中国几乎没有多大市场。

5．iOS

iOS 是苹果公司（Apple. Inc.）公司手机产品 iPhone 专用的智能手机系统。iPhone 由苹果公司首席执行官史蒂夫·乔布斯在 2007 年 1 月 9 日举行的 Macworld 宣布推出，2007 年 6 月 29 日在美国上市。iPhone 将创新的移动电话、可触摸宽屏 iPod 以及具有桌面级电子邮件、网页浏览、搜索和地图功能的因特网通信设备这三种产品完美地融为一体。iPhone 引入了基于大型多触点显示屏和领先性新软件的全新用户界面，让用户用手指即可控制 iPhone。iPhone 还开创了移动设备软件尖端功能的新纪元，重新定义了移动电话的功能。有人这样评价 iPhone“iPhone 是一款革命性的、不可思议的产品，比市场上的其他任何移动电话整整领先了五年。”苹果公司首席执行官史蒂夫·乔布斯说：“手指是我们与生俱来的终极定点设备，而 iPhone 利用它们创造了自鼠标以来最具创新意义的用户界面。”

6．Android

Android 一词的本义指“机器人”，同时也是 Google 公司于 2007 年 11 月 5 日宣布的基于 Linux 平台的开源手机操作系统的名称，该平台由操作系统、中间件、用户界面和应用软件组成，号称是首个为移动终端打造的真正开放和完整的移动软件。截止本书成形之时，Android 已经成为市场占有率最高的智能手机操作系统。

1.2　初识 Android

Android 平台采用了 WebKit 浏览器引擎，具备触摸屏、高级图形显示和上网功能，用户能够在手机上查看电子邮件、搜索网址并观看视频节目，同时 Android 还具有比 iPhone 等其他手机更强的搜索功能，可以说是一种融入全部 Web 应用的平台。

1.2.1　产生背景

Android 是 Google 公司开发的基于 Linux 平台的开源手机操作系统。Google 与开放手机联盟合作开发了 Android，这个联盟由包括中国移动、摩托罗拉、高通、宏达电（HTC）和 T-Mobile 在内的 30 多家技术和无线应用的领军企业组成。Google 通过与运营商、设备制造商、开发商和其他有关各方结成深层次的合作伙伴关系，希望借助建立标准化、开放式的移动电话软件平台，在移动产业内形成一个开放式的生态系统。

开放手机联盟的成立和 Android 的推出是对现状的重大改变，在带来初步效益之前，还需要不小的耐心和高昂的投入。但是，笔者认为如果意识到全球移动用户从中能获得的潜在利益，是值得付出这些努力的。

1.2.2　Android 手机介绍

2008 年 9 月 22 日，美国运营商 T-Mobile USA 在纽约正式发布第一款 Google 手机——T-Mobile G1。该款手机为中国台湾宏达电代工制造，是世界上第一部使用 Android 操作系统的手机，支持 WCDMA/HSPA 网络，理论下载速率 7.2Mbit/s，并支持 Wi-Fi。

摩托罗拉的首款 Android 手机 CLIQ，如图 1-1 所示。

搭载 Google Android 2.0 的摩托罗拉 Moto Droid 如图 1-2 所示。

图 1-1　摩托罗拉的首款 Android 手机

图 1-2　搭载 Google Android 2.0 的摩托罗拉 Moto Droid

1.3　Android 特性

Android 的主要特性如下。

- ❑ 应用程序框架，支持组建的重用与替换。
- ❑ Dalvik 虚拟机，专门为移动设备做了优化。
- ❑ 内部集成浏览器，该浏览器基于开源的 WebKit 引擎。
- ❑ 优化的图形库，包括 2D 和 3D 图形库，3D 图形库基于 OpenGL ES 1.0（硬件加速可选）。
- ❑ #SQLite，用做结构化的数据存储。
- ❑ 多媒体支持，包括常见的音频、视频和静态影像文件格式（如 MPEG4、H.264、MP3、AAC、AMR、JPG、PNG、GIF）。
- ❑ GSM 电话　（依赖于硬件）。
- ❑ 蓝牙 Bluetooth、EDGE、3G 和 WiFi（依赖于硬件）。
- ❑ 照相机、GPS、指南针和加速度计（依赖于硬件）。
- ❑ 丰富的开发环境，包括设备模拟器、调试工具、内存及性能分析图表，以及 Eclipse 集成开发环境插件。

1.4　Android 组件结构应用程序框架

和主流的开发工具一样，Android 也有自己的组件，在本节的内容中，将简要介绍 Android 组件结构应用程序框架的基本知识。

1.4.1　Android 组件结构

Android 采用了软件堆层（Software Stack，又名软件叠层）的架构，低层以 Linux 内核为基础，只提供基本功能，其他的应用软件则由各公司自行开发，以 Java 作为编写程序的一部分。

1.4.2 Android 应用程序框架

Android 会同一个核心应用程序包一起发布，该应用程序包包括 Email 客户端、SMS 短消息程序、日历、地图、浏览器及联系人管理等程序。所有的应用程序都是用 Java 编写的。

开发者也完全可以访问核心应用程序所使用的 API 框架。该应用程序架构用来简化组件软件的重用；任何一个应用程序都可以发布它的功能块，并且任何其他的应用程序都可以使用其所发布的功能块（不过得遵循框架的安全性限制）。该应用程序重用机制使得组件可以被用户替换。

以下所有的应用程序都由一系列的服务和系统组成。

- 一个可扩展的视图（Views）：可以用来建立应用程序，包括列表（Lists）、网格（Grids）、文本框（Text Boxes）、按钮（Buttons），甚至包括一个可嵌入的浏览器。
- 内容管理器（Content Providers）：使得应用程序可以访问另一个应用程序的数据（如联系人数据库），或者共享它们自己的数据。
- 一个资源管理器（Resource Manager）：提供非代码资源的访问，如本地字符串、图形和分层文件（Layout Files）。
- 一个通知管理器（Notification Manager）：使得应用程序可以在状态栏中显示客户通知信息。
- 一个活动类管理器（Activity Manager）：用来管理应用程序生命周期并提供常用的导航回退功能。

一个 Android 程序编译运行后的效果如图 1-3 所示。

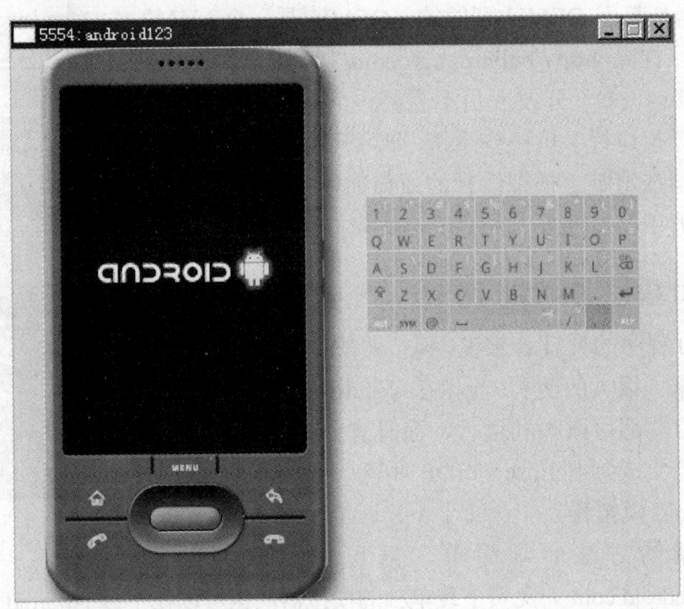

图 1-3　Android 程序运行效果

Android 系统提供给应用开发者一个框架，所有的应用开发都必须遵守这个框架的原则。在开发应用时就是在这个框架上进行扩展，下面来看 Android 框架都有些什么功能可供使用。

- ❑ android.app：提供高层的程序模型和基本的运行环境。
- ❑ android.content：包含对各种设备上的数据进行访问和发布。
- ❑ android.database：通过内容提供者浏览和操作数据库。
- ❑ android.graphics：底层的图形库，包含画布、颜色过滤、点及矩形等，可以将它们直接绘制到屏幕上。
- ❑ android.location：定位和相关服务的类。
- ❑ android.media：提供一些类管理多种音频、视频的媒体接口。
- ❑ android.net：提供帮助网络访问的类，超过通常的 java.net.* 接口。
- ❑ android.os：提供了系统服务、消息传输和 IPC 机制。
- ❑ android.opengl：提供 OpenGL 的工具。
- ❑ android.provider：提供访问 Android 内容提供者的类。
- ❑ android.telephony：提供与拨打电话相关的 API 交互。
- ❑ android.view：提供基础的用户界面接口框架。
- ❑ android.util：涉及工具性的方法，例如时间日期的操作。
- ❑ android.webkit：默认浏览器操作接口。
- ❑ android.widget：包含各种 UI（界面元素）元素在应用程序的布局中使用。

1.5 Android 模拟器

Android 中提供了一个模拟器来模拟 ARM 移动设备。Android 的模拟器是基于 QEMU 开发的，QEMU 是一个有名的开源虚拟机项目（详见 http://bellard.org/qemu/），它可以提供一个虚拟的 ARM 移动设备。开发人员不需要一个真实的手机，只需通过电脑即可模拟运行手机操作系统，即可开发出应用在手机上面程序。模拟器在电脑上模拟运行的效果如图 1-4 所示。

在本节的内容中，将简要介绍 Android 模拟器的基本知识。

图 1-4　模拟器模拟手机

1.5.1 Android 模拟器介绍

对于 Android 程序的开发者来说，模拟器的推出给开发者在开发和测试上带来了很大的便利。无论在 Windows 下还是 Linux 下，Android 模拟器都可以顺利运行，而且官方提供了 Eclipse 插件，可将模拟器集成到 Eclipse 的 IDE 环境。当然，也可以从命令行启动 Android 模拟器。

获取模拟器的方法非常简单，既可以从官方站点（http://developer.Android.com/）免费下载单独的模拟器，也可以在下载 Android SDK 后，解压后在其 SDK 的根目录下找到一个名为“tools”文件夹，此文件夹下包含了完整的模拟器和一些非常有用的工具。

Android SDK 中包含的模拟器的功能非常齐全，电话本、通话等功能都可正常使用，甚至其内置的浏览器和 Maps 都可以联网。用户可以使用键盘输入，鼠标单击模拟器按键输入，

也可以使用鼠标单击、拖动屏幕进行操纵。

1.5.2　模拟器和真机的区别

Android 模拟器和真机的不同之处如下。

- 不支持呼叫和接听实际来电，但可以通过控制台模拟电话呼叫（呼入和呼出）。
- 不支持 USB 连接。
- 不支持相机/视频捕捉。
- 不支持音频输入（捕捉），但支持输出（重放）。
- 不支持扩展耳机。
- 不能确定连接状态。
- 不能确定电池电量水平和交流充电状态。
- 不能确定 SD 卡的插入/弹出。
- 不支持蓝牙。

1.6　Android 的未来发展和市场前景

老牌智能手机软件平台制造商 Symbian 发言人表示，Google 的 Android 只不过是另一个 Linux，Symbian 对其他软件与其形成的竞争并不感到担心。除了北美之外，Symbian 在其它地区智能手机市场都占有大部分市场份额。但是事实是从 2011 年第一季度开始到现在，Android 已经成为市场占有率最高的智能手机操作系统。

1.6.1　Android 的未来发展

与 iPhone 相似，Android 采用 WebKit 浏览器引擎，具备触摸屏、高级图形显示和上网功能，用户能够在手机上查看电子邮件、搜索网址和观看视频节目等，比 iPhone 等其他手机更强调搜索功能，界面更强大，可以说是一种融入全部 Web 应用的单一平台。

但其最震撼人心之处在于 Android 手机系统的开放性和免费服务。Android 是一个对第三方软件完全开放的平台，开发者在为其开发程序时拥有更大的自由度，突破了 iPhone 等只能添加固定软件的枷锁；同时与 Windows Mobile、Symbian 等厂商不同，Android 操作系统免费向开发人员提供，这样可节省近三成成本。

Android 项目已经从手机运营商、手机厂商、开发者和消费者那里获得大力支持。Google 移动平台主管安迪·鲁宾（Andy Rubin）表示，与软件开发合作伙伴的密切接触正在进行中。在过去的四年间，市场上产品种类最多的智能手机产品便是 Android。几乎涵盖了所有的品牌，各种不同的配置。

1.6.2　Android 的市场前景

2011 年初，仅正式推出三年的 Android 已超越称霸十年的 Symbian 系统，采用 Android 系统的重要厂商包括美国 Motorola、韩国 Samsung、英国 Sony Ericsson；另外中国厂商如：HTC、联想、华为、中兴等，使之跃居全球最受欢迎的智能手机平台，Android 系统不但利用于智能手机，也在平板电脑市场急速扩大。

第2章　开始搭建 Android 开发环境

"工欲善其事，必先利其器"这句名言出自《论语》，意思是说要想高效地完成一件事，要有一个合适的工具。对于编程人员来说，开发工具至关重要。选择了 Android 这样一个高效的开发工具后，搭建一个开发平台则是首要的任务。在本章的内容中，将详细介绍搭建 Android 开发环境的基本知识。

2.1　开发 Android 应用前的准备

作为一门新兴技术，在进行开发前，首先要搭建一个对应的开发环境。而在搭建开发环境前，需要了解安装开发工具所需要的硬件和软件配置条件。在本节内容中，将首先讲解搭建 Android 开发环境的准备工作。

2.1.1　基本系统要求

开发基于 Android 的应用软件所需要的开发环境如表 2-1 所示。

表 2-1　开发系统所需参数

项　目	版 本 要 求	说　明	备　注
操作系统	Windows XP / Vista /7 Mac OS X 10.4.8+Linux Ubuntu Drapper	根据自己的电脑自行选择	选择自己最熟悉的操作系统
软件开发包	Android SDK	选择最新版本的 SDK	
IDE	Eclipse IDE+ADT	Eclipse3.3 (Europa), 3.4 (Ganymede)和 ADT(Android Development Tools)开发插件	选择"for Java Developer"
其他	JDK Apache Ant	Java SE Development Kit 5 或 6 Linux 和 Mac 上使用 Apache Ant 1.6.5+, Windows 上使用 1.7+版本	（单独的 JRE 不可以的，必须要有 JDK），不兼容 Gnu Java 编译器（gcj）

2.1.2　Android 软件开发包

Android 的软件开发包主要包括以下工具。

- ❑ JDK：可以到网址 http://www.oracle.com/technetwork/java/javase/downloads/index.html 下载。
- ❑ Eclipse（Europa）：可以到网址 http://www.eclipse.org/downloads/下载 Eclipse IDE for Java Developers。
- ❑ Android SDK：可以到网址 http://developer.android.com 下载。
- ❑ 配套的开发插件。

安装 Android 的 Eclipse 插件需注意，Eclipse 可以到对应的网站下载安装，如果通过网络

远程安装不成功，可以下载到本地安装。

2.2　Windows 开发环境搭建

因为当前主流的操作系统是 Windows，所以在此将首先详细讲解在 Windows 中搭建 Android 开发环境的基本知识。

2.2.1　安装 JDK、Eclipse、Android SDK

下面具体介绍 JDK 1.5、Eclipse 3.3、ADT1.5、Android SDK 的安装步骤，在配套的视频中也有详细的介绍。

1．安装 JDK

第 1 步：安装 Eclipse 的开发环境需要 JRE 的支持，在 Windows 上安装 JRE/JDK 非常简单，首先在 Oracle 官方网站下载，网址为 http://www.oracle.com/technetwork/java/javase/downloads/index.html，如图 2-1 所示。

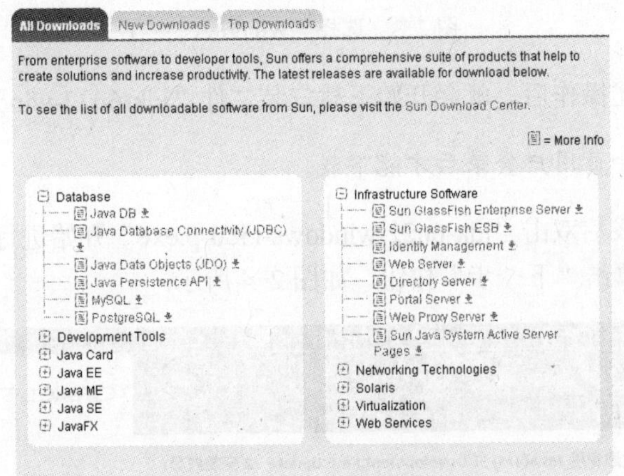

图 2-1　Sun 官方下载页面

第 2 步：在图 2-1 中可以看到有很多版本，运行 Eclipse 时虽然只需要 JRE 就可以了，但是在开发 Andriod 应用程序的时候，需要完整的 JDK（JDK 已经包含了 JRE），且要求其版本在 1.5+以上，这里选择 Java SE (JDK) 6，其下载页面如图 2-2 所示。

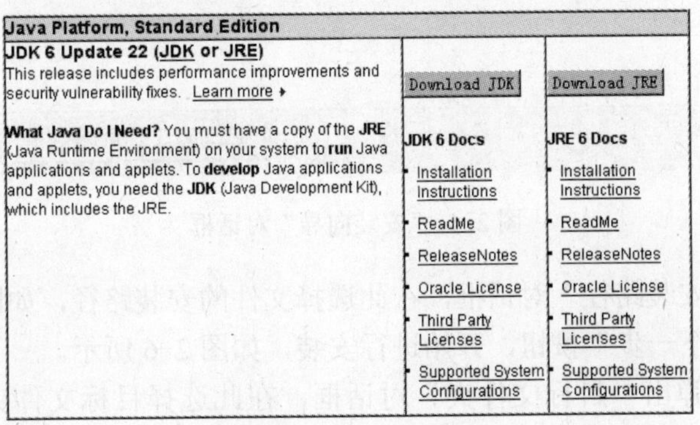

图 2-2　JDK 下载页面

第 3 步：在图 2-2 中选择“JDK 6 Update 22 (JDK or JRE)”，单击其右侧的“Download”按钮，出现让用户选择其操作系统和语言的界面，在此首先选择“Windows”，然后单击“Download”按钮，如图 2-3 所示。

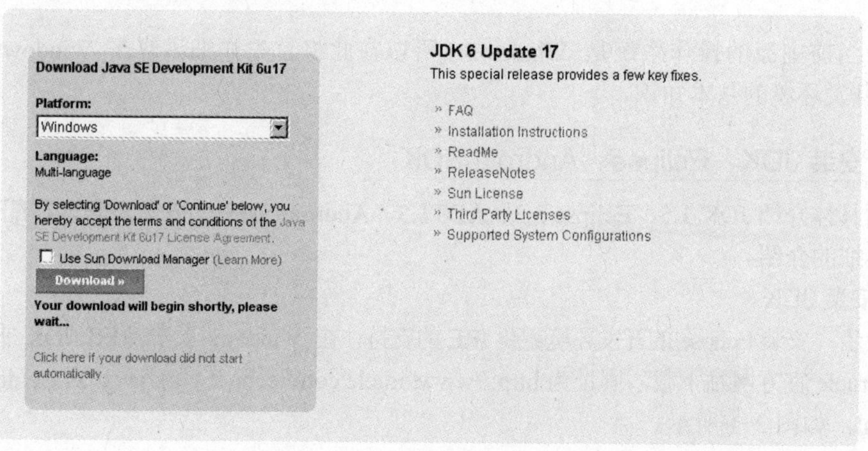

图 2-3　选择“Windows”

第 4 步：经过上述操作后，就会开始下载安装文件“jdk-6u22-windows-i586-p.exe”。

注意：在此需要会员用户登录后才能下载。

第 5 步：下载完成后双击“jdk-6u22-windows-i586-p.exe”开始进行安装。将弹出“安装向导”对话框，在此单击“下一步”按钮，如图 2-4 所示。

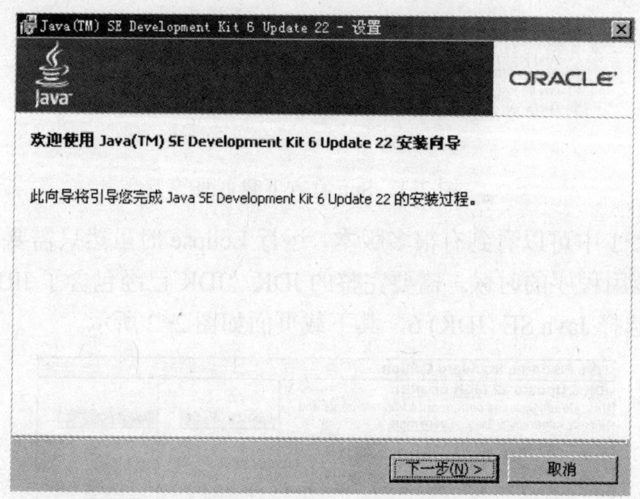

图 2-4　“安装向导”对话框

第 6 步：弹出“安装路径”对话框，在此选择文件的安装路径，如图 2-5 所示。

第 7 步：单击“下一步”按钮，开始进行安装，如图 2-6 所示。

第 8 步：完成后弹出“目标文件夹”对话框，在此选择目标文件夹的路径，如图 2-7 所示。

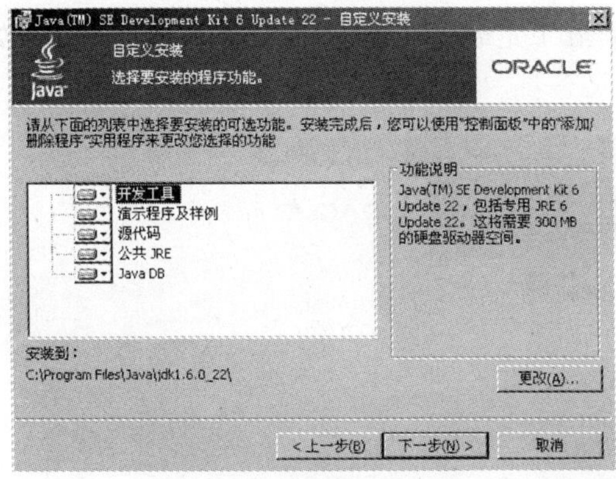

图 2-5 “安装路径”对话框

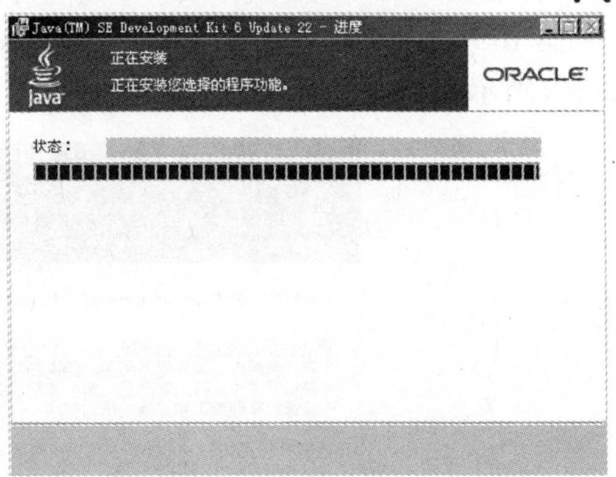

图 2-6 开始安装

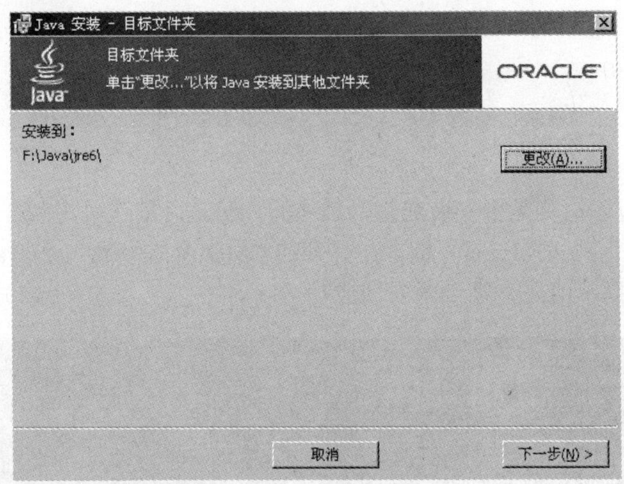

图 2-7 “目标文件夹”对话框

第 9 步：单击“下一步”按钮后，继续开始安装，如图 2-8 所示。

图 2-8 继续安装

第 10 步：完成后弹出“完成”对话框，单击“完成”按钮，完成整个安装过程，如图 2-9 所示。

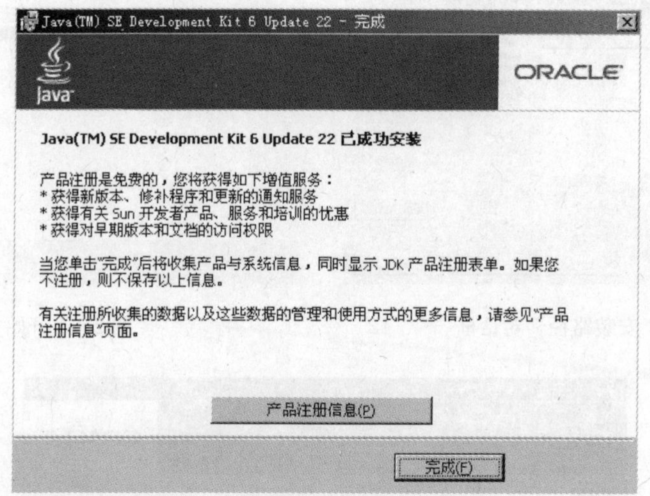

图 2-9　完成安装

完成安装后，可以检测是否安装成功，具体的方法是：依次单击“开始”→“运行”，在运行框中输入“cmd”并按〈Enter〉键，在打开的 CMD 窗口中输入“java –version”，如果显示如图 2-10 所示的提示信息，则说明安装成功。

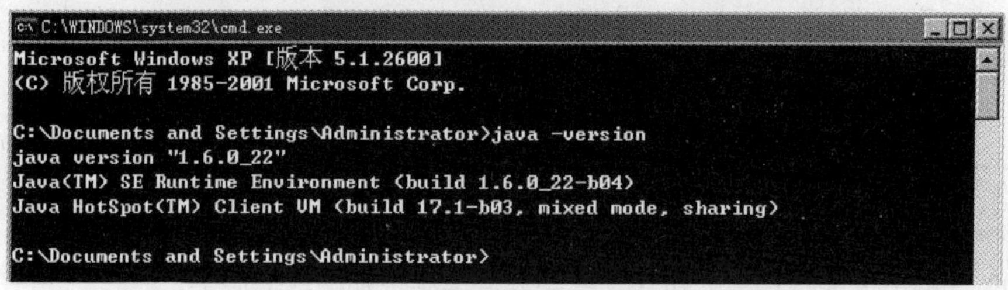

图 2-10　CMD 窗口

如果检测没有安装成功，则需要将它的目录的绝对路径添加到系统的路径（path）中，具体方法如下。

第 1 步：右键依次单击“我的电脑”→“属性”→“高级”，单击下面的“环境变量”，在下面的“系统变量”处选择新建在变量名处输入“JAVA_HOME”，变量值中输入刚才的目录，例如“C:\Program Files\Java\jdk1.5.0_14”，如图 2-11 所示。

编辑系统变量

变量名(N)：JAVA_HOME

变量值(V)：F:\Java\jdk1.6.0_22

确定　取消

图 2-11　设置系统变量

第 2 步：再新建一个变量，变量名为 classpath，变量值为“.;%JAVA_HOME%/lib/rt.jar;%JAVA_HOME%/lib/tools.jar”，单击“确定”后找到路径的变量，双击或单击编辑，在变量值最前面加上“%JAVA_HOME%/bin;”，如图 2-12 所示。

图 2-12　设置系统变量

第 3 步：再次依次单击“开始”→“运行”，在运行框中输入“cmd”并按〈Enter〉键，在打开的 CMD 窗口中输入“java –version”，如果显示如图 2-13 所示的提示信息，则说明安装成功。

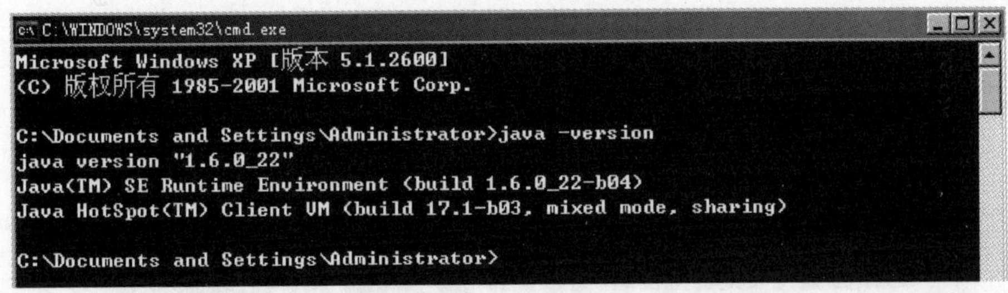

图 2-13　CMD 界面

注意：上述变量设置中，是按照笔者本人的安装路径设置的，笔者安装 JDK 的路径是 C:\Program Files\Java\jdk1.6.0_22。

2．安装 Eclipse

在安装好 JDK 后，就可以接着安装 Eclipse 了，具体过程如下。

第 1 步：打开 Eclipse 的下载页面 http://www.eclipse.org/downloads/，如图 2-14 所示。

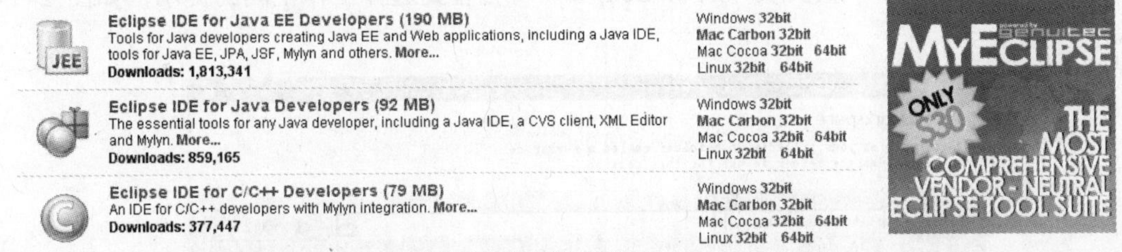

图 2-14　下载页面

第 2 步：在图 2-14 中选择“Eclipse IDE for Java Developers (92 MB)”，在其下载的镜像页面，选择离用户最近的镜像即可（一般推荐的下载速度就不错），如图 2-15 所示。

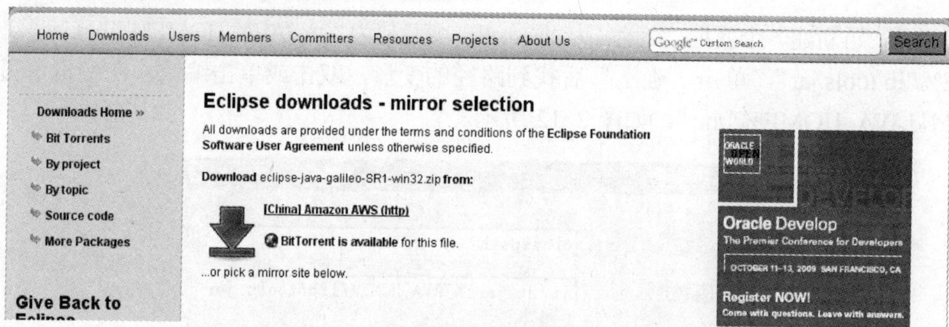

图 2-15　选择镜像

第 3 步：下载完成后，然后找到下载的压缩包“eclipse-java-galileo-SR1-win32.zip”。Eclipse 无须执行安装程序，解压此压缩文件就可以用，不过一定要先安装 JDK。在此假设 Eclipse 解压后存放的目录为“F:\eclipse”。

第 4 步：进入解压后的目录，此时可以看到一个名为“eclipse.exe”的可执行文件，双击此文件直接运行，Eclipse 能自动找到用户先期安装的 JDK 路径，启动界面如图 2-16 所示。

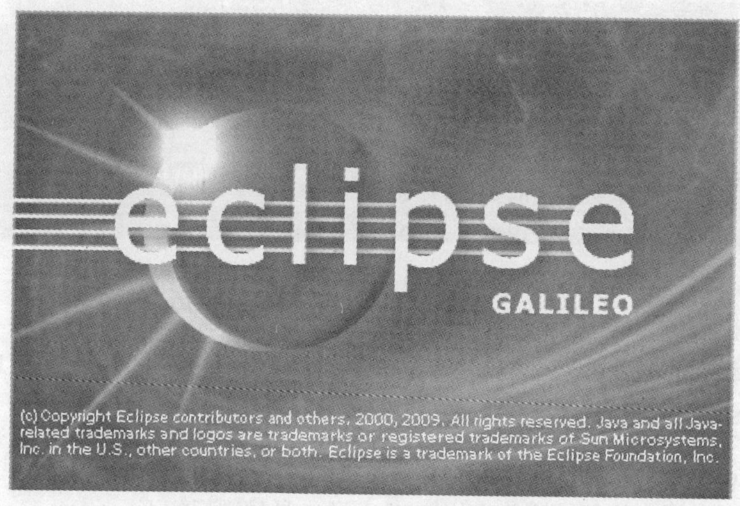

图 2-16　Eclipse 启动界面

第 5 步：因为是第一次安装、启动 Eclipse，将会看到选择工作空间的提示，如图 2-17 所示。

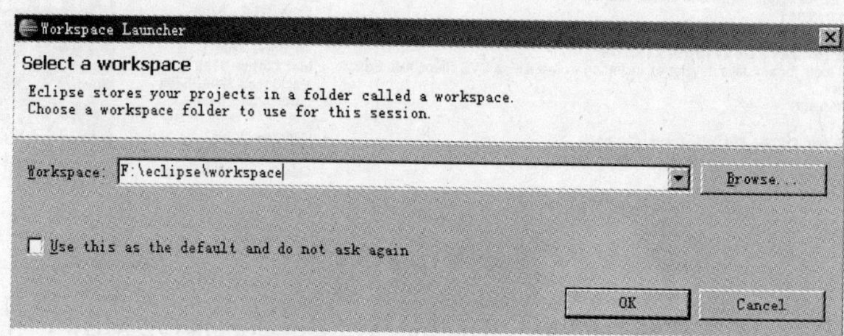

图 2-17　选择工作空间

此时单击“OK”按钮后，完成 Eclipse 的安装。系统进入初始欢迎界面，如图 2-18 所示。

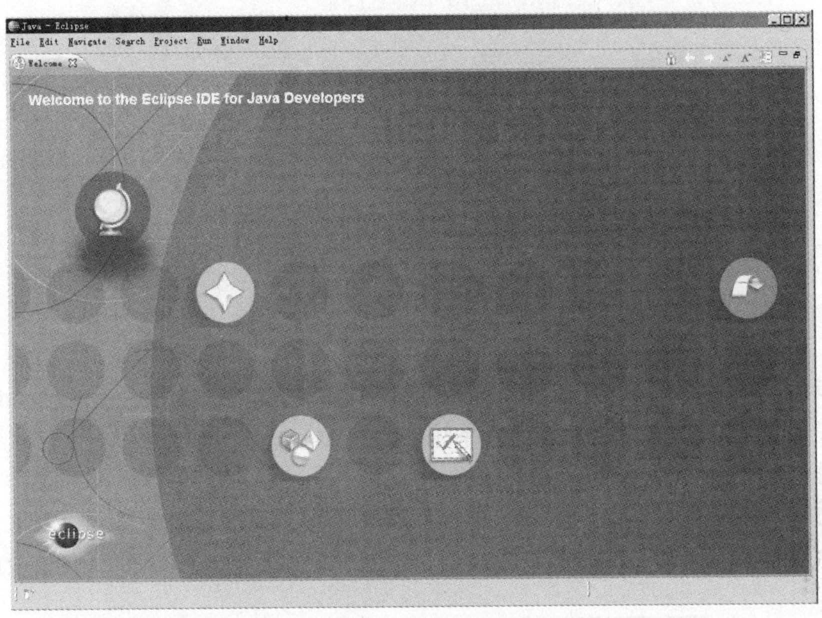

图 2-18　初始欢迎界面

3．安装 Android SDK

安装 JDK 和 Eclipse 完毕后，接下来需要下载安装 Andriod SDK，具体流程如下。

第 1 步：打开 Android 开发者社区网址 http://developer.android.com/，然后转到 SDK 下载页面下载，如图 2-19 所示。

Download the Android SDK

Welcome Developers! If you are new to the Android SDK, please read the steps below, for an overview of how to set up the SDK.

If you're already using the Android SDK, you should update to the latest tools or platform using the *Android SDK and AVD Manager*, starter package. See Adding SDK Components.

Platform	Package	Size	MD5 Checksum
Windows	android-sdk_r18-windows.zip	37448775 bytes	bfbfdf8b2d0fdecc2a621544d706fa98
	installer_r18-windows.exe (Recommended)	37456234 bytes	48b1fe7b431afe6b9c8a992bf75dd898
Mac OS X (intel)	android-sdk_r18-macosx.zip	33903758 bytes	8328e8a5531c9d6f6f1a0261cb97af36
Linux (i386)	android-sdk_r18-linux.tgz	29731463 bytes	6cd716d0e04624b865ffed3c25b3485c

图 2-19　SDK 下载页面

第 2 步：在此选择用于 Windows 平台的“android-sdk_r18-windows.zip”，单击后弹出下载界面，如图 1-20 所示。单击“保存文件”按钮，选择保存目录后开始下载。

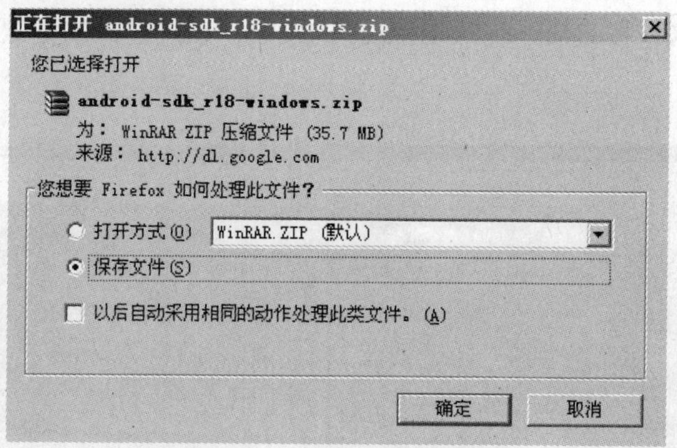

图 2-20 Android SDK 下载页面

第 3 步：下载完成后，解压压缩文件。假设下载后的文件解压存放在“F:\android\”目录下，并将其 tools 目录的绝对路径添加到系统的路径中，具体操作方法如下。

1）用鼠标右键单击“我的电脑”选择“属性”→“高级”，单击下面的“环境变量”按钮，在下面的“系统变量”选项组处选择“新建”选项，在变量名处输入“SDK_HOME”，变量值中输入刚才的目录，例如笔者的就是“F:\android-sdk-windows”，如图 2-21 所示。

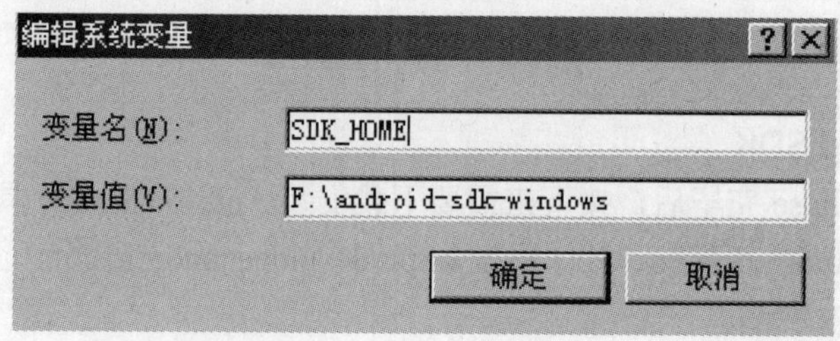

图 2-21 设置系统变量

2）找到 PATH 的变量，双击或单击编辑，在变量值最前面加上“%SDK_HOME%\tools;”，如图 2-22 所示。

图 2-22 设置系统变量

3）再次依次单击“开始”→“运行”，在运行框中输入“cmd”并按〈Enter〉键，在打开的 CMD 窗口中输入一个测试命令，例如 android-h，如果显示如图 2-23 所示的提示信息，则说明安装成功。

```
C:\WINDOWS\system32\cmd.exe
C:\Documents and Settings\Administrator>android -h

Usage:
  android [global options] action [action options]

Global options:
  -v --verbose  Verbose mode: errors, warnings and informational messages are pr
inted.
  -h --help     This help.
  -s --silent   Silent mode: only errors are printed out.

Valid actions are composed of a verb and an optional direct object:
-   list              : Lists existing targets or virtual devices.
-   list avd          : Lists existing Android Virtual Devices.
-   list target       : Lists existing targets.
- create avd          : Creates a new Android Virtual Device.
-   move avd          : Moves or renames an Android Virtual Device.
- delete avd          : Deletes an Android Virtual Device.
- update avd          : Updates an Android Virtual Device to match the folders of
 a new SDK.
- create project      : Creates a new Android Project.
- update project      : Updates an Android Project (must have an AndroidManifest.
xml).
- create test-project: Creates a new Android Test Project.
- update test-project: Updates an Android Test Project (must have an AndroidMani
```

图 2-23　设置系统变量

4．安装 ADT

Android 为 Eclipse 定制了一个插件，即 Android Development Tools（ADT）。这个插件为用户提供一个强大的综合环境用于开发 Android 应用程序。ADT 扩展了 Eclipse 的功能，可以让用户快速地建立 Android 项目，创建应用程序界面，在基于 Android 框架 API 的基础上添加组件，以及用 SDK 工具集调试应用程序，甚至导出签名（或未签名）的 APKs 以便发行应用程序。下面详细介绍安装配置 ADT 的基本方法。

在安装 ADT 插件前，需要首先打开 Eclipse 集成开发环境，接下来进行如下操作。

第 1 步：打开 Eclipse 后，依次单击菜单栏的“Help”→“Install New Software...”，如图 2-24 所示。

第 2 步：在弹出的对话框中，单击“Add”按钮，如图 2-25 所示。

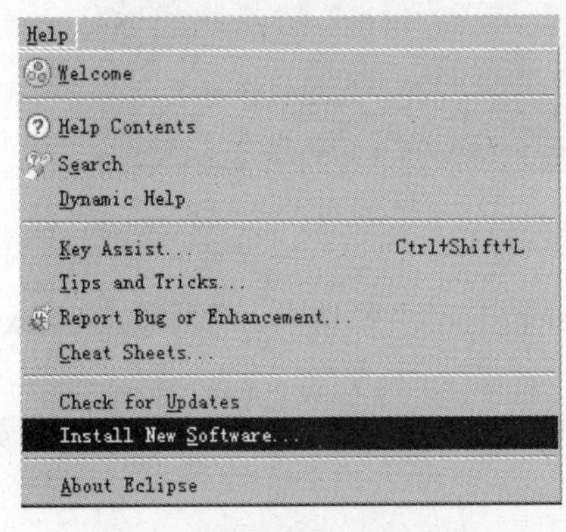

图 2-24　添加插件

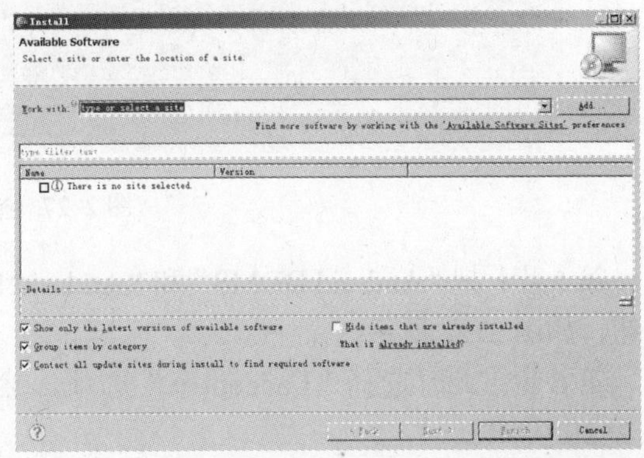

图 2-25　单击“Add”按钮

第 3 步：在弹出的“Add Site”对话框中分别输入名字和地址，名字可以自己命名，例如

“guan”，但是“Location”中必须输入插件的网络地址“http://dl-ssl.google.com/Android/eclipse/”，单击“OK”按钮，如图 2-26 所示。

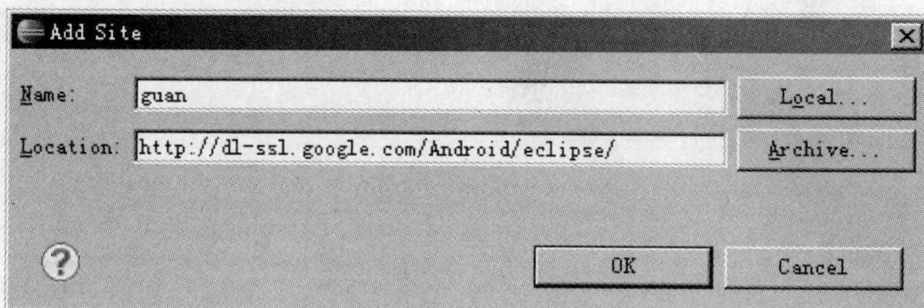

图 2-26　设置地址

第 4 步：单击“OK”按钮，完成设置，此时在“Install”界面将会显示可用的插件，如图 2-27 所示。

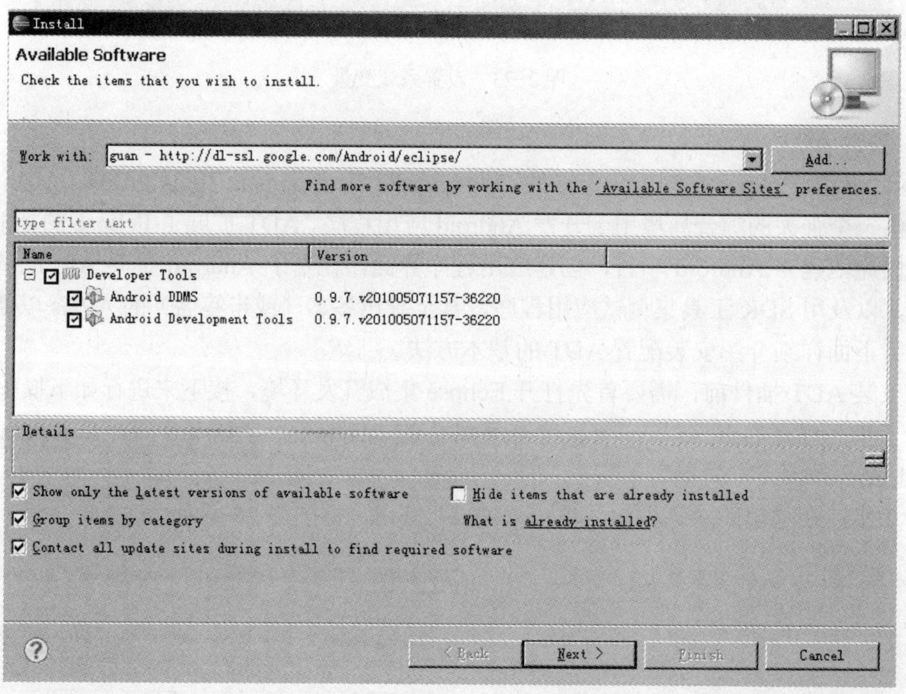

图 2-27　插件列表

第 5 步：把“Android DDMS”和“Android Development Tools”都选中，然后单击“Next”按钮，如图 2-28 所示。

第 6 步：此时选择“I accept…”选项，然后单击“Finish”按钮，开始进行安装，如图 2-29 所示。

注意：上述步骤可能会计算插件占用资源情况，过程有点慢，读者需要耐心，慢慢等待。完成后会提示重启 Eclipse 来加载插件，重启后就可以使用了。

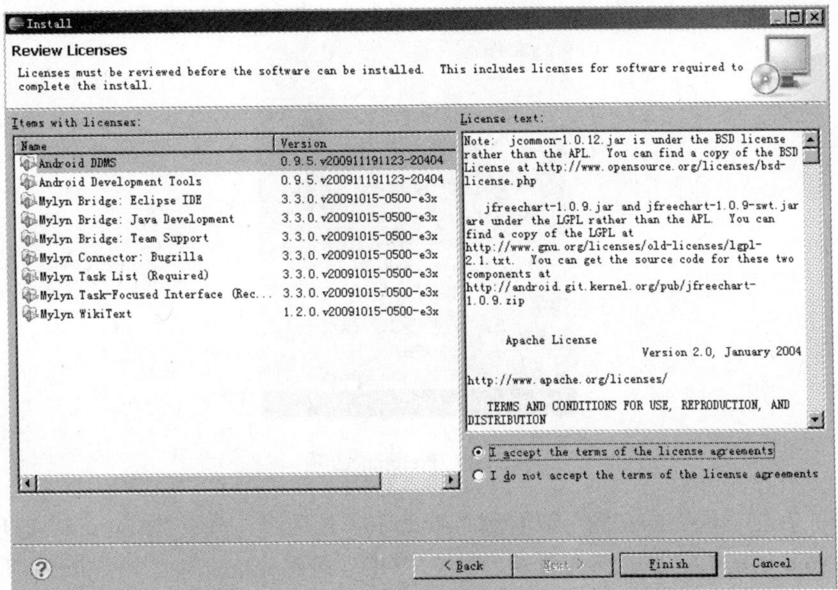

图 2-28　插件安装界面

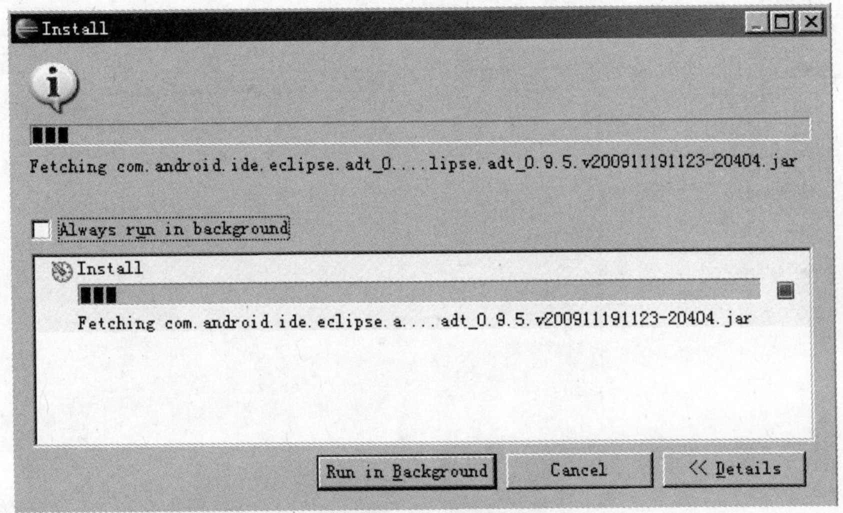

图 2-29　开始安装

另外，不同版本的 Eclipse 安装插件的方法和步骤是不同的，但是都大同小异，相信读者参照上面的操作能够自行解决。

2.2.2　设置 Android SDK Home

安装好插件后，还需要做如下配置才可以使用 Eclipse 创建 Android 项目，需要设置 Android SDK 的主目录。具体方法如下。

第 1 步：打开 Eclipse，在菜单中依次单击“Windows”→“Preferences”选项，如图 2-30 所示。

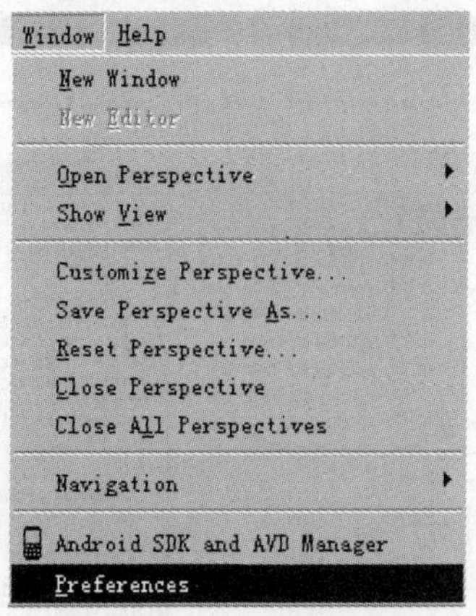

图 2-30　Preferences 项

第 2 步：在弹出的界面左侧可以看到“Android”选项组，选中“Android”选项后，在右侧设定 Android SDK Location 所在目录，单击“OK”按钮完成安装，如图 2-31 所示。

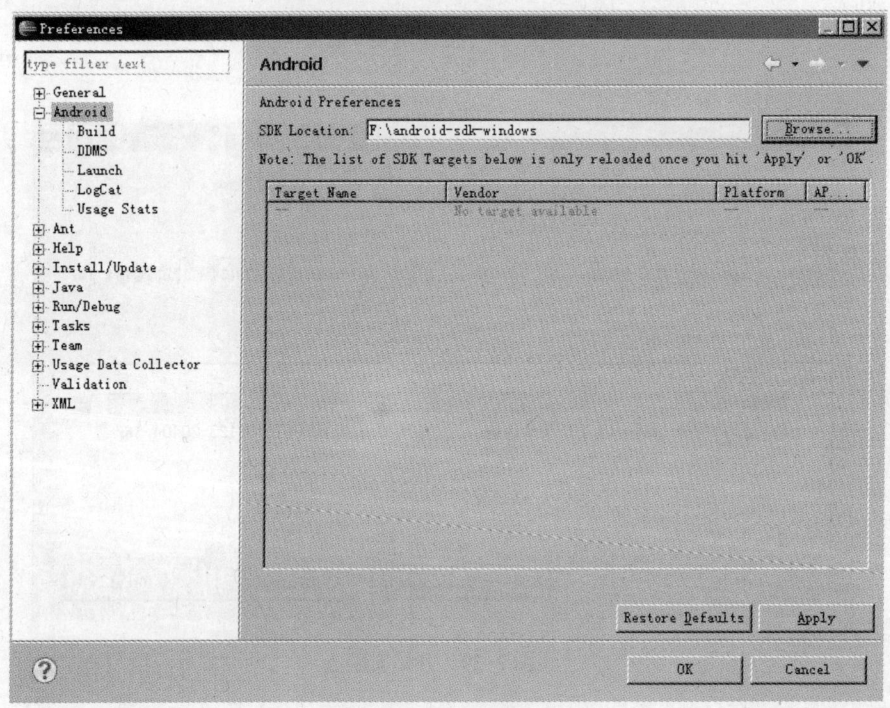

图 2-31　Preferences 项

2.2.3　验证开发环境

经过本章前面知识的讲解，Android 开发环境搭建完成。下面通过新建一个项目来验证当前的环境是否可以正常工作，具体流程如下。

第 1 步：打开 Eclipse，在菜单中依次选择“File”→“New”→“Project”选项，在弹出的对话框上可以看到“Android”选项，如图 2-32 所示。

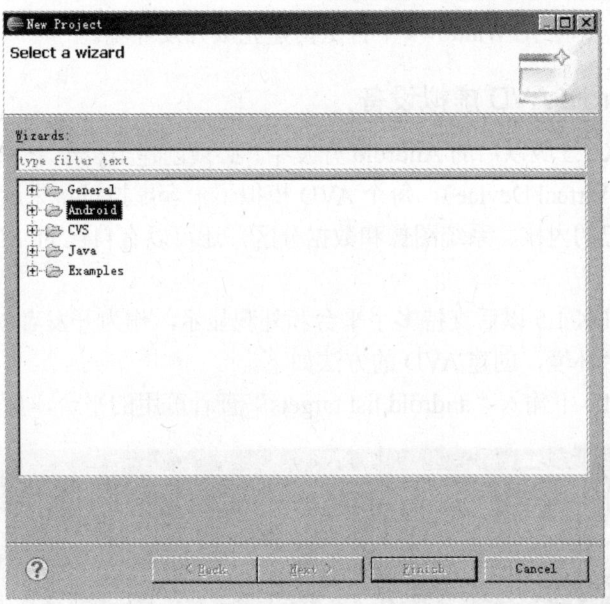

图 2-32 新建项目

第 2 步：在图 2-32 上选择“Android”选项，单击“Next”按钮，打开“New Android Project”对话框，在对应的文本框中输入必要的信息，如图 2-33 所示。

第 3 步：这里不再具体说明项目信息中各项的意义，后面章节会详细介绍。单击“Finish”按钮，会自动完成项目的创建工作，最后将可以看到如图 2-34 所示的项目结构。

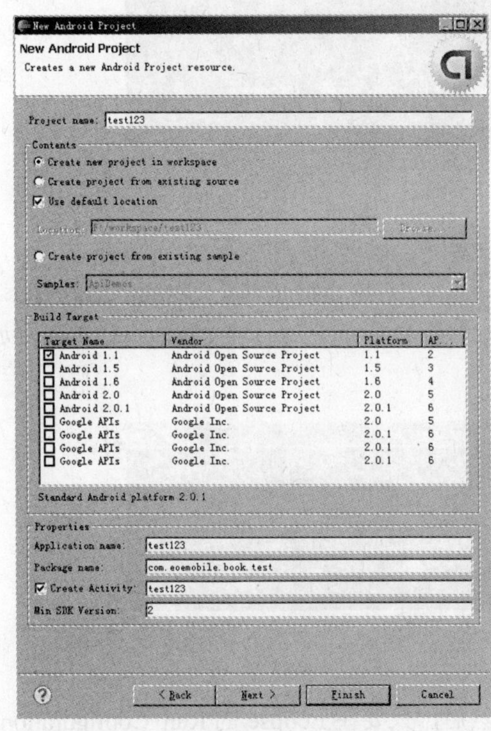

图 2-33 “New Android Project”对话框

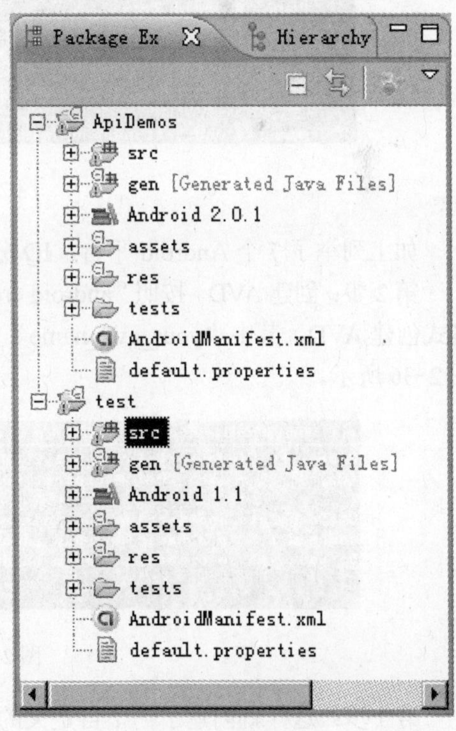

图 2-34 项目结构

通过上述操作，即可在 Windows 平台上搭建完成开发环境。

2.2.4 创建 Android AVD 虚拟设备

在 Android SDK 1.5 版以后的 Android 开发中，必须创建至少一个 AVD，AVD 即 Android 虚拟设备（Android Virtual Device），每个 AVD 模拟了一套虚拟设备来运行 Android 平台，这个平台至少要有自己的内核、系统图像和数据分区，还可以有自己的 SD 卡和用户数据以及外观显示等。

因为 Android SDK 1.5 以后支持多个平台和外观显示，作为开发者创建不同的 AVD 来模拟和测试不同的平台环境，创建 AVD 的方法如下。

第 1 步：在 CMD 下输入“android list targets”查看可用的平台，如图 2-35 所示。

```
C:\WINDOWS\system32\cmd.exe
     Name: Android 2.0
     Type: Platform
     API level: 5
     Revision: 1
     Skins: HVGA (default), QVGA, WQVGA400, WQVGA432, WVGA800, WVGA854
id: 5 or "android-6"
     Name: Android 2.0.1
     Type: Platform
     API level: 6
     Revision: 1
     Skins: HVGA (default), QVGA, WQVGA400, WQVGA432, WVGA800, WVGA854
id: 6 or "android-7"
     Name: Android 2.1-update1
     Type: Platform
     API level: 7
     Revision: 2
     Skins: HVGA (default), QVGA, WQVGA400, WQVGA432, WVGA800, WVGA854
id: 7 or "android-8"
     Name: Android 2.2
     Type: Platform
     API level: 8
     Revision: 1
     Skins: HVGA (default), QVGA, WQVGA400, WQVGA432, WVGA800, WVGA854

C:\Documents and Settings\Administrator>
```

图 2-35 CMD 界面

如上列举了 7 个 Android 平台，ID 分别是 1、2、3、4、5、6、7。

第 2 步：创建 AVD。按照“android create avd --name <your_avd_name> --target <targetID>”格式创建 AVD，其中“your_avd_name”是需要创建的 AVD 的名称，在 CMD 窗口界面中如图 2-36 所示。

```
C:\WINDOWS\system32\cmd.exe - android create avd --name mm --target 2
Microsoft Windows XP [版本 5.1.2600]
(C) 版权所有 1985-2001 Microsoft Corp.

C:\Documents and Settings\Administrator>android create avd --name mm --target 2
Android 1.5 is a basic Android platform.
Do you wish to create a custom hardware profile [no]
```

图 2-36 CMD 界面

第 3 步：这样就创建了一个自定义的 AVD，然后，只要在 Eclipse 的 Run Configurations 里面指定一个虚拟设备，即在 Target 下选中自己定义的这个 AVD，则 sdk_1_5_version 就可以

运行了，如图 2-37 所示。

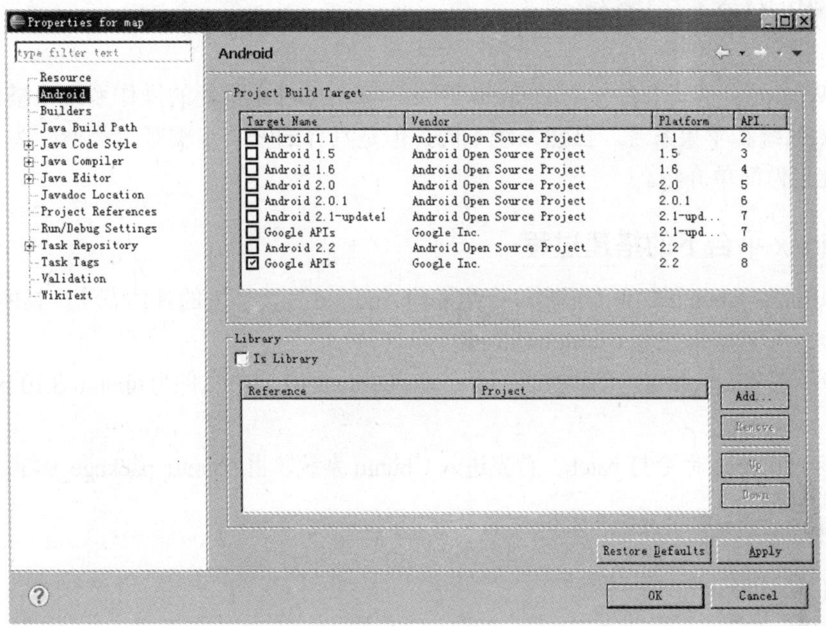

图 2-37　选择 AVD

第 4 步：选择“sdk_1_5_version”，单击“Apply”按钮后，然后单击“Start”按钮弹出“Launch”对话框，如图 2-38 所示。

第 5 步：此时单击“Launch”按钮后将会运行模拟器，如图 2-39 所示。

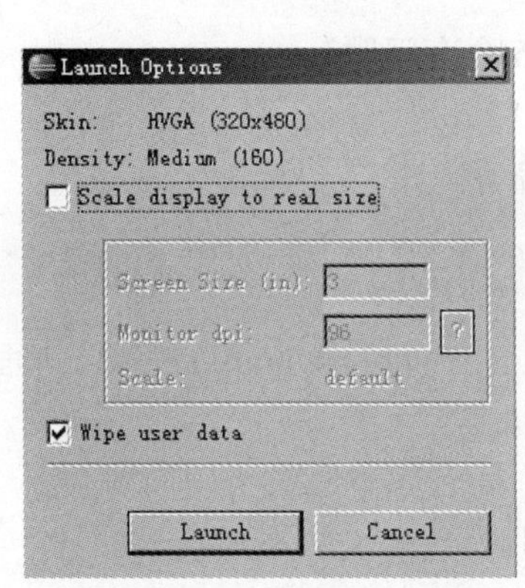

图 2-38　“Launch”对话框

图 2-39　模拟运行成功

关于应用结构分析和讲解，以及代码的调试部分内容会在本书后面的内容中进行详细介绍。至此，在 Windows 平台上的开发环境搭建完成，安装了运行环境 JDK、开发工具 Eclipse 和 Android SDK，并安装了 ADT，进行了 SDK Home 的配置，最后创建了一个 Android 虚拟设备。

2.3 其他平台下的搭建

除了 Windows 系统平台外，Android 开发环境还可以在其他的操作系统中搭建。例如常见的 Linux 系统和苹果系统。虽然系统不同，但是具体的安装步骤基本类似，为节省本书的篇幅，在此仅做简单介绍。

2.3.1 Linux 平台下的搭建过程

下面以 Linux Ubuntu 8.10 为平台，介绍搭建 Android 开发环境的具体方法。具体流程如下。

第 1 步：安装虚拟光驱 daemon400.exe。

第 2 步：在 Windows XP 下用虚拟光驱安装 Ubuntu 8.10, ISO 文件为 ubuntu-8.10-beta-desktop-i386.iso。

第 3 步：用 dpkg 命令打 patch。首先进入 Ubuntu 系统，将 ubuntu_package_0430.tar.gz 解压。

```
tar –zvxf ubuntu_package_0430.tar.gz
```

然后打开 patch。

```
sudo dpkg -i *.deb
```

如果存在没有成功的文件，则再依次执行。

```
sudo dpkg –i   filename.deb
```

注意：可能有些需要一起运行 dpkg，具体格式如下。

```
sudo dpkg –i   filename1.deb filename1.deb
```

另外，还需要重新将 Java5 执行 dpkg 命令（因为用 Java6 会有问题）。

第 4 步：编译原码和 Android SDK。

编译原码：解压原码到本地，进入原码目录，执行如下命令。

```
make
```

编译 SDK：在使用 make 命令之后，直接执行 make sdk，会在 out/host/linux-x86/sdk 下面生成 mdk 文件及文件夹，形如：android-sdk_eng.xxx_linux-x86。

第 5 步：安装 Eclipse。

只需直接解压 eclipse-jee-ganymede-SR2-linux-gtk.tar.gz 即可。

```
tar -zvxf eclipse-jee-ganymede-SR2-linux-gtk.tar.gz
```

第 6 步：Eclipse 下配置 Android。

第 7 步：测试刚才编译好的 SDK。

1）在 Eclipse 中将 Android SDK 目录设置成自己编译生成的 SDK 目录，例如 out/host/linux-x86/sdk/android-sdk_eng.xxx_linux-x86。选择“Window”→“preferences”→

“Android”选项中的 SDK Location，进行设置。

2）创建 AVD：在 Eclipse 中，选择“window”→“Android AVD Manager”选项，将 name、target、sdcard、skin 都填选好后，单击“Create AVD”即可。

3）在 CMD 窗口中进入到目录下，执行

```
emulator  - avd avdname
```

经过上述操作后，模拟器就运行起来了。

注意： 如果没有需要的 JDK、Eclipse 和 Android SDK，在 Linux 下也需要分别下载它们，只是在下载时选择 Linux 的资源即可，整个安装顺序和 Windows 下大同小异。

2.3.2　苹果平台下的搭建过程

苹果平台下搭建 Android 的具体过程和 Linux 下的类似，为节省本书篇幅，将不再进行详细介绍。只是在下载 Android SDK 的时候，注意选择 Mac OS 版本的资源即可。

2.4　安装过程中的常见问题

安装和搭建一个开发环境，常常会遇到一些意想不到的问题，这些问题可能是粗心造成的，也可能是使用的系统环境的差异造成的。在本节内容中，将简单介绍搭建 Android 开发环境中常见问题的解决方法。

1. Android 不能在线更新

在安装 Android 后，需要更新为最新的资源和配置。但是在启动 Android 后，经常会不能更新，弹出如图 2-40 所示的错误提示。

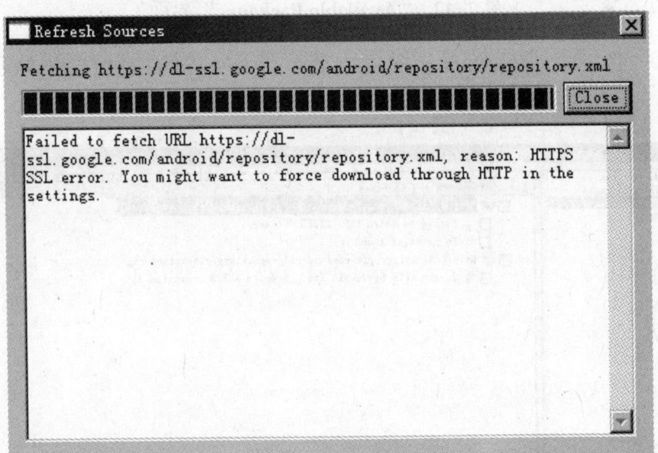

图 2-40　不能更新

Android 默认的在线更新地址是https://dl-ssl.google.com/android/eclipse/，但是经常会出现错误。如果此地址不能更新，可以自行设置更新地址，修改为 http://dl-ssl.google.com/android/repository/repository.xml。具体操作方法如下。

1）单击“Android SDK and AVD Manager”窗口左侧的“Available Packages”选项，然后单击下面的“Add Site...”按钮。如图 2-41 所示。

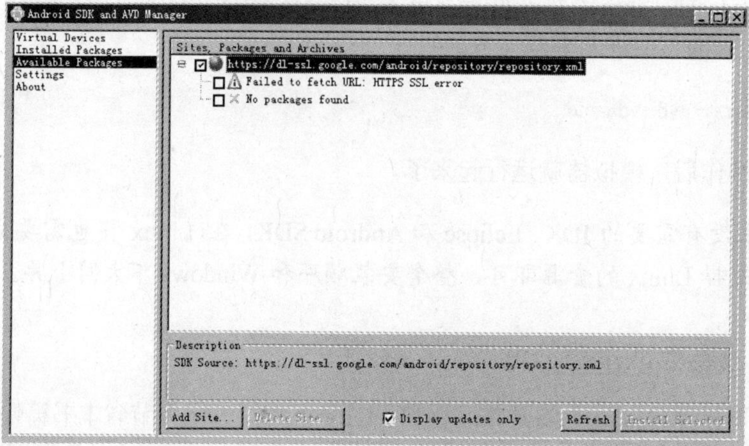

图 2-41 “Available Packages”界面

2）在弹出的“Add Site URL”对话框中输入修改后的地址 http://dl-ssl.google.com/android/repository/repository.xml，如图 2-42 所示。

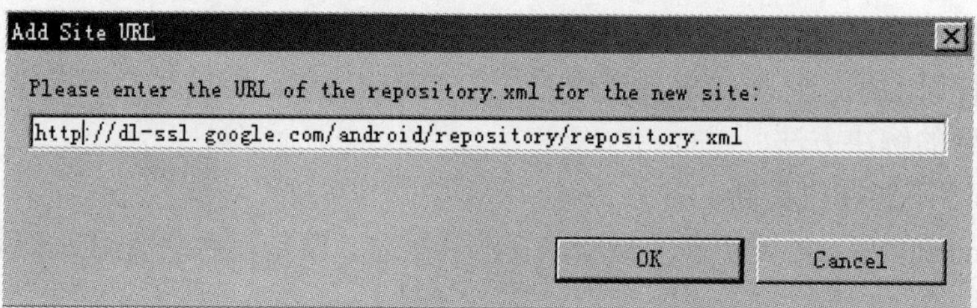

图 2-42 “Available Packages”界面

3）单击“OK”按钮，完成设置。经过上述操作后，就能够使用更新功能了，如图 2-43 所示。

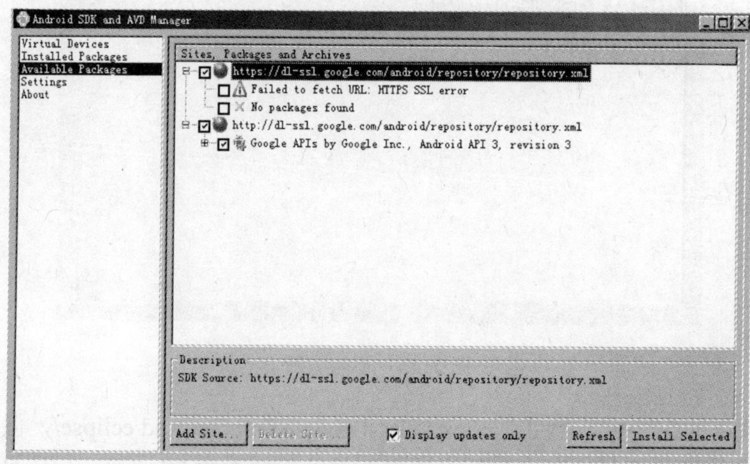

图 2-43 “Available Packages”界面

2．Eclipse 中新建 Android 工程时，一直显示“Project name must be specified”

一直显示“Project name must be specified”的情况如图 2-44 所示。

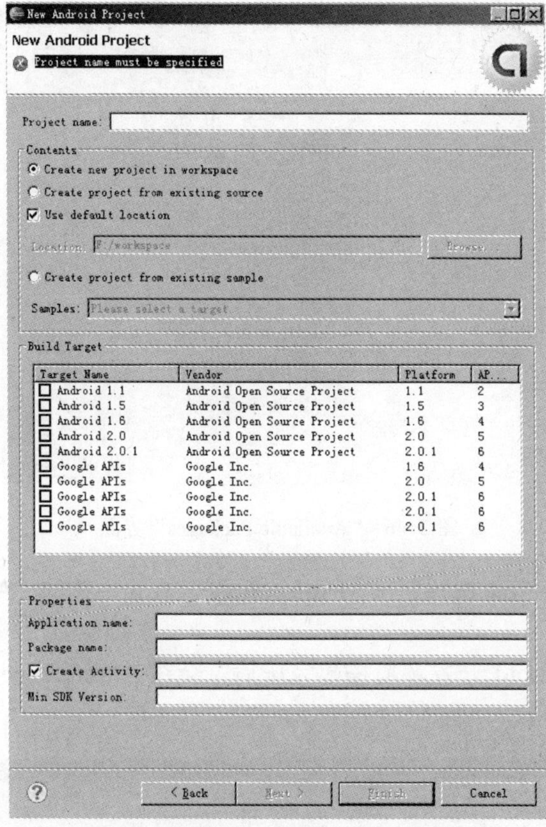

图 2-44　“Available Packages”界面

造成上述问题的原因是 Android 没有更新完成，需要进行完全更新。具体方法如下。

1）打开“Android SDK and AVD Manager”，选择左侧的“Installed Packages”，如图 2-45 所示。

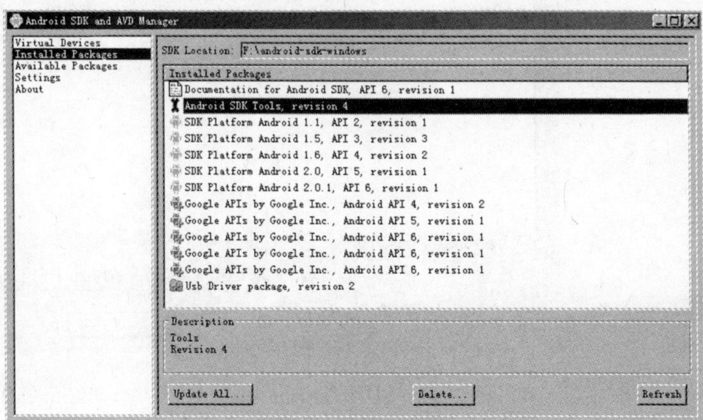

图 2-45　“Available Packages”界面

2）在右侧列表中选择“Android SDK Tools，revision4”，在弹出窗口中选择“Accept”，最后单击“Install Accepted”按钮后开始安装更新，如图 2-46 所示。

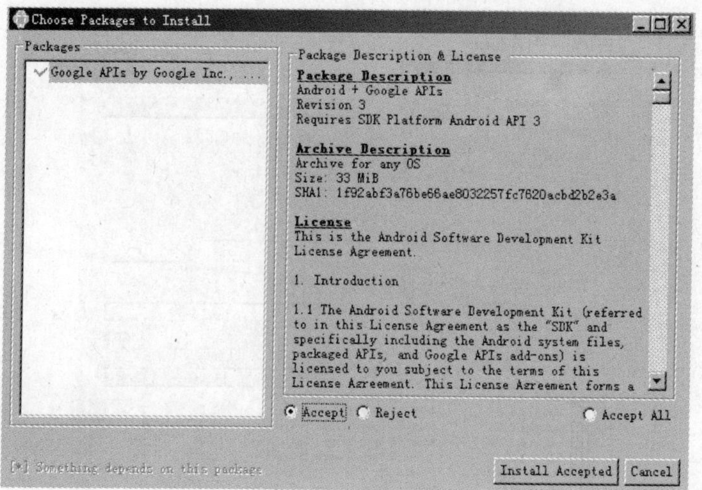

图 2-46 “Available Packages”界面

3. Eclipse 中单击“Window”→“Preference”，单击右侧的“Android”选项，Target 列表中没有 Target 选项

通常来说，当 Android 开发环境搭建完毕后，会在“Preference”中显示存在的 SDK Targets，如图 2-47 所示。

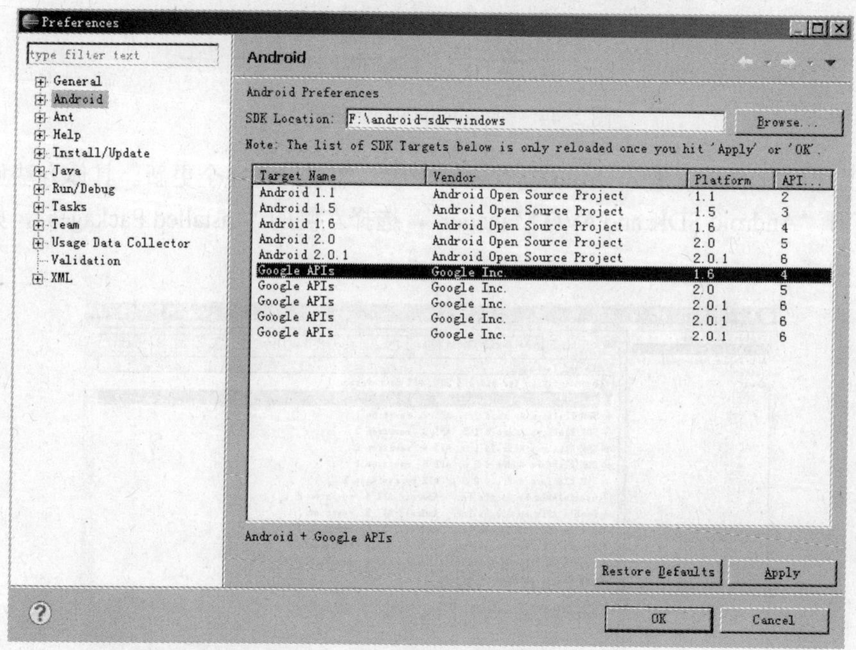

图 2-47 SDK Targets 列表

但是往往因为各种原因，会不显示 SDK Targets 列表，在界面中也不显示，并输出“Failed

to find an AVD compatible with target”提示的错误问题。上述问题的原因是 AVD 没有创建成功，如果出现上述问题，需要手动安装，当然前提是 Android 更新完毕。具体解决方法如下。

1）在“运行”中输入“cmd”，打开 CMD 窗口，如图 2-48 所示。

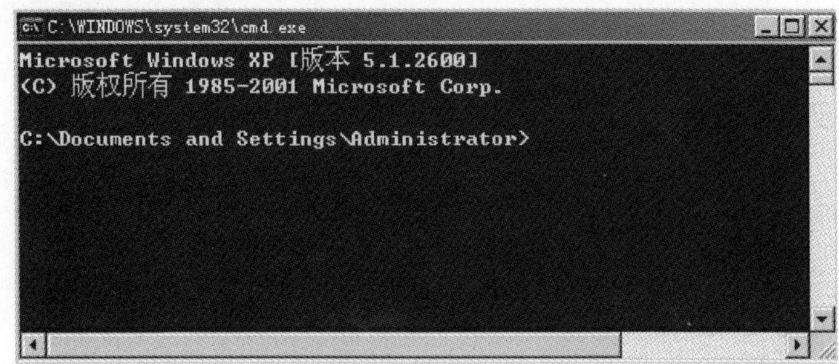

图 2-48　SDK Targets 列表

2）使用“android”命令创建一个 Targets，创建 AVD。按照“android create avd --name <your_avd_name> --target <targetID>”格式创建 AVD，其中“your_avd_name”是需要创建的 AVD 的名称，如图 2-49 所示。

图 2-49　CMD 界面

在如图 2-41 所示的窗口中创建了一个名为 aa，targetID 为 3 的 AVD，然后在 CMD 界面中输入“n”，即完成操作，如图 2-50 所示。

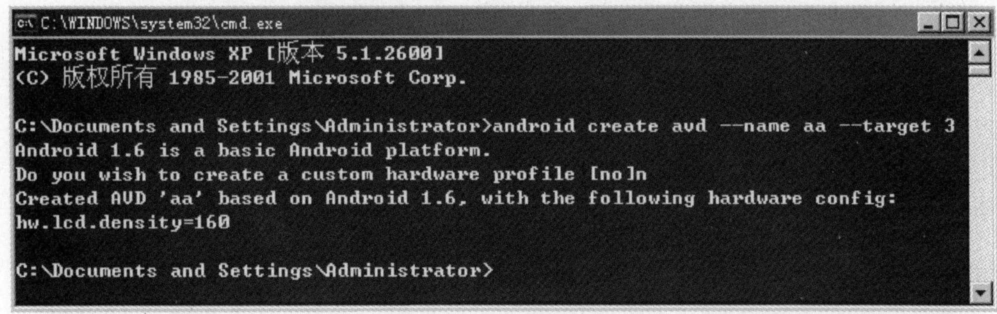

图 2-50　CMD 界面

4. 在线更新时，提示“Failed to rename directory d:\android-sdk-windows\tools to d:\android-sdk-windows\tools\temp\ToolPackage.old01”之类的错误

这种情况通常是因为杀毒软件或防火墙对这个文件进行了访问，建议读者在更新时关闭

杀毒软件和防火墙。如果还不行可进行如下操作。

1）在 android-sdk-windows 下复制 tools 文件夹，得到“复件 tools”文件夹

2）运行复件 tools 文件夹里的 android.bat，在出现的“Android SDK and AVD Manager”上执行之前的升级步骤即可成功。记得在此步之前关闭已打开的 Android SDK.. Manager（若有）。

3）删除“复件 tools”（非必要）。

第 3 章　用户人机界面设置

手机开发项目最终都会在手机屏幕上显示出来，因此，手机屏幕的显示效果往往会影响到消费者的购买欲望。本章将详细介绍，通过 Android 来设置用户人机界面的各种操作方法。希望读者能够举一反三，为进入本书后面知识的学习打下基础。

3.1　更改、显示文字标签

在实际应用中，经常需要在手机屏幕中显示一段简单地文字，有时为了满足现实的需求，还要对这行文字进行颜色和大小的设置修饰，以实现我们需要的绚丽效果。在本节的内容中，将通过一个具体实例来讲解在屏幕中更改显示文字标签的方法。

3.1.1　设计理念

实际上，可以在 Layout 中可以创建一个对象，并定义 strings.xml 中的字符串常数，然后通过文本组件 TextView 中的 setText 方法，在欲加载程序最初之际来更改 TextView 组件中显示的文字。

3.1.2　具体实现

本节实例保存在“光盘:\3\wenzi\”文件夹内，主文件 src/wenzi/wenzi.java 是此项目的主要文件，用于调用各个公用文件来实现具体的功能。具体代码如下。

```
package irdc.wenzi;
import android.app.Activity;
import android.os.Bundle;
/*必须引用 widget.TextView 接口才能在程序里声明 TextView 对象*/
import android.widget.TextView;
public class wenzi extends Activity
{
  /*必须引用 widget.TextView 才能在程序里声明 TextView 对象*/
  private TextView mTextView01;
  /** Called when the activity is first created. */
  @Override
  public void onCreate(Bundle savedInstanceState)
  {
    super.onCreate(savedInstanceState);
    /* 载入 main.xml Layout，此时 myTextView01:text 为 str_1 */
    setContentView(R.layout.main);
```

```
        /* 使用 findViewBtId 函数，利用 ID 找到该 TextView 对象 */
        mTextView01 = (TextView) findViewById(R.id.myTextView01);
        String str_2 = "欢迎来到 Android 世界...";
        mTextView01.setText(str_2);
    }
}
```

TextView 对象里的 setText 方法支持以下多态构造方法。

```
public final void setText(CharSequence text)
  public final void setText(int resid)
  public void setText(CharSequence text, TextView.BufferType type)
  public final void setText(int resid, TextView.BufferType type)
  public final void setText(char[] text, int start, int len)
```

在此，以最后 setText(char[] text, int start, int len) 为例，第一个参数 char 数组作为输出依据，第二个参数为从哪一个元素索引开始选取，第三个参数则为要取出多少个元素，再看下面的代码。

```
char char_1[] = new char[5];
char_1[0] = 'D';
char_1[1] = 'a';
char_1[2] = 'v';
char_1[3] = 'i';
char_1[4] = 'd';
mTextView01.setText(char_1,1,3);
```

如上述程序所示，输出的结果是“avi”，因为从第 1 个元素索引开始，共取 3 个元素；最后则要提醒读者，TextView.setTextView 不支持 HTML TAG 的输出，所以即便写成这样：

```
mTextView01.setText("<a ref=\"http://shop.
teac.idv.tw/MyBlog/\">我/a>");
```

实际输出时，也就是纯文本而已，并不会进行 HTML TAG 的转换。但若撇开 HTML TAG 之外（如"<"开头的标记），在 TextView 对象里加上了 android:autoLink="all"语句，那么正文中若有网址（http://），是可以被显示的，以下这个范例请读者实现。

```
<TextView xmlns:android="http://schemas.android.com/apk/res/android"
  android:layout_width="fill_parent"
  android:layout_height="wrap_content"
  android:autoLink="all"
  android:text="请访问：http://xxx.com
/>
```

经过上述操作设置，此实例的主要文件编程完毕。调试运行后的效果如图 3-1 所示。

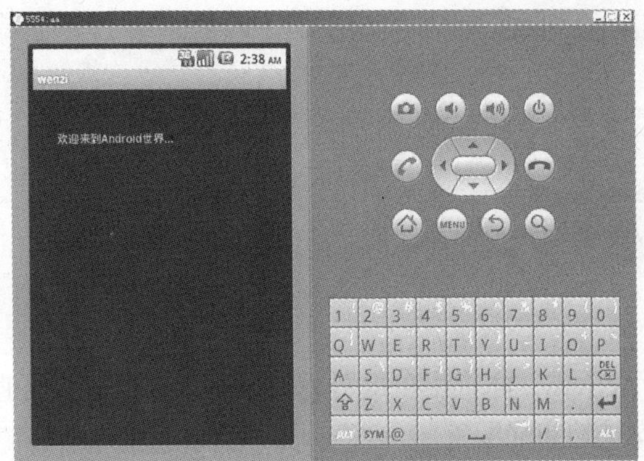

图 3-1　运行效果

3.2　更改屏幕背景颜色

为了满足特殊需求，也可以对屏幕画面的背景颜色进行设置。本节将通过一个简单的实例，来简要介绍更改屏幕窗口画面背景颜色的方法。

3.2.1　设计理念

如果是默认的运行结果，窗口的底色会是深黑色，这是 SDK 默认的颜色。要更改 Activity 类里的窗口底色有许多方法，最简单的方法就是将颜色色码事先定义在 drawable 当中，当程序 onCreate 创建的同时，加载预先定义的画面颜色。

本实例程序的设计方式是在 drawable 里指定布局页面的背景色（BackGround）为白色，但这里的“白色”（颜色色码为#FFFFFFFF）预先定义在 drawable 当中，当程序运行时，背景就会变为白色。

这是指定 Activity Layout 背景颜色最简单的方法，在范例最末，则将示范如何创建色彩板（color table），让 Android 手机程序可以像使用“常数”般直接取用，并反应在应用程序的运行阶段。

3.2.2　具体实现

本节实例保存在“光盘:\3\beijing\”文件夹内，下面简单介绍主要文件的具体含义。

主文件 src/beijing/beijing.java 是此项目的主要文件，调用各个公用文件来实现具体的功能。具体代码如下。

```
package irdc.beijing;

import android.app.Activity;
import android.os.Bundle;

public class beijing extends Activity
{
```

```
        /** Called when the activity is first created. */
        @Override
        public void onCreate(Bundle savedInstanceState)
        {
            super.onCreate(savedInstanceState);
            setContentView(R.layout.main);
        }
    }
```

上述代码继承自 Activity 类，并在重写方法 onCreate 之初，直接显示 R.layout.main（main.xml）这个页面安排的布局配置。

3.3 更改文字颜色

实际设计中最常用的定义颜色常数的方法，则是使用程序控制 TextView 对象上的文字或其他对象的背景色（setBackgroundDrawable 方法），如判断对象被单击时的背景色亮起，或当失去焦点时，又恢复成原来的背景色等。本节将通过一个简单的实例，介绍更改 TextView 文字颜色的方法。

3.3.1 设计理念

本实例扩展了前一个实例的实现，预先在布局页面中设计好两行 TextView 文本，并在运行 onCreate 同时，通过两种程序描述方法，实时更改原来布局页面里 TextView 文本的背景色以及文字颜色，最后学会使用 Android 默认的颜色常数（graphics.Color）来更改文字的前景色。

3.3.2 具体实现

本节实例保存在“光盘:\3\yanse\”文件夹内，下面简单介绍主要文件的具体含义。

1．主文件

主文件 src/yanse/yanse.java 是此项目的主要文件，调用各个公用文件来实现具体的功能。具体代码如下。

```
package irdc.yanse;

import irdc.yanse.R;
import android.app.Activity;
import android.content.res.Resources;
import android.graphics.Color;
import android.graphics.drawable.Drawable;
import android.os.Bundle;
import android.widget.TextView;

public class yanse extends Activity
{
  private TextView mTextView01;
```

```
    private TextView mTextView02;

    @Override
    public void onCreate(Bundle savedInstanceState)
    {
      super.onCreate(savedInstanceState);
      setContentView(R.layout.main);

      mTextView01 = (TextView) findViewById(R.id.myTextView01);
      mTextView01.setText("使用的是 Drawable 背景色文本。");

      Resources resources = getBaseContext().getResources();
      Drawable HippoDrawable = resources.getDrawable(R.drawable.white);
      mTextView01.setBackgroundDrawable(HippoDrawable);

      mTextView02 = (TextView) findViewById(R.id.myTextView02);
      mTextView02.setTextColor(Color.MAGENTA);
    }
  }
```

通过上述代码新建两个文本类成员变量：mTextView01 与 mTextView02，这两个变量在创建之初，以 findViewById 方法初始化为布局（main.xml）里的 TextView 对象。使用了 Resource 类以及 Drawable 类，分别创建了 resources 对象以及 HippoDrawable 对象，并将前一个范例中所创建的 R.drawable.white 以 getDrawable 方法加载，最后则调用了 setBackgroundDrawable 来更改 mTextView01 的文字底纹。更改 TextView 对象里的文字，则使用了 setText 方法。

在 mTextView02 当中，使用了 graphics.Color 里的颜色常数。接着，再利用 setTextColor 更改文字的前景色。

2．布局文件

在页面布局上，使用了 2 个 TextView 对象，关键在于下面的代码。

```
<?xml version="1.0" encoding="utf-8"?>
<LinearLayout xmlns:android="http://schemas.android.com/apk/res/android"
  android:orientation="vertical"
  android:layout_width="fill_parent"
  android:layout_height="fill_parent"
  >
  <TextView
  android:id="@+id/myTextView01"
  android:layout_width="fill_parent"
  android:layout_height="wrap_content"
  android:text="@string/str_textview01"
  />
  <TextView
  android:id="@+id/myTextView02"
  android:layout_width="fill_parent"
```

```
        android:layout_height="wrap_content"
        android:text="@string/str_textview02"
        />
</LinearLayout>
```

经过上述操作设置，此实例的主要文件编程完毕。调试运行后的效果如图 3-2 所示。

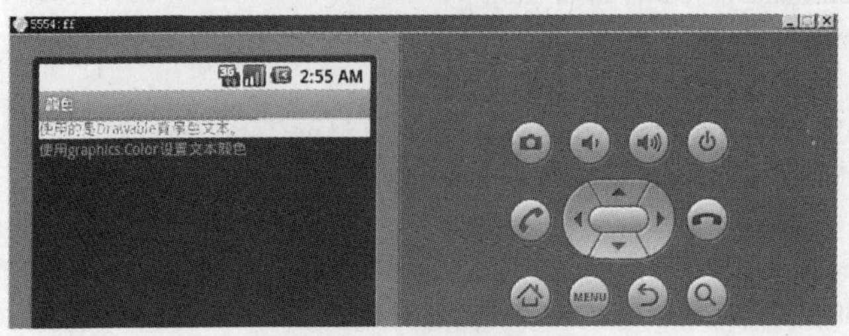

图 3-2　运行效果

3.4　置换 TextView 文字

前几个实例讲解了文本颜色和背景颜色的设置方法，其实也可以对 TextView 文字进行置换处理。本节将通过一个简单实例的实现过程，介绍置换 TextView 文字的方法。

3.4.1　设计理念

置换 TextView 文字可以通过 CharSequence 数据类型与 Resource ID 应用来实现。从一开始布局文件通过 Resource 初始化 TextView 对象的文字，到程序中动态更改 TextView 文字，但要如何在代码里取得 Resource 的字符串呢？在 Android 里，确实有一些方法可以直接以 R.string.*直接转换 ID 为 String，不过，这样的数据类型转换是非常规的，甚至是不妥的，正确的方法是利用 Context.getString 方法来取得存放在 global 里的 Resource ID。以下这个范例将示范如何在程序运行时（runtime），通过 CharSequence 依据 Resource ID 取出字符串，并正确更改 TextView 的文字。

3.4.2　具体实现

本节实例保存在“光盘:\3\zhihuan\”文件夹内，下面简单介绍主要文件的具体含义。

1．主文件

主文件 src/zhihuan/zhihuan.java 是此项目的主要文件，调用各个公用文件来实现具体的功能。此处主代码和前面程序的差异主要是：在更改 mTextView02 对象的文字时（setText 方法），合并了 str_3 与 str_2 这两个不同对象，由于 setText 方法同时支持 CharSequence 与 String 类型的参数，所以在此示范不同数据类型的字符串进行同步输出。具体代码如下。

```
package irdc.zhihuan;


import irdc.zhihuan.R;
import android.app.Activity;
```

```
import android.os.Bundle;
import android.widget.TextView;
public class zhihuan extends Activity
{
   private TextView mTextView02;

   @Override
   public void onCreate(Bundle savedInstanceState)
   {
      super.onCreate(savedInstanceState);
      setContentView(R.layout.main);

      mTextView02 = (TextView) findViewById(R.id.myTextView02);
      CharSequence str_2 = getString(R.string.str_2);

      String str_3 = "这是调用的 Resource";
      mTextView02.setText(str_3 + str_2);
   }
}
```

2．布局文件

在页面布局上，在 main.xml 里创建了两个 TextView，并采 LinearLayout 的方式配置，一上一下，在运行结果中 id 为 myTextView01 的 TextView 并没有任何文字的更改，维持一开始的 str_1（参考字符串常数里的文字），但在程序运行后，id 为 myTextView02 的 TextView 进行了文字的实时更改。具体代码如下。

```
<?xml version="1.0" encoding="utf-8"?>
<LinearLayout xmlns:android="http://schemas.
android.com/apk/res/android"
    android:orientation="vertical"
    android:background="@drawable/white"
    android:layout_width="fill_parent"
    android:layout_height="fill_parent"
    >
    <TextView
    android:id="@+id/myTextView01"
    android:layout_width="wrap_content"
    android:layout_height="wrap_content"
    android:text="@string/str_1"
    android:layout_x="30px"
    android:layout_y="50px"
    >
    </TextView>
    <TextView
    android:id="@+id/myTextView02"
    android:layout_width="wrap_content"
    android:layout_height="wrap_content"
```

```
    android:text="@string/str_2"
    android:layout_x="30px"
    android:layout_y="70px"
    >
    </TextView>
</LinearLayout>
```

经过上述操作设置，此实例的主要文件编程完毕。调试运行后的效果如图 3-3 所示。

注意：虽然在 values/strings.xml 里定义了默认的字符串常数，需留意若遇遇如“?”、“'”、“\”等符号时，必须使用转义字符“\”，例如，

```
\?
\'
\\
```

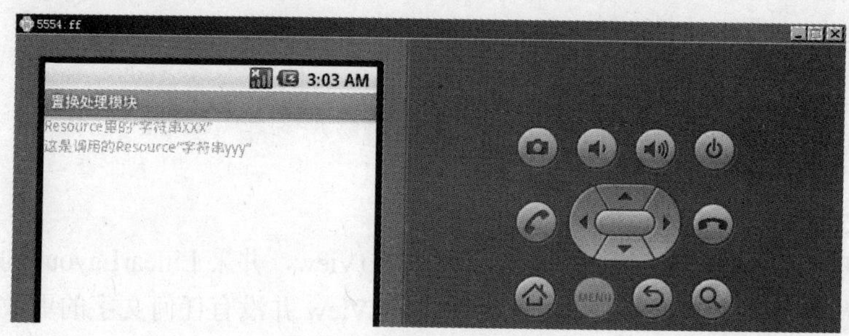

图 3-3　运行效果

3.5　获取手机分辨率的大小

不同的手机有不同的分辨率，分辨率越高屏幕越清晰。在当前手机市场中，屏幕的分辨率大小已经成为划分手机档次的一个标准。本节将通过一个简单实例的实现，介绍获取手机分辨率大小的方法。

3.5.1　设计理念

在开发手机应用程序时，除了对底层 API 的掌握度之外，最重要的仍是对屏幕的分辨率的把握，因各家手机厂商所采用的屏幕尺寸不同，用户接口呈现及布局自然也各异。

尽管 Android 可设置为随着窗口大小调整缩放比例，但即便如此，手机程序设计人员还是必须知道手机屏幕的边界，以避免缩放造成的布局变形问题。这个范例非常简短，只需几行程序即可取得手机的分辨率，其中的关键则是 DisplayMetrics 类的应用。

3.5.2　具体实现

本节实例保存在“光盘:\3\fenbian\”文件夹内，下面简单介绍主要文件的含义。

主文件 src/fenbian/fenbian.java 是此项目的主要文件，调用各个公用文件来实现具体的功能。具体代码如下。

```
package irdc.fenbian;

import irdc.fenbian.R;
import android.app.Activity;
import android.os.Bundle;
import android.util.DisplayMetrics;
import android.widget.TextView;
public class fenbian extends Activity
{
  private TextView mTextView01;

  /** Called when the activity is first created. */
  @Override
  public void onCreate(Bundle savedInstanceState)
  {
    super.onCreate(savedInstanceState);
    setContentView(R.layout.main);

    /* 必须引用 android.util.DisplayMetrics */
    DisplayMetrics dm = new DisplayMetrics();
    getWindowManager().getDefaultDisplay().getMetrics(dm);

    String strOpt = "当前这个手机屏幕分辨率是：" +
            dm.widthPixels + " × " + dm.heightPixels;

    mTextView01 = (TextView) findViewById(R.id.myTextView01);
    mTextView01.setText(strOpt);
  }
}
```

在 android.util 下有一个名为 DisplayMetrics 的对象，此对象记录了一些常用的信息，包含了显示信息、大小、维度、字体等；在使用时，请读者记得引用 android.util.DisplayMetrics。其中 DisplayMetrics 对象里的 widthPixels 和 heightPixels 字段是整数类型，在以下的程序当中，并没有对其作字符串类型的转换，因为字符串连接运算符的缘故，所以输出 strOpt 为字符串。

经过上述操作设置，此实例的主要文件编程完毕。调试运行后的效果如图 3-4 所示。

图 3-4　运行效果

上述程序一开始所创建的 DisplayMetrics 对象（程序中的 dm），不需要传递任何参数（构造时），但在调用 getWindowManager()方法之后，会取得现有的 Activity 组件中的 Handler，此时，调用 getDefaultDisplay 方法将取得的宽、高维度存放于 DisplayMetrics 对象 dm 中，而取得的宽、高维度以像素为单位（Pixel），“像素”所指的是“绝对像素”而非“相对像素”。

3.6 样式化处理对象

前几个实例讲解了某个单独对象的修饰、处理方法。如果一个个地指定文字的大小、颜色很浪费时间，实际上可以使用类似 CSS 样式的方法来指定颜色、大小。本节将通过一个简单实例的实现过程，介绍样式化处理对象的方法。

3.6.1 设计理念

在 Android 程序开发过程中，也可以通过设置样式（Style）的方式，初始化 TextView 的文本颜色、大小；当然这个范例只是抛砖引玉，在 Layout（布局）当中的任何对象都可以用样式化的方式来更改其外观。

在以下的范例中，将创建两个 TextView 对象作为对比，使其呈现两种不同的样式差异，而 Style 的写法与先前介绍到的颜色常数（color.xml）相同，同样是定义在 res/values 下，但其 XML 定义的方式不同。来看下面的这个实例。

3.6.2 具体实现

本节实例保存在“光盘:\3\yangshi\”文件夹内，下面简单介绍主要文件的具体含义。

1．主文件

主文件 src/yangshi/yangshi.java 是此项目的主要文件，调用各个公用文件来实现具体的功能。具体实现代码如下。

```
package irdc.yangshi;

import irdc.yangshi.R;
import android.app.Activity;
import android.os.Bundle;
public class yangshi extends Activity
{
  /** Called when the activity is first created. */
  @Override
  public void onCreate(Bundle savedInstanceState)
  {
    super.onCreate(savedInstanceState);
    setContentView(R.layout.main);
  }
}
```

由上述代码可知主程序十分简单，只是加载 R.layout.main 定义布局内容而已，但由于定

义在 main.xml 里的语句不同，自然也有不同的样貌。

2．布局文件

本实例的布局文件由 main.xml 和 style.xml 共同实现。其中文件 main.xml 是纯布局文件，设置了初始化 TextView 时，指定 Style 属性，使其应用 style.xml 里事先定义好的样式。

文件 style.xml 是纯修饰文件，犹如 CSS 样式文件一样，设置了各个元素的显示样式，并且是统一指定的。

style.xml 文件是这个实例的关键所在，定义了两个样式名称，分别为 DavidStyleText1 与 DavidStyleText2；<style>TAG 里以<item>描述的属性方式，与先前介绍 Drawable name 的描述类似。

经过上述操作设置，此实例的主要文件编程完毕。调试运行后的效果如图 3-5 所示。

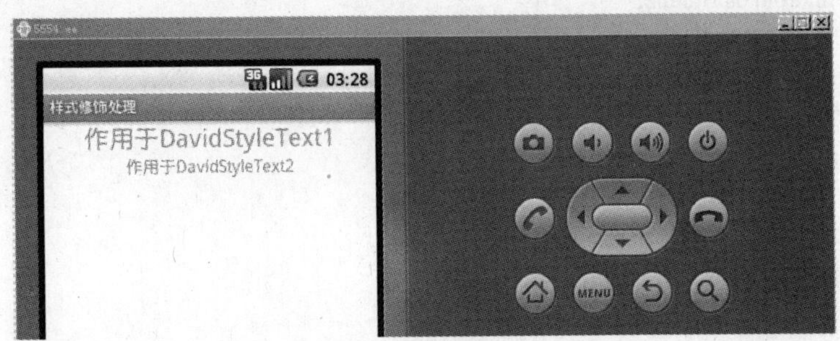

图 3-5　运行效果

注意： style 与 color 的 XML 语法相类似，都需要先声明 XML 的版本以及 encoding 为 utf-8，但其内的 resources 则需要以 stylename 作为样式名称，在最内层才是以 item 定义样式的范围，其具体的语法如下。

```
<style name=string [parent=string] >
    <item name=string>Hex value | string value | reference</item>+
</style>
```

3.7　响应按钮事件

前几个实例都是对显示样式的处理，从本节开始讲解和动态有关的知识。按钮在许多 Windows 窗口应用程序中，是最常见到的组件（Controls），此组件也常在网页设计里出现，诸如网页注册窗体、应用程序里等。本节将通过一个简单实例的实现，介绍 Android 中响应按钮事件的方法。

3.7.1　设计理念

按钮所触发的事件处理被称为 Event Handler（事件句柄），只不过在 Android 当中，按钮事件由系统的 Button.OnClickListener 所控制，熟悉 Java 程序设计的读者对 OnXxxListener 应该不陌生。下面的实例将展示如何在 Activity 组件中布局一个 Button（按钮），并设计这个按钮的事件处理函数，当单击按钮时，更改 TextView 里的文字。

3.7.2 具体实现

本节实例保存在“光盘:\3\anniu\”文件夹内，下面简单介绍主要文件的含义。

主文件 src/anniu/anniu.java 是此项目的主要文件，调用各个公用文件来实现具体的功能。在此文件开始，必须先在 Layout 当中加入一个 Button 对象及一个 TextView 对象，找不到这两个组件的话，系统会无法运行下去，在开发阶段会造成编译错误。具体实现代码如下。

```
package irdc.anniu;

import irdc.anniu.R;
import android.app.Activity;
import android.os.Bundle;
import android.view.View;
import android.widget.Button;
import android.widget.TextView;
public class anniu extends Activity
{
  private Button mButton1;
  private TextView mTextView1;

  /** Called when the activity is first created. */
  @Override
  public void onCreate(Bundle savedInstanceState)
  {
    super.onCreate(savedInstanceState);
    setContentView(R.layout.main);

    mButton1 = (Button) findViewById(R.id.myButton1);
    mTextView1 = (TextView) findViewById(R.id.myTextView1);
    mButton1.setOnClickListener(new Button.OnClickListener()
    {
      @Override
      public void onClick(View v)
      {
        mTextView1.setText("Hi, Everyone!!");
      }
    });
  }
}
```

在上述代码中，需要注意 onCreate 里创建的 Button.OnClickListener 事件，这也是触发按钮时会运行的程序段落；但由于 Eclipse 无法自动加载默认的传递参数（new Button.OnClickListener()），所以，在编写程序描述时，必须自行输入新创建的按钮所需的 OnClickListener() 事件。

经过上述操作设置，此实例的主要文件编程完毕。调试运行后的效果如图 3-6 所示。当单击屏幕中的按钮后，屏幕上的文本将会发生变化，由原来的“anniu”变为“Hi,Everyone”，

如图 3-7 所示。

图 3-6　运行效果

读者应该注意到，上述代码中只有一个按钮，但在 Activity 里，其实可以布局多个按钮，只需要在 Layout 里多配置一个按钮对象即可。

图 3-7　运行效果

3.8　页面的转换处理

在上网冲浪时，人们经常去不同的页面，来浏览浩瀚的网络资源。用手机上网时，也经常需要访问不同的页面。本节将通过一个简单实例的实现，介绍页面转换处理的方法。

3.8.1　设计理念

在 Android 应用中，是通过 setContentView 来实现页面的转换处理的。在网页的世界里，想要在两个网页间做转换，只要利用超链接（HyperLink）就可以实现，但在手机的世界里，要如何实现手机页面之间的转换呢？最简单的方式就是改变 Activity 的 Layout！在这个范例中，将布局两个 Layout，分别为 Layout1（main.xml）与 Layout2（mylayout.xml），默认载入的 Layout 为 main.xml，且在 Layout1 中创建一个按钮，当单击按钮时，显示第二个 Layout（mylayout.xml）；同样地，在 Layout2 里也设计一个按钮，当单击第二个 Layout 的按钮之后，则显示回原来的 Layout1，现在就来示范如何在两个页面之间互相切换。

3.8.2　具体实现

本节实例保存在“光盘:\3\zhuanhuan\”文件夹内，下面简单介绍主要文件的含义。

1．主文件

主文件 src/zhuanhuan/zhuanhuan.java 是此项目的主要文件，调用各个公用文件来实现具体的功能。具体实现代码如下。

```
package irdc.zhuanhuan;

/*加载相关类*/
import irdc.zhuanhuan.R;
import android.app.Activity;
import android.os.Bundle;
import android.view.View;
import android.widget.Button;
public class zhuanhuan extends Activity
{
  /** Called when the activity is first created. */
  @Override
  public void onCreate(Bundle savedInstanceState)
  {
    super.onCreate(savedInstanceState);
    /* 载入布局文件 main.xml */
    setContentView(R.layout.main);

    /* 通过方法 findViewById()取得 Button 对象，并添加 onClickListener（单击事件） */
    Button b1 = (Button) findViewById(R.id.button1);
    b1.setOnClickListener(new Button.OnClickListener()
    {
      public void onClick(View v)
      {
        jumpToLayout2();
      }
    });
  }

  /* method jumpToLayout2：将 layout 由 main.xml 切换成 mylayout.xml */
  public void jumpToLayout2()
  {
    /* 将布局改成 mylayout.xml */
    setContentView(R.layout.mylayout);

    /* 以 findViewById()取得 Button 对象，并添加 onClickListener */
    Button b2 = (Button) findViewById(R.id.button2);
    b2.setOnClickListener(new Button.OnClickListener()
    {
      public void onClick(View v)
      {
        jumpToLayout1();
      }
    });
```

```
    }

    /* method jumpToLayout1：将 layout 由 mylayout.xml 切换成 main.xml */
    public void jumpToLayout1()
    {
      /* 将布局改成 main.xml */
      setContentView(R.layout.main);

      /* 以 findViewById()取得 Button 对象，并添加 onClickListener */
      Button b1 = (Button) findViewById(R.id.button1);
      b1.setOnClickListener(new Button.OnClickListener()
      {
        public void onClick(View v)
        {
          jumpToLayout2();
        }
      });
    }
  }
```

在上述代码中，预加载的布局文件是 main.xml，屏幕上显示的是黑色背景的“This is Layout 1!!”，在第一个界面上的按钮被单击的同时，改变 Activity 的布局为 mylayout.xml，屏幕上显示变为白色背景的“This is Layout 2!!”，并利用按钮单击时，调用方法的不同，做两个不同布局间的切换。

2．修饰文件

本实例的布局文件由 main.xml 和 mylayout.xml 共同实现。其中，文件 main.xml 是为了突出显示不同布局界面间切换的效果，改变两个布局的背景色及输出文字。在 main.xml 中定义其背景色为黑色，输出文字为“This is Layout 1!!”。

而在 mylayout.xml 中，定义了其背景色为白色，输出文字为“This is Layout 2!!”。

经过上述操作设置，此实例的主要文件编程完毕。调试运行后的效果如图 3-8 所示。当单击屏幕中的按钮后，屏幕上的文本将会发生变化，由原来的“anniu”变为“zhunahuan”，如图 3-9 所示。

图 3-8　运行效果

在现实应用中，基本都是采用改变 Activity Layout（界面布局）来实现手机页面转换的效果，当然也可以搭配之前介绍过的 Style 与 Theme，进行更加灵活的布局配置运用。例如，让用户自行决定要使用的系统样式、背景及文字颜色等，接着直接应用来改变布局。

图 3-9　转换后效果

3.9　调用另一个 Activity

在上一个实例中介绍了如何运用切换 Layout 布局界面的方式，进行手机页面间的转换。如果要转换的页面并不单只是背景、颜色或文字等内容，而是不同 Activity 之间的置换，那就不是仅仅改变 Layout 就能完成的，尤其是需要传递的变量不像网页可以通过 Cookie 或 Session 来实现，在程序里要移交主控权到另外一个 Activity，光靠先前的 Layout 技巧是办不到的。本节将通过一个简单实例的实现过程，介绍调用另一个 Activity 的方法。

3.9.1　设计理念

在 Android 的程序设计中，可在主程序里使用 startActivity() 这个方法来调用另一个 Activity（主程序本身即是一个 Activity），但其中的关键并不在于 startActivity 这个方法，而是 Intent 这个特有的对象，Intent 就如同其英文字义，是“想要”或“意图”之意，在主 Activity 当中，告诉程序自己是什么，并想要前往哪里，这就是 Intent 对象所处理的事了。本范例并没有特别的 Layout 布局，而是直接在主 Activity（Activity1）当中部署一个按钮，当单击按钮的同时，告诉主 Activity 前往 Activity2，并在 Activity2 里创建一个回到 Activity1 的按钮，本范例将利用此简易的程序描述，示范如何在一个 Activity 中调用另一个 Activity 的方法。

3.9.2　具体实现

本节实例保存在“光盘:\3\zhihuan1\”文件夹内，下面简单介绍主要文件的含义。

1. 主文件

主文件 src/zhihuan1/zihuan.java 和 src/zhihuan1/zihuan_1.java 是此项目的主要文件，调用各个公用文件来实现具体的功能。文件 zihuan.java 的具体实现代码如下。

```
package irdc.zhihuan;

/* 引入相关类 */
import irdc.zhihuan.R;
import android.app.Activity;
import android.os.Bundle;
```

```
import android.view.View;
import android.widget.Button;
import android.content.Intent;
public class zhihuan extends Activity
{
  /** Called when the activity is first created. */
  @Override
  public void onCreate(Bundle savedInstanceState)
  {
    super.onCreate(savedInstanceState);
    /* 载入 mylayout.xml Layout */
    setContentView(R.layout.main);

    /* 以 findViewById()取得 Button 对象，并添加 onClickListener */
    Button b1 = (Button) findViewById(R.id.button1);
    b1.setOnClickListener(new Button.OnClickListener()
    {
      public void onClick(View v)
      {
        /* 新建一个 Intent 对象，并指定要启动的类 */
        Intent intent = new Intent();
          intent.setClass(zhihuan.this, zhihuan_1.class);

          /* 调用一个新的 Activity */
          startActivity(intent);
          /* 关闭原本的 Activity */
          zhihuan.this.finish();
      }
    });
  }
}
```

在上述代码中加载的 Layout 是 main.xml，屏幕上显示的是黑色背景的“Activity 1!!”，在按钮被单击时调用另一个 Activity（zhihuan_1），并将主 Activity 关闭，接着将主控权交给下一个 Activity，即 Activity2。

文件 zihuan_1.java 的具体实现代码如下。

```
package irdc.zhihuan;

/* 引入相关类 */
import irdc.zhihuan.R;
import android.app.Activity;
import android.os.Bundle;
import android.content.Intent;
import android.view.View;
import android.widget.Button;
```

```
public class zhihuan_1 extends Activity
{
  /** Called when the activity is first created. */
  @Override
  public void onCreate(Bundle savedInstanceState)
  {
    super.onCreate(savedInstanceState);
    /* 载入布局文件 main.xml */
    setContentView(R.layout.mylayout);

    /* 以 findViewById()取得 Button 对象，并添加 onClickListener */
    Button b2 = (Button) findViewById(R.id.button2);
    b2.setOnClickListener(new Button.OnClickListener()
    {
      public void onClick(View v)
      {
        /* 新建一个 Intent 对象，并指定要启动的类 */
        Intent intent = new Intent();
       intent.setClass(zhihuan_1.this, zhihuan.class);
       /* 调用一个新的 Activity */
        startActivity(intent);
        /* 关闭原本的 Activity */
        zhihuan_1.this.finish();
      }
    });

  }
}
```

上述程序程序是第二个 Activity 的主程序，其加载的布局文件为 mylayout.xml，屏幕上所显示的是白色背景的“This is Activity 2!!”，当主 Activity（Activity1）调用这个 Activity（Activity2）后，同样为 Button 添加 onClickListener()，使 Button 被点击时，重新调用 Activity1（zhihuan），并将 Activity2（zhihuan_1）关闭（finish()）。

2．修饰文件

本实例的布局文件由 main.xml 和 mylayout.xml 共同实现。其中文件 main.xml 是为了突出显示 Layout 间切换的效果，特意将两个 Layout 的背景及输出文字有所区别。在 main.xml 中定义其背景色为黑色，输出文字为“Activity 1!!”。

在文件 mylayout.xml 中定义其背景色为白色，输出文字为“Activity 2!!”。

3．配置文件

因为在该实例中添加了一个 Activity，所以必须在 AndroidManifest.xml 中定义一个新的 activity，并给予名称 name，否则程序将无法编译运行。具体实现代码如下所示。

```
<?xml version="1.0" encoding="utf-8"?>
  <manifest xmlns:android="http://schemas.android.com/apk/res/android"
        package="irdc.zhihuan"
```

```
        android:versionCode="1"
        android:versionName="1.0.0">
    <application android:icon="@drawable/icon" android:label="@string/app_name">
        <activity android:name=".zhihuan"
                  android:label="@string/app_name">
            <intent-filter>
                <action android:name="android.intent.action.MAIN" />
                <category android:name="android.intent.category.LAUNCHER" />
            </intent-filter>
        </activity>
        <activity android:name="zhihuan_1"></activity>
    </application>
</manifest>
```

经过上述操作设置，此实例的主要文件编程完毕。调试运行后的效果如图 3-10 所示。当单击屏幕中的按钮后，屏幕上的文本将会发生变化。如图 3-11 所示。

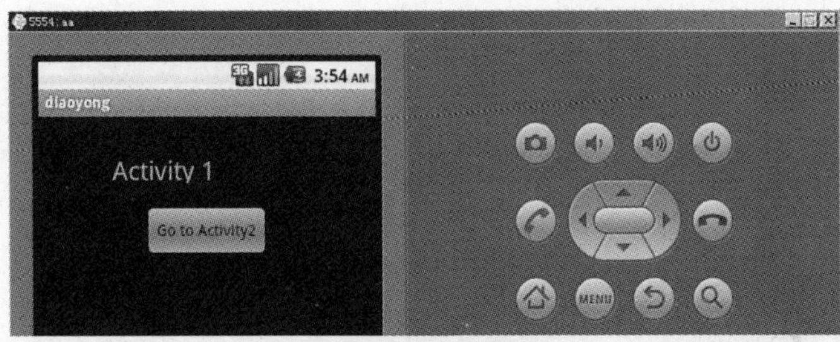

图 3-10　运行效果

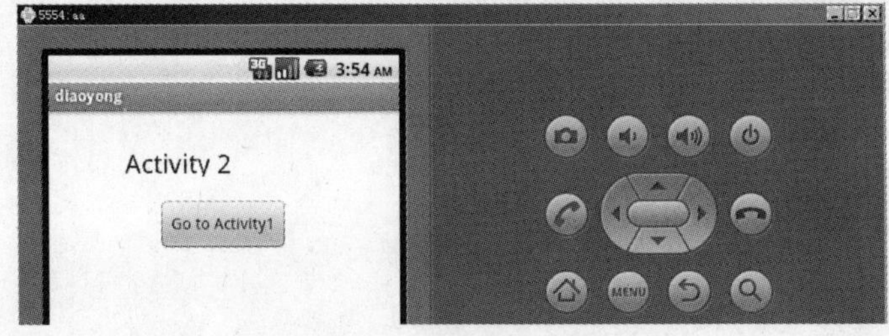

图 3-11　转换后效果

系统中新添加 Activity 时，必须在 AndroidManifest.xml 里定义一个新的 activity。

```
<activity android:name="zhihuan_1"></activity>
```

否则，系统将会因为找不到 Activity 而发生编译错误。

3.10　不同 Activity 之间的数据传递

在上节的实例中，介绍了如何在 Activity 中调用另一个 Activity，但是在现实应用中，经常需要在不同的 Activity 之间传递数据。本节将通过一个简单实例，介绍在不同 Activity 之间传递数据的方法。

3.10.1 设计理念

如果需要在调用另外一个 Activity 的同时传递数据，需要利用 android.os.Bundle 对象封装数据的能力，将欲传递的数据或参数，通过 Bundle 来传递不同 Intent 之间的数据。本实例将设计一个简易表单类型，在 Activity1 中收集 User 输入的数据，在离开 Activity1 的同时，将 User 选择的结果传递至下一个 Activity2，以一个简单“身材计算器”来演示如何传递数据到下一个 Activity 里。

3.10.2 具体实现

本节实例保存在“光盘:\3\butong\”文件夹内，下面简单介绍主要文件的含义。

1. 主文件

主文件 src/butong/butong.java 和 src/butong/butong_1.java 是此项目的主要文件，调用各个公用文件来实现具体的功能。文件 butong.java 的具体实现代码如下。

```
package irdc.butong;

/*  加载相关类 */
import irdc.butong.R;
import android.app.Activity;
import android.content.Intent;
import android.os.Bundle;
import android.view.View;
import android.widget.Button;
import android.widget.EditText;
import android.widget.RadioButton;
public class butong extends Activity
{
  /** Called when the activity is first created. */
  @Override
  public void onCreate(Bundle savedInstanceState)
  {
    super.onCreate(savedInstanceState);
    /*  载入布局文件 main.xml */
    setContentView(R.layout.main);

    /*  以 findViewById()取得 Button 对象，并添加 onClickListener */
    Button b1 = (Button) findViewById(R.id.button1);
    b1.setOnClickListener(new Button.OnClickListener()
    {
      public void onClick(View v)
      {
        /*取得输入的身高*/
        EditText et = (EditText) findViewById(R.id.height);
```

```
                double height=Double.parseDouble(et.getText().toString());
                /*取得选择的性别*/
                String sex="";
                RadioButton rb1 = (RadioButton) findViewById(R.id.sex1);
                if(rb1.isChecked())
                {
                  sex="M";
                }
                else
                {
                  sex="F";
                }
                /*新建一个 Intent 对象，并指定类*/
                Intent intent = new Intent();
                intent.setClass(butong.this,butong_1.class);

                /*新建一个 Bundle 对象，并将要传递的数据传入*/
                Bundle bundle = new Bundle();
                bundle.putDouble("height",height);
                bundle.putString("sex",sex);

                /*将 Bundle 对象分派给 Intent*/
                intent.putExtras(bundle);

                /*调用 Activity butong_1*/
                startActivity(intent);
              }
            });
          }
        }
```

在上述代码中，定义了“性别”选项的单选按钮组以及输入身高的“EditText”控件，并运用 Intent 及 Bundle 对象，再调用 Activity2（butong_1）时，同时将数据传入。

文件 src/butong/butong_1.java 的具体代码如下。

```
package irdc.butong;
/* 加载相关类 */
import irdc.butong.R;
import java.text.DecimalFormat;
import java.text.NumberFormat;
import android.app.Activity;
import android.os.Bundle;
import android.widget.TextView;

public class butong_1 extends Activity
{
  /** Called when the activity is first created. */
```

```
@Override
public void onCreate(Bundle savedInstanceState)
{
  super.onCreate(savedInstanceState);
  /* 加载布局文件 main.xml */
  setContentView(R.layout.myalyout);

  /* 取得 Intent 中的 Bundle 对象 */
  Bundle bunde = this.getIntent().getExtras();

  /* 取得 Bundle 对象中的数据 */
  String sex = bunde.getString("sex");
  double height = bunde.getDouble("height");

  /* 判断性别 */
  String sexText="";
  if(sex.equals("M"))
  {
    sexText="男性";
  }
  else
  {
    sexText="女性";
  }

  /* 取得标准体重 */
  String weight=this.getWeight(sex, height);

  /* 设置输出文字 */
  TextView tv1= (TextView) findViewById(R.id.text1);
  tv1.setText("你是一位"+sexText+"\n 你的身高是"
              +height+"厘米\n 你的标准体重是"+weight+"公斤");
}

/* 四舍五入的方法 */
private String format(double num)

{
  NumberFormat formatter = new DecimalFormat("0.00");
  String s=formatter.format(num);
  return s;
}
```

在 Activity2（butong_1）中需要接收来自 Activity1（butong）传递来的数据。在 Activity1 是以 Bundle 封装对象，所以在 Activity2 也是以 Bundle 的方式解开封装的数据；程序中以 getIntent().getExtras() 方法取得随着 Bundle 对象传递过来的性别与身高，经过计算之后，显示

在屏幕上。

2．配置文件

由于本范例中有两个 Activity，所以文件中必须有两个 Activity 的声明，否则系统将无法运行。配置文件 AndroidManifest.xml 的具体实现代码如下所示。

```
<?xml version="1.0" encoding="utf-8"?>
<manifest xmlns:android="http://schemas.android.com/apk/res/android"
    package="irdc.butong"
    android:versionCode="1"
    android:versionName="1.0.0">
  <application android:icon="@drawable/icon" android:label="@string/app_name">
    <activity android:name=".butong"
              android:label="@string/app_name">
      <intent-filter>
        <action android:name="android.intent.action.MAIN" />
        <category android:name="android.intent.category.LAUNCHER" />
      </intent-filter>
    </activity>
    <activity android:name="butong_1"></activity>
  </application>
</manifest>
```

经过上述操作设置，此实例的主要文件编程完毕。调试运行后的效果如图 3-12 所示。当输入身高、选择性别并单击“计算”按钮后，将显示标准体重。如图 3-13 所示。

图 3-12 运行效果

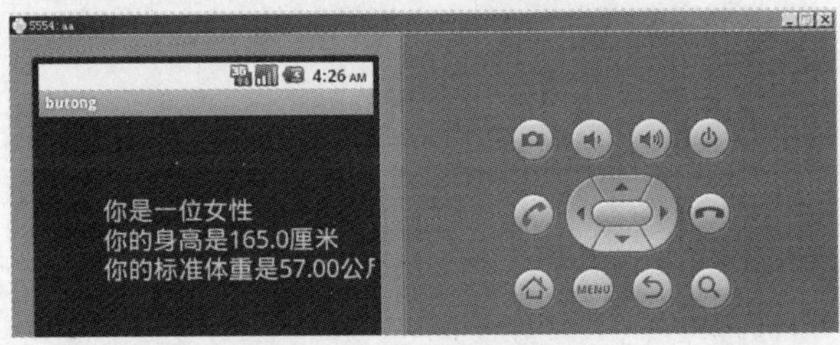

图 3-13 转换后效果

3.11 返回数据到前一个 Activity

在现实应用中，经常需要将数据返回到前一个 Activity。本节将通过一个简单实例，介绍将数据返回到前一个 Activity 的方法。

3.11.1 设计理念

上个实例描述，将数据从 Activity1 传递至 Activity2，但是如果要再回到 Activity1，数据是否再封装一次呢？而且前一个 Activity1 早就被程序销毁了，倘若在 Activity1 最后以 finish()方法结束程序，再通过 Activity2 将数据采用 Bundle 的方式通过新打开 Activity1 传递参数，这样的做法虽然也可以恢复 User 输入的数据，但是并不符合我们的期待。

如果要在第二个页面加上一个“回上页”的按钮，而并不是通过模拟器中的回复键来实现，且回上页后又能保留之前输入的相关信息，那么就必须使用 startActivityForResult()来唤起一个 Activity。利用这个方法，前一个 Activity1 便会有一个等待第二个 Activity2 的返回，而返回的数据就可以实现想要的效果。

3.11.2 具体实现

本节实例保存在“光盘:\3\qian\”文件夹内，下面简单介绍主要文件的含义。

1．主文件

主文件 src/qian/qian.java 和 src/qian/qian_1.java 是此项目的主要文件，调用各个公用文件来实现具体的功能。文件 qian.java 的具体实现代码如下。

```
package irdc.qian;

/* 加载相关类 */
import irdc.qian.R;
import android.app.Activity;
import android.content.Intent;
import android.os.Bundle;
import android.view.View;
import android.widget.Button;
import android.widget.EditText;
import android.widget.RadioButton;

public class qian extends Activity
{
  private EditText et;
  private RadioButton rb1;
  private RadioButton rb2;

  /** Called when the activity is first created. */
  @Override
  public void onCreate(Bundle savedInstanceState)
```

```
{
  super.onCreate(savedInstanceState);
  /*  载入布局文件 main.xml */
  setContentView(R.layout.main);

  /*  以 findViewById()取得 Button 对象，并添加 onClickListener */
  Button b1 = (Button) findViewById(R.id.button1);
  b1.setOnClickListener(new Button.OnClickListener()
  {
    public void onClick(View v)
    {
      /*取得输入的身高*/
      et = (EditText) findViewById(R.id.height);
      double height=Double.parseDouble(et.getText().toString());
      /*取得选择的性别*/
      String sex="";
      rb1 = (RadioButton) findViewById(R.id.sex1);
      rb2 = (RadioButton) findViewById(R.id.sex2);
      if(rb1.isChecked())
      {
        sex="M";
      }
      else
      {
        sex="F";
      }

      /*新建一个 Intent 对象，并指定类*/
      Intent intent = new Intent();
      intent.setClass(qian.this,qian_1.class);

      /*新建一个 Bundle 对象，并将要传递的数据传入*/
      Bundle bundle = new Bundle();
      bundle.putDouble("height",height);
      bundle.putString("sex",sex);

      /*将 Bundle 对象分派给 Intent 对象*/
      intent.putExtras(bundle);

      /*调用 Activity qian_1*/
      startActivityForResult(intent,0);
    }
  });
}

/*  覆盖  onActivityResult()*/
```

```
    @Override
    protected void onActivityResult(int requestCode, int resultCode,
                                    Intent data)
    {
      switch (resultCode)
      {
        case RESULT_OK:
          /* 取得来自 Activity2 的数据，并显示于画面上 */
          Bundle bunde = data.getExtras();
          String sex = bunde.getString("sex");
          double height = bunde.getDouble("height");

          et.setText(""+height);
          if(sex.equals("M"))
          {
            rb1.setChecked(true);
          }
          else
          {
            rb2.setChecked(true);
          }
          break;
        default:
          break;
      }
    }
}
```

对于上述代码，在 Activity1 主程序中，将调用 Activity 的方法更改成 startActivityForResult (intent,0)，其中 0 为下一个 Activity 要返回值的依据，可指定为自行定义的参考标识符（Identifier）。程序覆盖了 onActivityResult() 这个方法，使程序在收到 result 后，再重新加载写回原本输入的值。

对于文件 qian_1.java，在 Activity2 的主程序中，设计当按钮被单击时，将 Bundle 对象与结果返回给前一个 Activity1。具体代码如下。

```
package irdc.qian;

/* 加载相关类 */
import irdc.qian.R;

import java.text.DecimalFormat;
import java.text.NumberFormat;
import android.app.Activity;
import android.content.Intent;
import android.os.Bundle;
import android.view.View;
```

```
import android.widget.Button;
import android.widget.TextView;

public class qian_1 extends Activity
{
  Bundle bunde;
  Intent intent;
  /** Called when the activity is first created. */
  @Override
  public void onCreate(Bundle savedInstanceState)
  {
    super.onCreate(savedInstanceState);
    /* 载入 mylayout.xml Layout */
    setContentView(R.layout.myalyout);

    /* 取得 Intent 中的 Bundle 对象 */
    intent=this.getIntent();
    bunde = intent.getExtras();

    /* 取得 Bundle 对象中的数据 */
    String sex = bunde.getString("sex");
    double height = bunde.getDouble("height");

    /* 判断性别 */
    String sexText="";
    if(sex.equals("M"))
    {
      sexText="男性";
    }
    else
    {
      sexText="女性";
    }

    /* 取得标准体重 */
    String weight=this.getWeight(sex, height);

    /* 设置输出文字 */
    TextView tv1=(TextView) findViewById(R.id.text1);
    tv1.setText("你是一位"+sexText+"\n 你的身高是"+height+
                  "厘米\n 你的标准体重是"+weight+"公斤");

    /* 以 findViewById()取得 Button 对象，并添加 onClickListener 单击监听事件 */
    Button b1 = (Button) findViewById(R.id.button1);
    b1.setOnClickListener(new Button.OnClickListener()
    {
```

```
            public void onClick(View v)
            {
              /* 返回结果回上一个 activity */
              qian_1.this.setResult(RESULT_OK, intent);

              /* 结束这个 activity */
              qian_1.this.finish();
            }
          });
        }

        /* 四舍五入的 method */
        private String format(double num)
        {
          NumberFormat formatter = new DecimalFormat("0.00");
          String s=formatter.format(num);
          return s;
        }

        /* 以 findViewById()取得 Button 对象，并添加 onClickListener */
        private String getWeight(String sex,double height)
        {
          String weight="";
          if(sex.equals("M"))
          {
            weight=format((height-80)*0.7);
          }
          else
          {
            weight=format((height-70)*0.6);
          }
          return weight;
        }
      }
```

2．配置文件

由于本范例中有两个 Activity，所以文件中必须有两个 Activity 的声明，否则系统将无法运行。配置文件 AndroidManifest.xml 的具体实现代码如下。

```
<?xml version="1.0" encoding="utf-8"?>
  <manifest xmlns:android="http://schemas.android.com/apk/res/android"
      package="irdc.qian"
      android:versionCode="1"
      android:versionName="1.0.0">
    <application android:icon="@drawable/icon" android:label="@string/app_name">
      <activity android:name=".qian"
```

```
                android:label="@string/app_name">
            <intent-filter>
                <action android:name="android.intent.action.MAIN" />
                <category android:name="android.intent.category.LAUNCHER" />
            </intent-filter>
        </activity>
        <activity android:name="qian_1"></activity>
    </application>
</manifest>
```

经过上述操作设置，此实例的主要文件编程完毕。调试运行后的效果如图 3-14 所示。当输入身高、选择性别并单击“计算”按钮后，将显示标准体重。如图 3-15 所示。

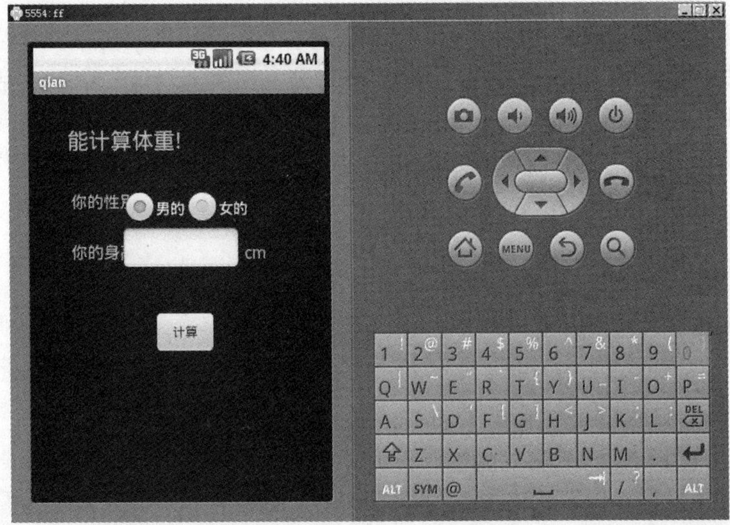

图 3-14　运行效果

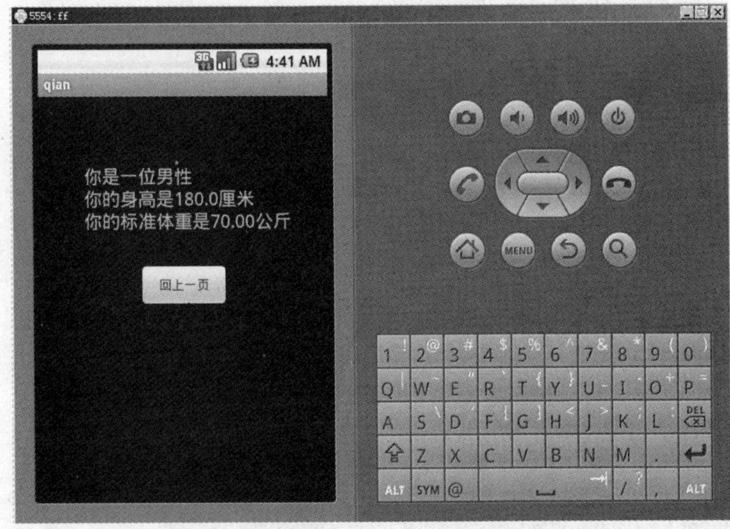

图 3-15　转换后效果

在上述实例中，为了在回到上一页时，能够显示之前所输入的数据，所以将原本传递次Activity 的 Intent（里面包含了有数据的 Bundle 对象）再重新返回给主 Activity1。如果要在次Activity2 中返回其它的数据，例如，经过计算后的结果、数据，此时只需将要返回的数据再放入 Bundle 对象中即可。

3.12 实现交互对话框

在现实应用中，经常需要在手机屏幕中实现交互对话框功能。本节将通过一个简单实例的实现，介绍在手机中实现交互对话框功能的方法。

3.12.1 设计理念

在网络中读者会经常遇到“提示”、“警告”或“确认”之类的对话框，当按下按钮后，会显示出一些提示信息。在手机开发中，同样需要开发出类似功能的交互对话框。

在 Android SDK 中，有很多的窗口，其中具有交互功能的是 AlertDialog 窗口。本实例的设计理念就是基于 AlertDialog 组件的，仅仅通过一个简单的“确认”按钮来作为演示，当单击这个按钮后，会产生 AlertDialog 对话框。

3.12.2 具体实现

本节实例保存在“光盘:\3\duihua\”文件夹内，下面简单介绍主要文件的具体含义。

主文件 src/duihua/duihua.java 是此项目的主要文件，调用各个公用文件来实现具体的功能。文件 duihua.java 的具体实现代码如下。

```
package irdc.duihua;

import irdc.duihua.R;
import android.app.Activity;
import android.app.AlertDialog;
import android.content.DialogInterface;
import android.os.Bundle;
import android.view.View;
import android.widget.Button;
public class duihua extends Activity
{
  private Button mButton1;

  /** Called when the activity is first created. */
  @Override
  public void onCreate(Bundle savedInstanceState)
  {
    super.onCreate(savedInstanceState);
    setContentView(R.layout.main);
    mButton1 = (Button) findViewById(R.id.myButton1);
    mButton1.setOnClickListener(new Button.OnClickListener()
```

```
        {

          @Override
          public void onClick(View v)
          {
            new AlertDialog.Builder(duihua.this)
            .setTitle(R.string.app_about)
            .setMessage(R.string.app_about_msg)
            .setPositiveButton(R.string.str_ok,
                new DialogInterface.OnClickListener()
                 {
                public void onClick(DialogInterface dialoginterface, int i)
                  {
                  }
                  }
                 )
             .show();
          }
        });
      }
    }
```

在上述代码中，设计了按钮事件，当按下按钮时，触发 onClick 单击事件，以 new 方式打开 AlertDialog 对象 AlertDialog.Builder 函数，生成一个具有 Title、Message 和确认按钮的对话框窗口。

调试运行后的效果如图 3-16 所示。当单击“单击这里”按钮后，将弹出一个对话框。如图 3-17 所示。

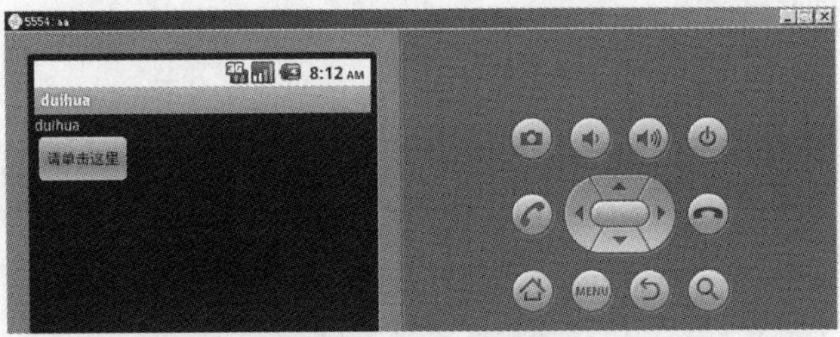

图 3-16　运行效果

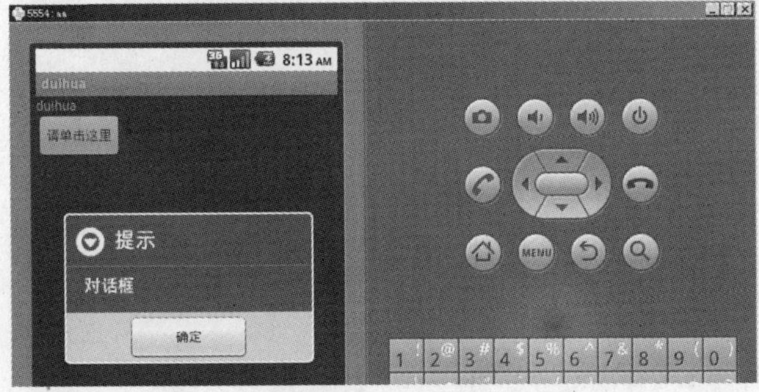

图 3-17　弹出对话框

3.13 置换文字颜色

在现实应用中，经常需要置换手机屏幕中文字的颜色。本节将通过一个简单实例，介绍置换手机屏幕中文字颜色的方法。

3.13.1 设计理念

在前面的实例中，读者了解了多种与 TextView 文字相关的处理方法，并且也了解了与按钮相关的处理方法。本实例将对前面的知识进行整合，通过按钮的 setOnClickListener 和 onClick 方法的方式，在单击按钮后触发 setTextColor 方法来改变文本颜色，并且创建一个字形定义的颜色数组 mColor，当单击按钮时，会根据这个颜色数组的索引值的变化来置换 TextView 的文字颜色。

3.13.2 具体实现

本节实例保存在“光盘:\3\wenyan\”文件夹内，下面简单介绍主要文件的具体含义。

主文件 src/wenyan/wenyan.java 是此项目的主要文件，调用各个公用文件来实现具体的功能。文件 wenyan.java 的具体实现代码如下。

```
package irdc.wenyan;

import irdc.wenyan.R;
import android.app.Activity;
/*必须引用 graphics.Color 才能使用 Color.*的对象*/
import android.graphics.Color;

import android.os.Bundle;
import android.view.View;

/*必须引用 widget.Button 才能声明使用 Button 对象*/
import android.widget.Button;

/*必须引用 widget.TextView 才能声明使用 TestView 对象*/
import android.widget.TextView;
public class wenyan extends Activity
{
  private Button mButton;
  private TextView mText;
  private int[] mColors;
  private int colornum;

  /** Called when the activity is first created. */
  @Override

  public void onCreate(Bundle savedInstanceState)
  {
```

```
        super.onCreate(savedInstanceState);
        setContentView(R.layout.main);

        /*通过 findViewById 构造器来使用 main.xml 与 string.xml
        中 button 与 textView 的参数*/
        mButton=(Button) findViewById(R.id.mybutton);
        mText= (TextView) findViewById(R.id.mytext);

        /*声明并构造一整数 array 来存储欲使用的文字颜色*/
        mColors = new int[]
                        {
        Color.BLACK, Color.RED, Color.BLUE,
        Color.GREEN, Color.MAGENTA, Color.YELLOW
                        };
        colornum=0;

        /*使用 setOnClickListener 让按钮聆听事件*/
        mButton.setOnClickListener(new View.OnClickListener()
        {
          /*使用 onClick 让用户点下按钮来驱动变动文字颜色*/
          public void onClick(View v)
          {
            if (colornum < mColors.length)
            {
              mText.setTextColor(mColors[colornum]);
              colornum++;
            }
            else
              colornum=0;
          }
        });
    }
}
```

在上述代码中，使用了 setTextColor 方法来改变文字的颜色，驱动 setTextColor 的事件是通过按钮上的 onClick 方法实现的。

调试运行后的效果如图 3-18 所示；当单击“单击”按钮后，文本将会改变颜色，如图 3-19 所示；再次单击“单击”按钮后，文本还会改变颜色。如图 3-20 所示。

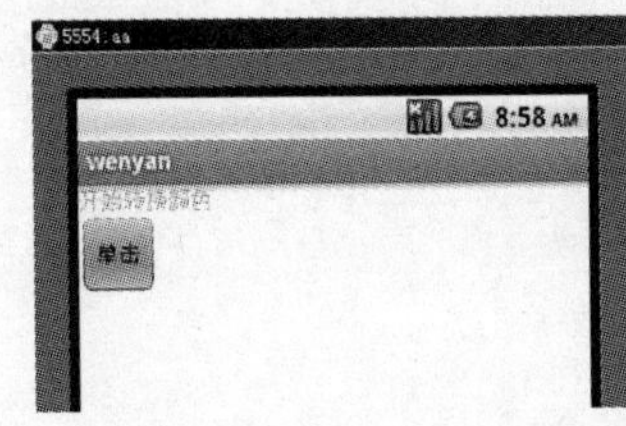

图 3-18　运行效果

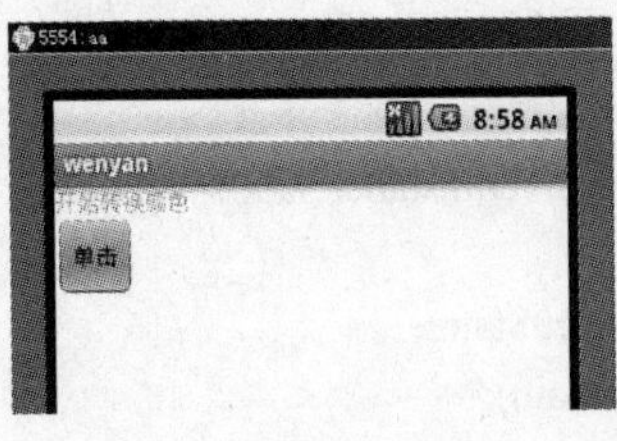

图 3-19　改变颜色

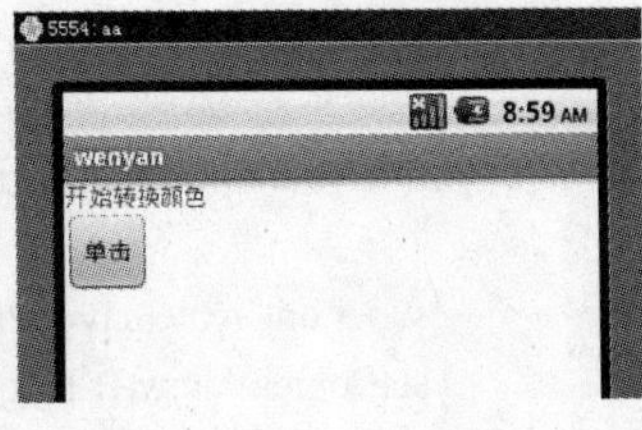

图 3-20　改变颜色

3.14 设置文字字体

在现实应用中，经常需要置换手机屏幕中文字的字体格式。本节将通过一个简单实例的实现，介绍置换手机屏幕中文字字体的方法。

3.14.1 设计理念

文字的字体格式主要包括文字大小（Size）和字体（Font）。改变字体和改变颜色的原理一样，即通过按钮对象的 Button.onClickListener 单击监听事件来改变 TextView 控件对象的字体大小和字体。

在 TextView 对象中有许多和字体有关的方法。例如，使用 setTextSize 改变文字的大小，使用 setTypeface 方法来指定字体。在本实例中将涉及两个按钮，一个控制文本大小，一个控制文本字体，并通过外部资源 assets，引用外部字体文件 True Type Font，再通过 Typeface 类的 creatFromAsset 方法，让 TextView 通过 SetTypeface 方法来改变字体。

3.14.2 具体实现

本节实例保存在“光盘:\3\ziti\”文件夹内，主文件 src/zitin/ziti.java 是此项目的主要文件，通过调用各个公用文件来实现具体的功能。文件 ziti.java 的具体实现代码如下。

```
package irdc.ziti;

import irdc.ziti.R;
import android.app.Activity;
/*必须引用 graphics.Typeface 才能使用 creatFromAsset()来改变字体*/
import android.graphics.Typeface;
import android.os.Bundle;
import android.view.View;
import android.widget.Button;
import android.widget.TextView;

public class ziti extends Activity
{
  /** Called when the activity is first created. */
  private TextView mText;
  private Button sizeButton;
  private Button fontButton;
  @Override

  public void onCreate(Bundle savedInstanceState)
  {
    super.onCreate(savedInstanceState);
    setContentView(R.layout.main);
```

```
        mText=(TextView)findViewById(R.id.mytextview);
        sizeButton=(Button) findViewById(R.id.sizebutton);
        fontButton=(Button) findViewById(R.id.fontbutton);
        /*设置 onClickListener 与按钮对象连接*/
        sizeButton.setOnClickListener(new View.OnClickListener()
        {
          public void onClick(View v)
          {
            /*使用 setTextSize()来改变字体大小 */
            mText.setTextSize(20);
          }
        }
        );
        fontButton.setOnClickListener(new View.OnClickListener()
        {
          public void onClick(View v)
          {
            /*必须事先在 assets 底下创建一 fonts 文件夹
             * 并放入要使用的字体文件(.ttf)
             * 并提供相对路径给 creatFromAsset()来创建 Typeface 对象*/
            mText.setTypeface
            (Typeface.createFromAsset(getAssets(),
            "fonts/HandmadeTypewriter.ttf"));
          }
        }
        );
      }
    }
```

在上述代码中，通过 setTextSize 方法改变文字的大小，setTypeface 方法来指定文本的字体。

调试运行后的效果如图 3-21 所示；当单击“转换字体”按钮和“变大”按钮后，文本样式将会改变。如图 3-22 所示。

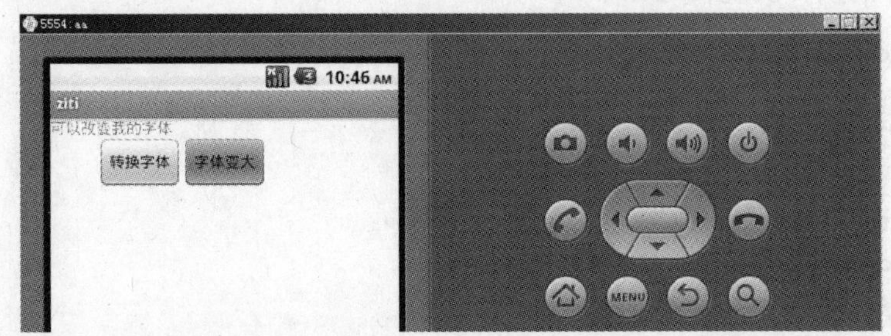

图 3-21　运行效果

在本实例中，使用了类 Typeface，并使用了外部字体文件来改变文字的字体。但是在当前的 Android 中，缺乏对字体的支持，即使使用了它不支持的字体，也不会报错，而是以 Droid

Type 来替换。所以当使用外部字体但字体没有变换时，通常是因为这个字体在 Android 中不被支持。

图 3-22　改变样式

3.15　拖动相片特效

在很多手机应用中，可以拖动一幅图片，这样的效果很吸引用户的眼球。本节将通过一个简单实例，介绍在手机屏幕中拖动图片特效的方法。

3.15.1　设计理念

在 Android 中，拖动图片特效可以通过 Android.content.Context、Android.widget.BaseAdapter 和 Android.widget.ImageView 等来实现，这些通常被应用到相册和图片类型选择器上。要理解实例，需要了解 Context 子类和 widget 里的 BaseAdapter 类。在 Activity 中，Context 犹如画布，随时会被处理覆盖。Context 是作为 Android.content 的子类。在本实例中，在 Layout 中布局一个 Gallery（切换组件），再通过 widget.BaseAdapter 作为容器来存放 Gallery 所需要的图片，为了快速掌握 Gallery 的使用方法，在实例中使用了 Android 的 Icon 图标。

3.15.2　具体实现

本节实例保存在“光盘:\3\texiao\”文件夹内，下面简单介绍主要文件的具体含义。

主文件 src/texiao/texiao.java 是此项目的主要文件，调用各个公用文件来实现具体的功能。文件 texiao.java 的具体实现代码如下。

```
package irdc.texiao;

import irdc.texiao.R;
import android.app.Activity;
import android.graphics.Color;
import android.os.Bundle;
import android.widget.TextView;

/*欲在 Layout 里使用 Gallery widget，必须引用这些模块*/
import android.content.Context;
import android.widget.Gallery;
import android.view.View;
```

```
import android.view.ViewGroup;
import android.widget.BaseAdapter;
import android.widget.ImageView;

public class texiao extends Activity
{
  private TextView mTextView01;

  /** Called when the activity is first created. */
  @Override
  public void onCreate(Bundle savedInstanceState)
  {
    super.onCreate(savedInstanceState);
    setContentView(R.layout.main);

    mTextView01 = (TextView) findViewById(R.id.myTextView01);
    mTextView01.setText(getString(R.string.str_txt1));
    mTextView01.setTextColor(Color.BLUE);

    ((Gallery) findViewById(R.id.myGallery1))
                    .setAdapter(new ImageAdapter(this));
  }

  public class ImageAdapter extends BaseAdapter
  {
    /* 类成员 myContext 为 Context 父类 */
    private Context myContext;

    /*使用 android.R.drawable 里的图片作为图库来源，类型为整数数组*/
    private int[] myImageIds =
                    {
                      android.R.drawable.btn_minus,
                      android.R.drawable.btn_radio,
                      android.R.drawable.ic_lock_idle_low_battery,
                      android.R.drawable.ic_menu_camera
                    };
    /* 构造器只有一个参数，即要存储的 Context */
    public ImageAdapter(Context c) { this.myContext = c; }

    /* 返回所有已定义的图片总数量 */
    public int getCount() { return this.myImageIds.length; }

    /* 利用 getItem 方法，取得目前容器中图像的数组 ID */
    public Object getItem(int position) { return position; }
    public long getItemId(int position) { return position; }
```

```
        /* 取得目前欲显示的图像 View，传入数组 ID 值使之读取与成像 */
        public View getView(int position, View convertView,
                            ViewGroup parent)
        {
          /* 创建一个 ImageView 对象 */
          ImageView i = new ImageView(this.myContext);

          i.setImageResource(this.myImageIds[position]);
          i.setScaleType(ImageView.ScaleType.FIT_XY);

          /* 设置这个 ImageView 对象的宽高，单位为 dip */
          i.setLayoutParams(new Gallery.LayoutParams(120, 120));
          return i;
        }

        /*依据距离中央的位移量 利用 getScale 返回 views 的大小(0.0f to 1.0f)*/
        public float getScale(boolean focused, int offset)
        {
          /* Formula: 1 / (2 ^ offset) */
          return Math.max(0,1.0f/(float)Math.pow(2,Math.abs(offset)));
        }
      }
    }
```

在上述代码中，创建了一个继承于类 BaseAdapte 的 ImageAdapte 方法。此 ImageAdapte 方法的功能是暂时保存要显示的图片，作为 Gallery 组件图片的引用。

调试运行后的效果如图 3-23 所示；当单击某个图片后，会以特效滚动的样式显示。如图 3-24 所示。

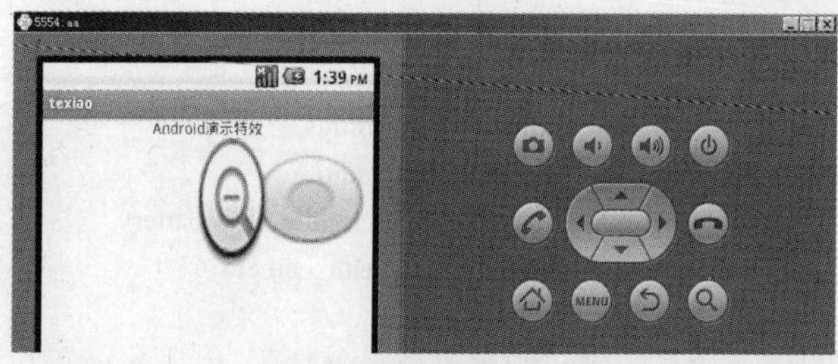

图 3-23　运行效果

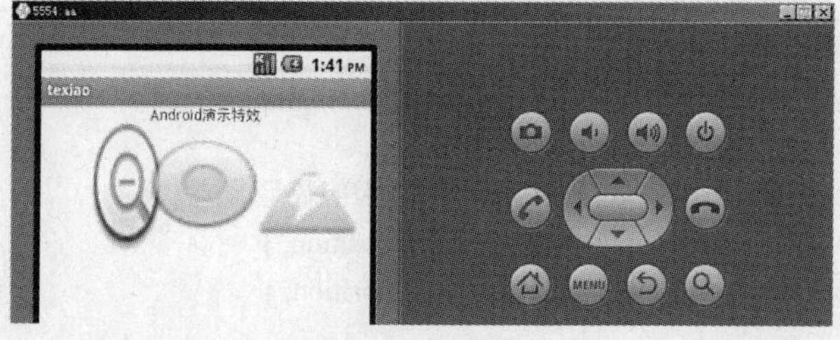

图 3-24　改变显示

上述实例是使用了 Android 的内置图标，另外也可以从外部导入图片，只需在 res\drawable 文件夹下导入图形文件即可。这些图形文件会在部署阶段同应用程序一起打包成.apk。

3.16 制作一个计算器

计算器是手机中的最常用程序之一，通过计算器，可以方便人们对日常计算的处理。本节将通过一个简单实例，介绍在手机上制作计算器的方法。

3.16.1 设计理念

在 Android 中，我们可以使用 Button 组件作为计算器中的数字按键和运算符号按键，并将计算结果显示在 TextView 当中。每次单击一个 Button 组件，都会响应执行一个处理事件，此功能是通过 onClick(View v)实现的，为每个 Button 按钮设置了监听事件方法 setOnClickListener()。

3.16.2 具体实现

本节实例保存在“光盘:\3\jisuanqi\”文件夹内，下面简单介绍主要文件的具体含义。

主文件 src/jisuanqi/jisuanqi.java 是此项目的主要文件，调用各个公用文件来实现具体的功能。文件 jisuanqi.java 的具体实现代码如下。

```
package irdc.jisuanqi;
……………………………………………………
public class jisuanqi extends Activity {
    /** Called when the activity is first created. */

  public Button mButton2;
  public Button mButton3;
  public Button mButton4;
  public Button mButton5;
  public EditText mEditText1;
  public EditText mEditText2;
  public TextView mTextView2;
  public TextView mTextView4;

      @Override
     public void onCreate(Bundle savedInstanceState)
 {
      super.onCreate(savedInstanceState);
      setContentView(R.layout.main);

      mTextView2 = (TextView) findViewById(R.id.mTextView2);
      mTextView4 = (TextView) findViewById(R.id.mTextView4);
      mButton2 = (Button) findViewById(R.id.mButton2);
      mButton3 = (Button) findViewById(R.id.mButton3);
      mButton4 = (Button) findViewById(R.id.mButton4);
```

```
        mButton5 = (Button) findViewById(R.id.mButton5);
        mEditText1 = (EditText) findViewById(R.id.mText1);
        mEditText2 = (EditText) findViewById(R.id.mText2);

        mButton2.setOnClickListener(new Button.OnClickListener()
    {
        @Override
        public void onClick(View v)
        {
          mTextView2.setText("+");
          String strRet = Integer.toString( Integer.parseInt(mEditText1.
                  getText().toString())+ Integer.parseInt
                  (mEditText2.getText().toString()) );
          mTextView4.setText(strRet);

        }
      });

      mButton3.setOnClickListener(new Button.OnClickListener()
    {
          @Override
          public void onClick(View v)
          {
            mTextView2.setText("-");
            String strRet = Integer.toString( Integer.parseInt(mEditText1.
                    getText().toString())- Integer.parseInt
                    (mEditText2.getText().toString()) );
            mTextView4.setText(strRet);

          }
    });

      mButton4.setOnClickListener(new Button.OnClickListener()
    {
          @Override
          public void onClick(View v)
          {
            mTextView2.setText("*");
            String strRet = Integer.toString( Integer.parseInt(mEditText1.
                    getText().toString())* Integer.parseInt
                    (mEditText2.getText().toString()) );
            mTextView4.setText(strRet);

          }
    });
```

```
        mButton5.setOnClickListener(new Button.OnClickListener()
    {
            @Override
            public void onClick(View v)
            {
                mTextView2.setText("/");
                String strRet = Integer.toString( Integer.parseInt(mEditText1.
                        getText().toString())/ Integer.parseInt
                        (mEditText2.getText().toString()) );
                mTextView4.setText(strRet);

            }
    });
    }
}
```

在上述代码中，特别要注意对象的名称，一定要创建按钮对象。因为需要用到逻辑运算，所以如果要输出数值，就需要以 Java Integer.toString()或 Float.toString()来处理这些数值。

调试运行后的效果如图 3-25 所示；当输入数值，选择一个算法后，将会计算出对应的结果。如图 3-26 所示。

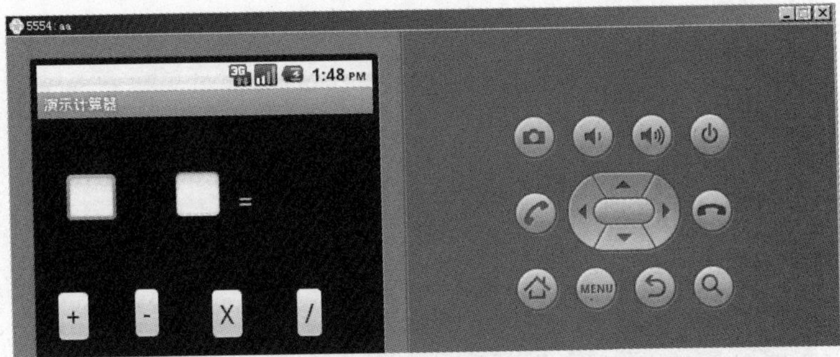

图 3-25　运行效果

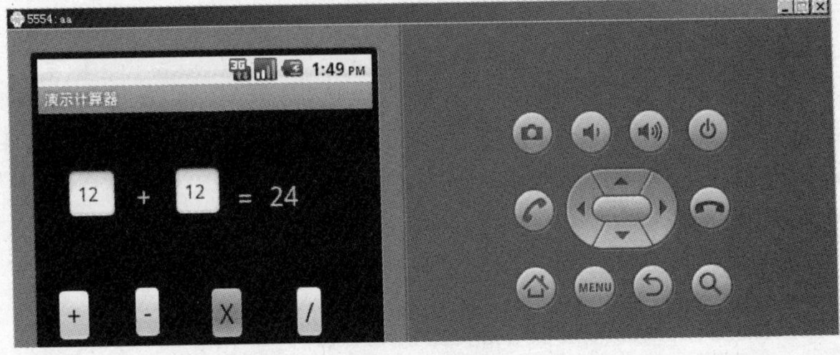

图 3-26　计算结果

上述实例只能计算整数，不能计算小数类型。可以对其进行修改，将 Integer 改为 Float 数据类型，这样就可以计算小数了。

3.17 设置 About（关于）信息

关于信息的作用是说明当前软件或硬件的基本信息，常见于计算机领域。例如，网站中的“关于我们”。本节将通过一个简单实例的实现，介绍在手机中实现设置 About（关于）信息的方法。

3.17.1 设计理念

在 Android 中，手机接口是 Menu Shotcut，即所谓的 Menu Key。本实例将讲解 Android Menu Key 的设计方法，演示“关于”对话框、“离开”对话框的语法。在程序中除了默认覆盖的 onCreate 外，还需要建立 2 个类函数 onCreateMenu()和 onOptionItemSelected()。其中，前者是创建 Menu 菜单项目，后者则是处理菜单被选择运行后的事件。在实例最后当用户单击“关于”菜单后，会弹出 AlertDialog，显示出“关于”的信息。

3.17.2 具体实现

本节实例保存在“光盘:\3\guanyu\”文件夹内，下面简单介绍主要文件的具体含义。

主文件 src/guanyu/guanyu.java 是此项目的主要文件，调用各个公用文件来实现具体的功能。文件 guanyu.java 的具体实现代码如下。

```
package irdc.guanyu;
………………………………………………
public class guanyu extends Activity
{
  /** Called when the activity is first created. */
  @Override
  public void onCreate(Bundle savedInstanceState)
  {
    super.onCreate(savedInstanceState);
    setContentView(R.layout.main);
  }

  public boolean onCreateOptionsMenu(Menu menu)
  {
    menu.add(0, 0, 0, R.string.app_about);
    menu.add(0, 1, 1, R.string.str_exit);
    return super.onCreateOptionsMenu(menu);
  }

  public boolean onOptionsItemSelected(MenuItem item)
  {
    super.onOptionsItemSelected(item);
    switch(item.getItemId())
    {
```

```
        case 0:
            openOptionsDialog();
            break;
        case 1:
            finish();
            break;
    }
    return true;
}

private void openOptionsDialog()
{
    new AlertDialog.Builder(this)
    .setTitle(R.string.app_about)
    .setMessage(R.string.app_about_msg)
    .setPositiveButton(R.string.str_ok,
        new DialogInterface.OnClickListener()
        {
         public void onClick(DialogInterface dialoginterface, int i)
         {
         }
         }
        )
    .show();
}
}
```

在上述代码中，创建了一个 onCreateOptionsMenu(Menu menu)类函数，用于添加 Menu 菜单项，然后再利用 onOptionItemSelected()选择事件获取菜单选择项目，处理对应的事件，即 getItemId()=0 是“关于”，getItemId()=1 是“离开”。

调试运行后的效果如图 3-27 所示；当单击“Menu”菜单后会弹出两个子菜单，如图 3-28 所示；当单击“关于”菜单后会弹出一个对话框，这个对话框就是“关于”的说明信息。如图 3-29 所示。

图 3-27　运行结果

图 3-28　单击 Menu 后

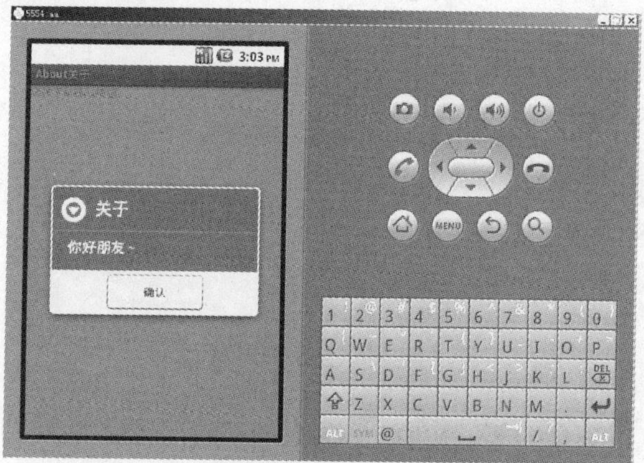

图 3-29 “关于”对话框

3.18 程序加载中

“程序加载中”常见于各种应用软件，表示程序还没有加载完成，请用户慢慢等待。例如，经常显示“程序正在加载中，请稍后.....”。本节将通过一个简单实例，介绍在手机中实现“程序加载中，请稍后.....”效果的方法。

3.18.1 设计理念

我们经常在 Windows 程序中看到“程序正在加载中”的提示。在 Android 中，此功能是通过 Progress Dialog 类来运行，此类被封装在 Android.app.ProgressDialog 类里，但需要注意的是 Android 中的 Progress Dialog 类必须在后台程序运行完毕前，使用 dismiss()方法来关闭取得焦点的对话框，否则程序将会陷入无法终止的无穷循环中；或者在线程里不可有任何更改 Context 或 parent View 的任何状态、文字输出等事件。因为线程里的 Context 或 View 并不属于 parent，两者之间没有关联。在本实例中，将以线程来模拟后台程序的运行，再通过线程运行完毕时，关闭这个加载中的动画对话框。在实例中，将设计一个按钮，单击按钮后开始线程的周期，在运行过程中显示 ProgressDialog 对话框，当线程运行完毕后，结束 ProgressDialog 对话框。

3.18.2 具体实现

本节实例保存在“光盘:\3\dengdai\”文件夹内，下面简单介绍主要文件的具体含义。

主文件 src/dengdai /dengdai.java 是此项目的主要文件，调用各个公用文件来实现具体的功能。文件 guanyu.java 的具体实现代码如下。

```
package irdc.dengdai;
..............................................
public class dengdai extends Activity
{
  private Button mButton1;
  private TextView mTextView1;
```

```
public ProgressDialog myDialog = null;

/** Called when the activity is first created. */
@Override
public void onCreate(Bundle savedInstanceState)
{
  super.onCreate(savedInstanceState);
  setContentView(R.layout.main);

  mButton1 =(Button) findViewById(R.id.myButton1);
  mTextView1 = (TextView) findViewById(R.id.myTextView1);
  mButton1.setOnClickListener(myShowProgressBar);
}

Button.OnClickListener myShowProgressBar =
new Button.OnClickListener()
{
  public void onClick(View arg0)
  {
    final CharSequence strDialogTitle =
          getString(R.string.str_dialog_title);
    final CharSequence strDialogBody =
          getString(R.string.str_dialog_body);

    /* 显示 Progress 对话框 */
    myDialog = ProgressDialog.show
              (
                dengdai.this,
                strDialogTitle,
                strDialogBody,
                true
              );

    mTextView1.setText(strDialogBody);

    new Thread()
    {
      public void run()
      {
        try
        {
          /* 在这里写上要运行的程序片段 */
          /* 为了明显看见效果，以暂停 3 秒作为示范 */
          sleep(3000);
        }
        catch (Exception e)
```

```
                {
                    e.printStackTrace();
                }
                finally
                {
                    /* 卸载所创建的 myDialog 对象 */
                    myDialog.dismiss();
                }
            }
        }.start(); /* 开始运行运行线程 */
    } /*End: public void onClick(View arg0)*/
  };
}
```

在上述代码中，创建了一个按钮，当单击按钮后，会触发 myShowProgressBar 事件，在事件中更改了 TextView 里的文字，并将焦点传递给前台的 ProgressDialog.show 方法。当运行 3s 后，再将焦点传递返回给原来的 Activity。

调试运行后的效果如图 3-30 所示；当单击“按下后”按钮后会弹出“请稍后”提示信息，如图 3-31 所示。

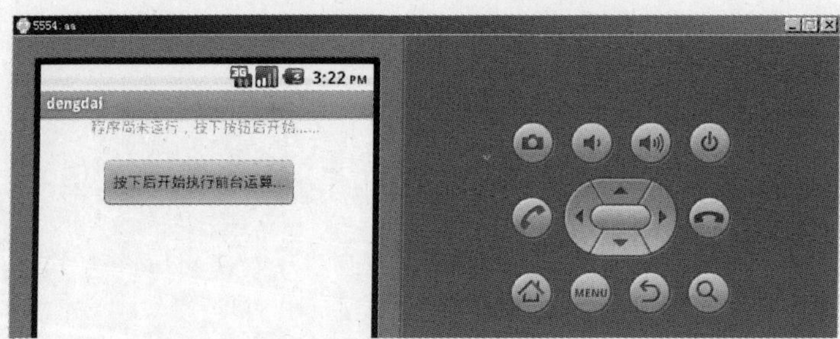

图 3-30　运行结果

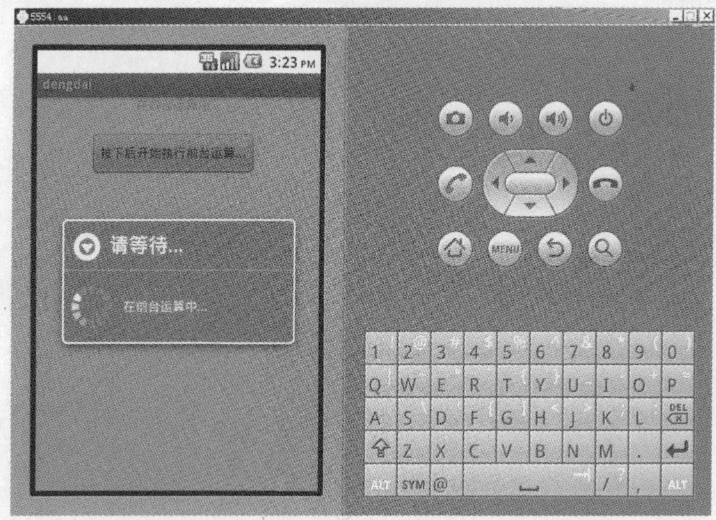

图 3-31　显示提示信息

上述功能是基于 Android.app.ProgressDialog 类实现的，Android.app.ProgressDialog 中包括了几种重要的构造方法，具体信息如表 3-1 所示。

表 3-1 Android.app.ProgressDialog 类的构造函数

方 法	格 式
static	show(Contextcontext,CharSequencetitle,CharSequencemessage)
static	show(Contextcontext,CharSequencetitle,CharSequencemessage, booleanindeterminate)
static	show(Contextcontext,CharSequencetitle,CharSequencemessage, booleanindeterminate,booleancancelable)
static	show(Contextcontext,CharSequencetitle,CharSequencemessage, booleanindeterminate,booleancancelable, DialogInterface.OnCancelListenercancelListener)

有关 Android.app.ProgressDialog 类更加详细的信息，请读者参阅下面的网址。

```
http://www.chinaup.org/docs/reference/android/app/ProgressDialog.html
```

3.19 可选择的对话框

对话框是网络中常见的窗口，手机软件中也必不可少。前面内容中已经介绍了对话框的生成方法，本节将进一步介绍一种更为复杂的对话框——内容可选的对话框。本节将通过一个简单实例的实现，介绍实现可选择对话框的方法。

3.19.1 设计理念

在前面的内容中，介绍了 AlertDialog.Builder 对话框，其实在这个对话框内还可以包含对话窗口，即含有多个子对话框。在 Android 应用中，可以使用 AlertDialog.Builder 对话框，开发出有多个选项的对话框。在本实例中，将使用一个按钮事件，触发按钮事件后，将通过类似列表项目的方式呈现在 Alert Dialog 中。上述做法可以实现常见的投票处理、事物选择处理和遥控器等应用。

3.19.2 具体实现

本节实例保存在“光盘:\3\xuanze\”文件夹内，下面简单介绍主要文件的具体含义。

主文件 src/xuanze/xuanze.java 是此项目的主要文件，调用各个公用文件来实现具体的功能。文件 xuanze.java 的具体实现代码如下。

```
public class xuanze extends Activity
{
   public Button mButton1;
   public TextView mTextView1;

   /** Called when the activity is first created. */
   @Override
```

```
public void onCreate(Bundle savedInstanceState)
{
  super.onCreate(savedInstanceState);
  setContentView(R.layout.main);

  mButton1 =(Button) findViewById(R.id.myButton1);
  mTextView1 = (TextView) findViewById(R.id.myTextView1);
  mButton1.setOnClickListener(myShowAlertDialog);
}

Button.OnClickListener myShowAlertDialog = new Button.OnClickListener()
{
  public void onClick(View arg0)
  {
    new AlertDialog.Builder(xuanze.this)
       .setTitle(R.string.str_alert_title)
       .setItems(R.array.items_irdc_dialog,
       new DialogInterface.OnClickListener()
       {
         public void onClick(DialogInterface dialog, int whichcountry)
         {
           CharSequence strDialogBody = getString(R.string.str_alert_body);
           String[] aryShop =
           getResources().getStringArray(R.array.items_irdc_dialog);
             new AlertDialog.Builder(xuanze.this)
             .setMessage(strDialogBody + aryShop[whichcountry])
             .setNeutralButton(R.string.str_ok, new DialogInterface.OnClickListener()
             {
               public void onClick(DialogInterface dialog, int whichButton)
               {

               }
             })
             .show();
         }
     })
     .setNegativeButton("      ", new DialogInterface.OnClickListener()
     {
       @Override
       public void onClick(DialogInterface d, int which)
       {
```

```
                d.dismiss();
            }
        })
        .show();
    } /*结束 onClick 方法*/
  };
}
```

在上述代码中，单击按钮后会弹出选择。调试运行后的效果如图 3-32 所示；当单击“单击后”开始选择按钮后会弹出选择菜单，如图 3-33 所示；选择一个选项后会弹出一个对话框，如图 3-34 所示。再次单击“确认”按钮，返回图 3-32 所示的初始界面。

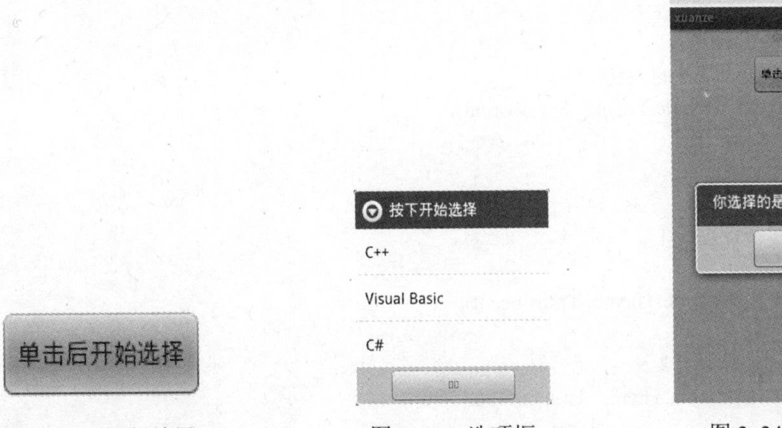

图 3-32　运行结果　　图 3-33　选项框　　图 3-34　选择提示

3.20　主题变换

主题变换是手机程序中的常见应用之一，在一些手机中可以选择自己需要的风格主题。本节将通过一个简单实例的实现，介绍 Android 实现主题变换的方法。

3.20.1　设计理念

在前面的内容中，介绍过 Style 的使用方法。通过使用 Style，可以快速开发出不同外观的效果。通过使用 Style，还可以大大方便对程序的维护。Style 的引入，改善了程序员和视觉设计人员之间存在的沟通问题，解决了 J2EE 和 Windows Mobile 中存在已久的问题。在 Android 中，除了可以使用 Style 定制外，还可以针对每个 Activity、前景、背景和透明度等进行设置规划。本节下面介绍的实例，是基于 Style 的简易 Theme（主题）应用。

3.20.2　具体实现

本节实例保存在“光盘:\3\zhuti\”文件夹内，下面简单介绍主要文件的具体含义。

1．主文件

主文件 src/zhuti/zhuti.java 是此项目的主要文件，调用各个公用文件来实现具体的功能。

文件 zhuti.java 的具体实现代码如下。

```
package irdc.zhuti;

import irdc.zhuti.R;
import android.app.Activity;
import android.os.Bundle;
public class zhuti extends Activity
{
  /** Called when the activity is first created. */
  @Override
  public void onCreate(Bundle savedInstanceState)
  {
    super.onCreate(savedInstanceState);

    /*
     * 应用透明背景的主题
     * setTheme(R.style.Theme_Transparent);
     */

    /*
     * 应用布景主题 1
     */
      setTheme(R.style.Theme_Translucent);
    /*
     * 应用布景主题 2
    * setTheme(R.style.Theme_Translucent2);
    */
    setContentView(R.layout.main);
  }
}
```

在上述代码中，利用 setTheme 方法指定了 Activity 的主题，其中主题设置文件在 Style.xml 中。

2．主题设置文件

主题设置文件 Style.xml 中已经预先设置好了 Theme、ThemeTranslucent、ThemeTransparent、TextAppearance.Theme.PlaneText 四种主题样式。

调试运行后的效果如图 3-35 所示。

图 3-35　运行效果

第 4 章 玩转 Android 组件

手机项目开发几乎都需要通过组件来实现。组件就如同一个模块，通过调用这些组件能够实现对应的功能效果。在本章的内容中，将通过具体的实例来详细讲解 Android 中各个主要组件的基本知识。希望读者能够在学习中举一反三，为学习本书后面知识打下基础。

4.1 EditText 和 setOnKeyListener 事件实现文本处理

在本节的实例中，将演示使用 EditText 文本编辑组件和 setOnKeyListener 监听按钮事件实现文本处理的功能。本实例将以 EditText 组件和 TextView 组件来演示如何在捕捉用户键盘输入文字的同时，实时取得文字，同步显示于 TextView 中，类似手机版的 Ajax 效果，实时输入实时输出。

本实例实现代码保存在“光盘:\daima\4\example1”，下面开始讲解本实例的具体实现流程。

在主程序文件 example1.java 中，关键之处是利用 EditText.OnKeyListener 来拦截 EditText 的键盘输入事件，只需在其中重写 onKey()方法即可实现。在 onKey()方法中，将 EditText.getText() 取出的文字，显示于 TextView 当中，是一个简单易懂的范例练习。文件 example1.java 的具体实现代码如下。

```
package irdc.example1;
import irdc.example1.R;
import android.app.Activity;
import android.os.Bundle;
import android.view.KeyEvent;
import android.view.View;
import android.widget.EditText;
import android.widget.TextView;

public class example1 extends Activity
{
  /*声明 TextView、EditText 对象*/
  private TextView mTextView01;
  private EditText mEditText01;

  /** Called when the activity is first created. */
    @Override
  public void onCreate(Bundle savedInstanceState)
  {
```

```
        super.onCreate(savedInstanceState);
        setContentView(R.layout.main);

        /*取得 TextView、EditText*/
        mTextView01 = (TextView)findViewById(R.id.myTextView);
        mEditText01 = (EditText)findViewById(R.id.myEditText);

        /*设置 EditText 用 OnKeyListener 事件来启动*/
        mEditText01.setOnKeyListener(new EditText.OnKeyListener()
        {
          @Override
          public boolean onKey(View arg0, int arg1, KeyEvent arg2)

          {
            /*设置 TextView 显示 EditText 所输入的内容*/
            mTextView01.setText(mEditText01.getText());
            return false;
          }
        });
    }
}
```

执行后的效果如图 4-1 所示，当在文本框输入字符后，在下方会即时显示文本框内输入的字符，如图 4-2 所示。

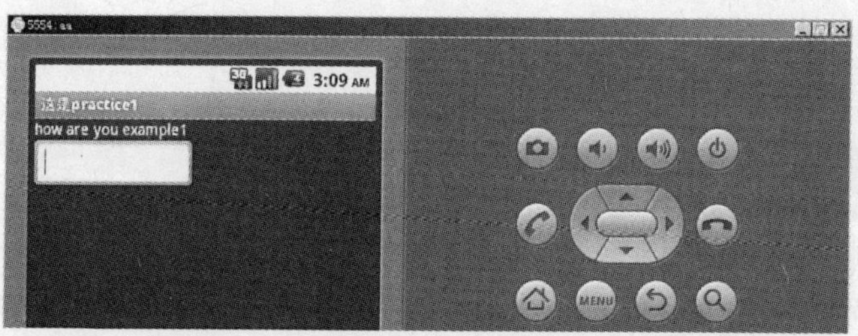

图 4-1　初始效果

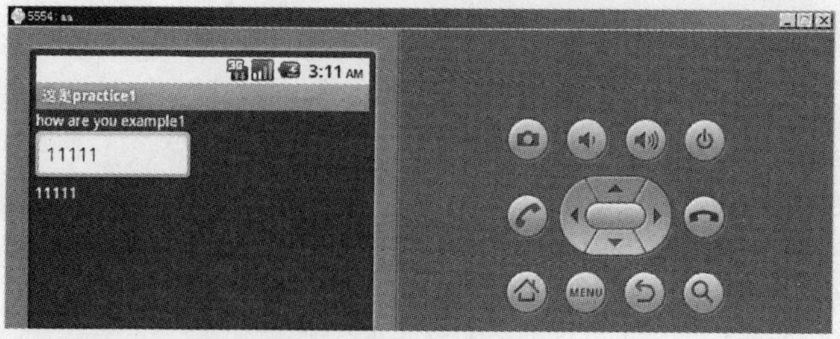

图 4-2　即时显示提示信息

上述的实时输入实时显示效果可以扩展在许多手机应用程序中，读者可以在方法 OnKeyListener()里做实时文字过滤效果。例如，当用户输入不雅的文字时，可以提示用户不

接受部分关键字，以输入“Shit”为例，在 TextView 就会出现：Sh*t，此种做法可以过滤掉不雅文字。

此外，不仅可以重写 Widget 中组件的 setOnKeyListener()方法，而且也可以重写 View（视图）组件中的 View.setOnKeyListener()方法，这个方法能够捕捉用户点击键盘时的事件。在此需要特别注意，在使用这些方法时需要拦截这个事件，也就是说当 View 要取得 Focus（焦点）时才能触发 onKeyDown（按键按下时）事件。

4.2 实现背景图片按钮

在现实应用中，有时为了特殊需要，会需要设计一个具有背景图的按钮，本节的实例是基于 ImageButton（图像按钮）来实现的。

实现原理如下。首先，将按钮背景图预先加载至 Drawable 文件夹里（*.png 格式的图形文件），利用这些图片，作为图片按钮的背景图；然后，在布局中配置一个“一般按钮”，读者可以查看两者的对照效果，在运行效果中，可以明显看出图片按钮与一般按钮在外观上的差异。

要设置图片按钮背景图有许多方法，此程序使用的方法是 ImageButton.setImageResource()，需要传递的参数即是“res/drawable/”目录下面的 Resource ID，除了设置背景图片的方法外，程序需要用到 onFocusChange 焦点变化监听与 onClick（单击）等作为按钮单击之后的处理事件，最后通过 TextView 来显示目前图片按钮的状态为 onClick（单击）、onFocus（事件在对象获得焦点时发生）或 offFocus（事件在对象离开焦点时发生），并且同步更新按钮的背景图，让用户有动态交互的感觉。

在本节的实例中，将演示使用 ImageButton 组件实现背景图片按钮的功能。本实例将以 EditText 和 TextView 来演示如何在捕捉用户键盘输入文字的同时，实时取得文字，同步显示于 TextView，类似手机版的 Ajax 效果，实时输入实时输出。

本实例实现代码保存在“光盘:\daima\4\example2”，下面开始讲解本实例的具体实现流程。

在主程序文件 example2.java 中，在图片按钮上设置两个监听事件：onFocusChangeListener 与 onClickListener 函数，并实现 Image Button 图片的置换。文件 example2.java 的具体实现代码如下。

```
package irdc.example2;
import irdc.example2.R;
import android.app.Activity;
import android.os.Bundle;
import android.view.View;
/*使用 OnClickListener 与 OnFocusChangeListener 来区分按钮的状态*/
import android.view.View.OnClickListener;
import android.view.View.OnFocusChangeListener;
import android.widget.Button;
import android.widget.ImageButton;
import android.widget.TextView;
```

```
public class example2 extends Activity
{
  /*声明三个对象变量(图片按钮,按钮,与 TextView)*/
  private ImageButton mImageButton1;
  private Button mButton1;
  private TextView mTextView1;

  /** Called when the activity is first created. */
  @Override
  public void onCreate(Bundle savedInstanceState)
  {
    super.onCreate(savedInstanceState);
    setContentView(R.layout.main);

    /*通过 findViewById 建构三个对象*/
    mImageButton1 =(ImageButton) findViewById(R.id.myImageButton1);
    mButton1=(Button)findViewById(R.id.myButton1);
    mTextView1 = (TextView) findViewById(R.id.myTextView1);

    /*通过 OnFocusChangeListener 来应答 ImageButton 的 onFous 事件*/
    mImageButton1.setOnFocusChangeListener(new OnFocusChangeListener()
    {
      public void onFocusChange(View arg0, boolean isFocused)
      {

        /*若 ImageButton 状态为 onFocus（离开焦点）改变 ImageButton 的图片
         * 并改变 textView 的文字*/
        if (isFocused==true)
        {
          mTextView1.setText("图片按钮状态为:Got Focus");
          mImageButton1.setImageResource(R.drawable.iconfull);
        }
        /*若 ImageButton 状态为 offFocus 改变 ImageButton 的图片
         *并改变 textView 的文字*/
        else
        {
          mTextView1.setText("图片按钮状态为:Lost Focus");
          mImageButton1.setImageResource(R.drawable.iconempty);
        }
      }
    });

   /*通过 onClickListener 来应答 ImageButton 的 onClick 事件*/
```

```
        mImageButton1.setOnClickListener(new OnClickListener()
        {
          public void onClick(View v)
          {
            /*若图片按钮状态为 onClick 改变图片按钮的图片
             * 并改变 textView 的文字*/
            mTextView1.setText("图片按钮状态为:Got Click");
            mImageButton1.setImageResource(R.drawable.iconfull);
          }
        });

        /*通过 onClickListener 来应答 Button 的 onClick 事件*/
        mButton1.setOnClickListener(new OnClickListener()
        {
          public void onClick(View v)
          {
            /*若 Button 状态为 onClick 改变 ImageButton 的图片
             * 并改变 textView 的文字*/
            mTextView1.setText("图片按钮状态为:Lost Focus");
            mImageButton1.setImageResource(R.drawable.iconempty);
          }
        });
      }
    }
```

通过上述代码，实现了图片样式的按钮效果，执行后的显示效果如图 4-3 所示，单击“普通按钮”后的效果如图 4-4 所示。

除了在运行时用 onFocus() 与 onClick() 事件来设置按钮背景图片外，Android 的 MVC 设计理念，可以让程序运行之初就以 XML 定义的方式来初始化图片按钮的背景图，仅需先将图片导入“res/drawable”文件夹即可。

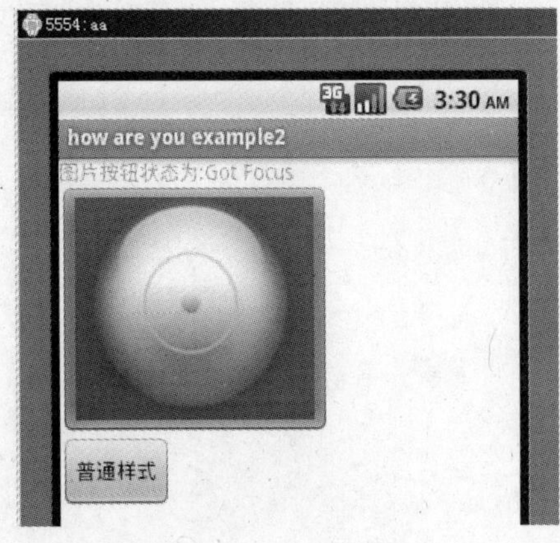

图 4-3　初始效果

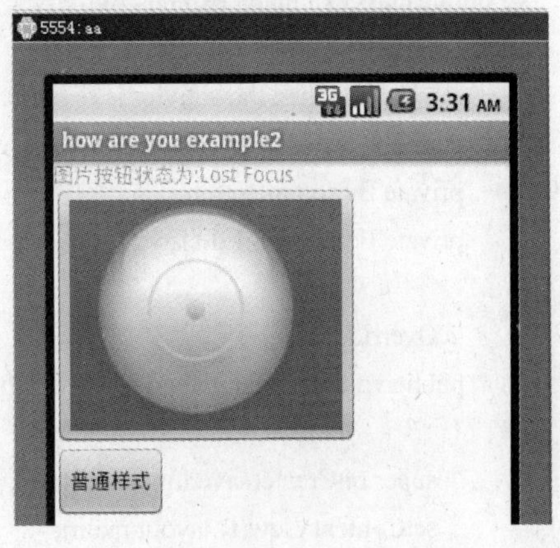

图 4-4　单击后的效果

4.3 Toast 实现温馨提示

Toast 是 Android 中的提示对象，它是一个简短的小信息，将要告诉用户的信息，以一个浮动在最上层的 View 显示。显示 Toast 之后，静待几秒后便会自动消失，最常见的应用就是音量大小的调整，当单击音量调整钮之后，会看见跳出的音量指示 Toast 对象，等待调整完之后便会消失。

通过 Toast 的特性，可以在不影响用户通话或聆听音乐情况下，显示要给用户的信息。这样可以在任何程序运行时，通过 Toast 的方式，显示运行变量或手机环境的概况。

在本节的实例中，将使用一个 EditText 组件来接受用户输入的文字，当单击按钮时，将 EditText 里的文字，以方法 Toast.makeText()让文字显示于 Toast 对象中，这段文字会在显示一段时间后自动消失。

本实例实现代码保存在“光盘:\daima\4\example3”，下面开始讲解本实例的具体实现流程。

在主程序文件 example3.java 中，构建 2 个组件，EditText 组件与 Button 组件，并在 Button 组件的 onClick() 方法中使用 Toast 对象的 makeText()方法来显示输入的文字。文件 example3.java 的具体实现代码如下。

```
package irdc.example3;

import irdc.example3.R;
import android.app.Activity;
import android.os.Bundle;
import android.text.Editable;
import android.view.View;
import android.view.View.OnClickListener;
import android.widget.Button;
import android.widget.EditText;
import android.widget.Toast;

public class example3 extends Activity
{
  /** Called when the activity is first created. */
  /*声明两个对象变量(按钮与编辑文字)*/
  private Button mButton;
  private EditText mEditText;

  @Override
  public void onCreate(Bundle savedInstanceState)
  {
    super.onCreate(savedInstanceState);
    setContentView(R.layout.main);
```

```
    /*通过 findViewById()取得对象 */
    mButton=(Button)findViewById(R.id.myButton);
    mEditText=(EditText)findViewById(R.id.myEditText);

    /*设置 onClickListener 给 Button 对象聆听 onClick 事件*/
    mButton.setOnClickListener(new OnClickListener()
    {
      @Override
      public void onClick(View v)
      {

      /*声明字符串变量并取得用户输入的 EditText 字符串*/
      Editable Str;
      Str=mEditText.getText();

      /*使用系统标准的 makeText()方法来产生 Toast 信息*/
      Toast.makeText(
        example3.this,
        "你的愿望　"+Str.toString()+"已送达宝贝信箱",
        Toast.LENGTH_LONG).show();

      /*清空 EditText*/
      mEditText.setText("");
      }
    });
  }
}
```

通过上述代码，实现了温馨祝福提示的效果，执行后的显示效果如图 4-5 所示，用户可以输入祝福语句，单击“example3”按钮后的效果如图 4-6 所示。

图 4-5　初始效果

Toast 提示信息在显示一定时间后会消失，在 Toast 构造参数中的第二个参数为显示的时间常数，可设置为 LENGTH_LONG 或 LENGTH_SHORT，前者提示时间较长，后者较短，作为传递 makeText()方法的参数使用。

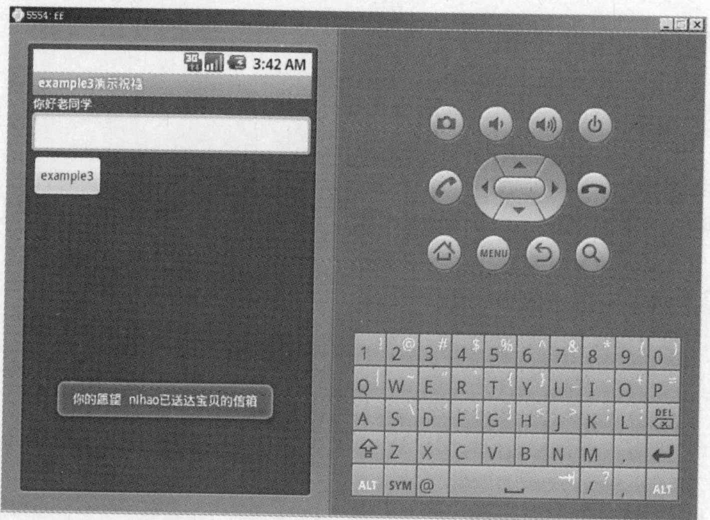

图 4-6　温馨提示效果

4.4　CheckBox 实现一个简单物品清单

CheckBox 是一个复选框组件，可以供用户选择。在本节实例中，通过 CheckBox.setOnCheckedChangeListener 在程序中设计了 3 个 CheckBox 选取项，分别表示 3 种物品列表，当用户勾选其中一个物品，就在 TextView 里显示已选择的物品列表。

本实例实现代码保存在“光盘:\daima\4\example4”，下面开始讲解本实例的具体实现流程。

在主程序文件 example4.java 中，分别创建了 3 个 CheckBox 对象和一个 TextView 对象，通过监听 setOnCheckedChangeListener（当选项组中的按钮的勾选状态发生改变时的监听事件）事件，利用 onCheckedChanged()方法来更新 TextView 文字。文件 example4.java 的具体实现代码如下。

```
package irdc.example4;

import irdc.example4.R;
import android.app.Activity;
import android.os.Bundle;
import android.widget.CheckBox;
import android.widget.CompoundButton;
import android.widget.TextView;

public class example4 extends Activity
{
  /*声明对象变量*/
  private TextView mTextView1;
  private CheckBox mCheckBox1;
  private CheckBox mCheckBox2;
  private CheckBox mCheckBox3;
```

```
/** Called when the activity is first created. */
@Override
public void onCreate(Bundle savedInstanceState)
{

  super.onCreate(savedInstanceState);
  setContentView(R.layout.main);

  /*通过 findViewById 取得 TextView 对象并调整文字内容*/
  mTextView1 = (TextView) findViewById(R.id.myTextView1);
  mTextView1.setText("你所选择的项目有: ");

  /*通过 findViewById 取得三个 CheckBox 对象*/
  mCheckBox1=(CheckBox)findViewById(R.id.myCheckBox1);
  mCheckBox2=(CheckBox)findViewById(R.id.myCheckBox2);
  mCheckBox3=(CheckBox)findViewById(R.id.myCheckBox3);

  /*设置 OnCheckedChangeListener 给三个 CheckBox 对象*/
  mCheckBox1.setOnCheckedChangeListener(mCheckBoxChanged);
  mCheckBox2.setOnCheckedChangeListener(mCheckBoxChanged);
  mCheckBox3.setOnCheckedChangeListener(mCheckBoxChanged);
}

/*声明并建构 onCheckedChangeListener 对象*/
private CheckBox.OnCheckedChangeListener mCheckBoxChanged
= new CheckBox.OnCheckedChangeListener()
{
  /*implement onCheckedChanged 方法*/
  @Override
  public void onCheckedChanged(CompoundButton buttonView,
                               boolean isChecked)
  {
    /*通过 getString()取得 CheckBox 的文字字符串*/
    String str0="所选的项目为: ";
    String str1=getString(R.string.str_checkbox1);
    String str2=getString(R.string.str_checkbox2);
    String str3=getString(R.string.str_checkbox3);
    String plus=";";
    String result="但是超过预算啰!!";
    String result2="还可以再多买几本喔!!";

    /*任一 CheckBox 被勾选后,该 CheckBox 的文字会改变 TextView 的文字内容
     * 三个对象总共八种情况*/
```

```
            if(mCheckBox1.isChecked()==true & mCheckBox2.isChecked()==true
                & mCheckBox3.isChecked()==true)
            {
              mTextView1.setText(str0+str1+plus+str2+plus+str3+result);
            }
            else if(mCheckBox1.isChecked()==false & mCheckBox2.isChecked()==true
                & mCheckBox3.isChecked()==true)
            {
              mTextView1.setText(str0+str2+plus+str3+result);
            }
            else if(mCheckBox1.isChecked()==true & mCheckBox2.isChecked()==false
                & mCheckBox3.isChecked()==true)
            {
              mTextView1.setText(str0+str1+plus+str3+result);
            }
            else if(mCheckBox1.isChecked()==true & mCheckBox2.isChecked()==true
                & mCheckBox3.isChecked()==false)
            {
              mTextView1.setText(str0+str1+plus+str2+result);
            }
            else if(mCheckBox1.isChecked()==false & mCheckBox2.isChecked()==false
                & mCheckBox3.isChecked()==true)
            {
              mTextView1.setText(str0+str3+plus+result2);
            }
            else if(mCheckBox1.isChecked()==false & mCheckBox2.isChecked()==true
                & mCheckBox3.isChecked()==false)
            {
              mTextView1.setTcxt(str0+str2);
            }
            else if(mCheckBox1.isChecked()==true & mCheckBox2.isChecked()==false
                & mCheckBox3.isChecked()==false)
            {
              mTextView1.setText(str0+str1);
            }
            else if(mCheckBox1.isChecked()==false & mCheckBox2.isChecked()==false
                & mCheckBox3.isChecked()==false)
            {
              mTextView1.setText(str0);
            }
        }
    };
}
```

在上述代码中，首先设置一共拥有 3600 块钱，然后提供了一个货物清单供用户选择要买的商品。当用户选择商品后，会显示对应的购买物品，如果超出了 3600 这个额度，会显示出对应的提示。

实例执行后的初始效果如图 4-7 所示，当选择一种商品后，在下方会显示出对应的提示，如图 4-8 所示。如果超出了预算 3600，也会显示对应的提示，如图 4-9 所示。

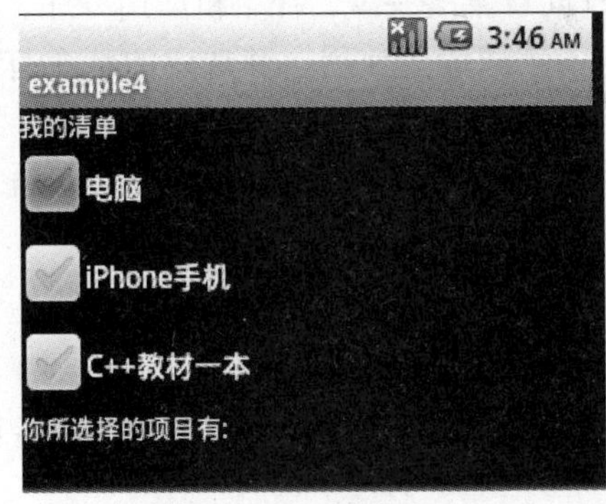

图 4-7　初始效果

图 4-8　显示提示效果

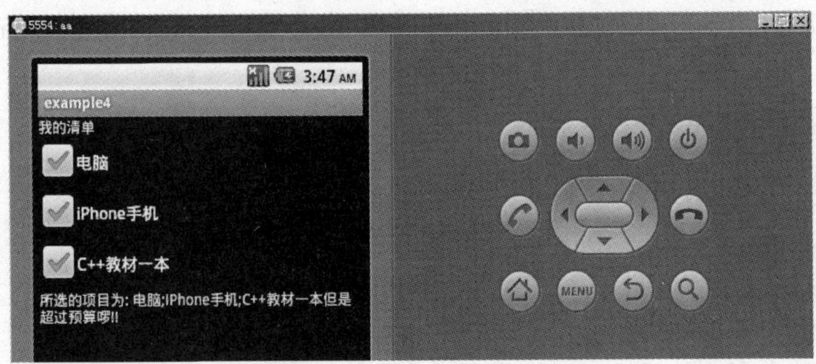

图 4-9　超出预算提示效果

为了满足特殊需求，可以将 OnCheckedChangeListener 改为 OnTouchListener（屏幕触控事件），具体实现代码如下。

```
private CheckBox.OnTouchListener mCheckBoxTouch =
new CheckBox.OnTouchListener()
{
  @Override
  public boolean onTouch(View v, MotionEvent event)
  {
    /* 判断在触控笔指压此组件时的状态 */
    if(mCheckBox1.isChecked()==false)
    {
      /*当触控笔放开后的动作*/
    }
    else if(mCheckBox1.isChecked()==true)
    {
```

```
            /*当触控笔压下后的动作*/
        }
        return false;
    }
};
```

当用户使用某些网络服务时，系统会要求用户同意某些条款。在手机应用程序、手机游戏的设计过程中，也会常遇见 CheckBox 在同意条款情境的使用，其选取的状态有两种：isChecked=true 与 isChecked=false。

在本节实例中，使用 TextView 放入条款文字，在下方配置一个 CheckBox Widget 作为选取项，通过 Button.onClickListener 按钮事件处理，取得用户同意条款的状态。

当 CheckBox.isChecked 为 true，更改 TextView 的文字内容为“你已接受同意!!”，当未选取 CheckBox 时，Button 是不可以选择的（被 Disabled）。

本实例实现代码保存在“光盘:\daima\4\example5”，下面开始讲解本实例的具体实现流程。

在主程序文件 example5.java 中，利用 CheckBox.OnClickListener（单击监听）里的事件来判断 Button 该不该显示，其方法就是判断 Button.Enabled（按钮是否选择）的值；在一开始时，默认参数为 false，当单击 CheckBox 时，Button 参数就修改为 true。文件 example5.java 的具体实现代码如下。

```
package irdc.example5;

import irdc.example5.R;
import android.app.Activity;
import android.os.Bundle;
import android.view.View;
import android.widget.Button;
import android.widget.CheckBox;
import android.widget.TextView;

public class example5 extends Activity
{
    /** Called when the activity is first created. */

  /*声明 TextView、CheckBox、Button 对象*/
  public TextView myTextView1;
  public TextView myTextView2;
  public CheckBox myCheckBox;
  public Button myButton;

    @Override
  public void onCreate(Bundle savedInstanceState)
```

```
{
  super.onCreate(savedInstanceState);
  setContentView(R.layout.main);

  /*取得 TextView、CheckBox、Button*/
  myTextView1 = (TextView) findViewById(R.id.myTextView1);
  myTextView2 = (TextView) findViewById(R.id.myTextView2);
  myCheckBox = (CheckBox) findViewById(R.id.myCheckBox);
  myButton = (Button) findViewById(R.id.myButton);

  /*将 CheckBox、Button 默认为未选择状态*/
  myCheckBox.setChecked(false);
  myButton.setEnabled(false);

  myCheckBox.setOnClickListener(new CheckBox.OnClickListener()
  {
    @Override
    public void onClick(View v)
    {
      if(myCheckBox.isChecked())
      {
        /*设置按钮为不能选择对象*/
        myButton.setEnabled(true);
        myTextView2.setText("");
      }
      else
      {
        /*设置按钮为可以选择对象*/
        myButton.setEnabled(false);
        myTextView1.setText(R.string.text1);
        /*在 TextView2 里显示出"请勾选我同意"*/
        myTextView2.setText(R.string.no);
      }
    }
  });

  myButton.setOnClickListener(new Button.OnClickListener()
  {
    @Override
    public void onClick(View v)
    {
      if(myCheckBox.isChecked())
      {
        myTextView1.setText(R.string.ok);
```

```
            }else
            {
            }
        }
    });

  }
}
```

实例执行后的初始效果如图 4-10 所示，当选择“同意”选项并单击“确定”按钮后，会输出对应的提示界面，如图 4-11 所示。

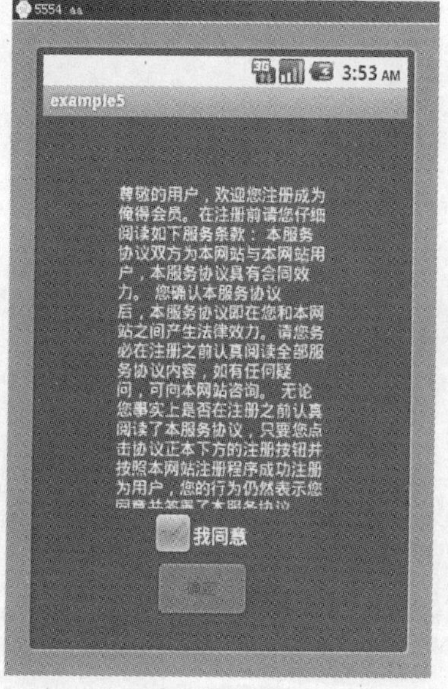

图 4-10　初始效果

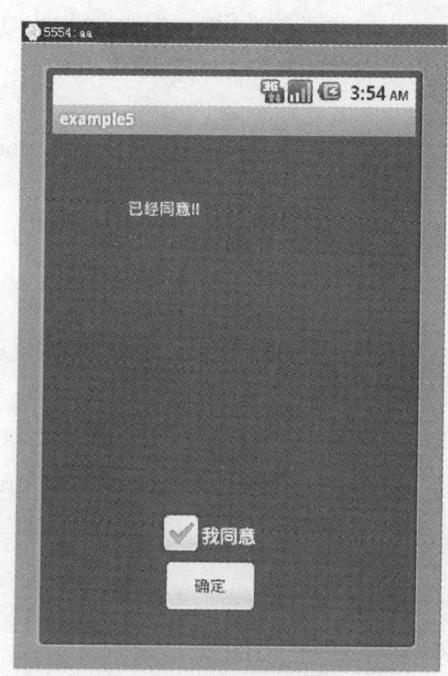

图 4-11　输出提示

当 CheckBox 在默认内容为空白时（没有任何默认的提示文字下），可设置提示用户的文字，其调用的方法为 CheckBox.setHint()方法。

4.5　单选按钮组实现选择处理

单选按钮组能够将不同的单选按钮布置于同一个单选按钮组，同属一个单选按钮组里的按钮，只能做出单一选择，虽然前一章曾经介绍过单选按钮组与单选按钮，但当时使用的是按钮事件，在此要示范“单击”的同时就运行事件处理，不再需要按钮的辅助了。

在本节实例中，先设计一个 TextView Widget 和一个单选按钮组，并在单选按钮组内放置 2 个单选按钮，默认为不选择。在程序运行阶段，利用 onCheckedChanged（选择改变事件）作为启动事件装置，让用户选择其中一个按钮时，显示被选择的内容，最后将 RadioGroup（单选按钮）的选项文字显示于 TextView 当中。

本实例实现代码保存在“光盘:\daima\4\example6”，下面开始讲解本实例的具体实现流程。

在主程序文件 example6.java 中，使用 OnCheckedChangeListener 来启动 RadioGroup 的事件，随后将被选择的单选按钮（mRadio1.getText()）的文字显示于 TextView。文件 example6.java 的具体实现代码如下。

```
package irdc.example6;

import irdc.example6.R;
import android.app.Activity;
import android.os.Bundle;
import android.widget.RadioButton;
import android.widget.RadioGroup;
import android.widget.TextView;

public class example6 extends Activity
{
  public TextView mTextView1;
  public RadioGroup mRadioGroup1;
  public RadioButton mRadio1,mRadio2;

  /** Called when the activity is first created. */
  @Override
  public void onCreate(Bundle savedInstanceState)
  {
    super.onCreate(savedInstanceState);
    setContentView(R.layout.main);

    /*取得 TextView、RadioGroup、RadioButton 对象的内容*/
    mTextView1 = (TextView) findViewById(R.id.myTextView);
    mRadioGroup1 = (RadioGroup) findViewById(R.id.myRadioGroup);
    mRadio1 = (RadioButton) findViewById(R.id.myRadioButton1);
    mRadio2 = (RadioButton) findViewById(R.id.myRadioButton2);

    /*单选按钮组用 OnCheckedChangeListener 来运行*/
    mRadioGroup1.setOnCheckedChangeListener(mChangeRadio);
  }

  private RadioGroup.OnCheckedChangeListener mChangeRadio = new
            RadioGroup.OnCheckedChangeListener()
  {
    @Override
    public void onCheckedChanged(RadioGroup group, int checkedId)
    {
      if(checkedId==mRadio1.getId())
      {
        /*把 mRadio1 对象的内容传到 mTextView1*/
```

```
                mTextView1.setText(mRadio1.getText());
            }
            else if(checkedId==mRadio2.getId())
            {
                /*把 mRadio2 对象的内容传到 mTextView1*/
                mTextView1.setText(mRadio2.getText());
            }
        }
    };
}
```

实例执行后的初始效果如图 4-12 所示，当选择一个选项后会显示出选择的值，如图 4-13 所示。

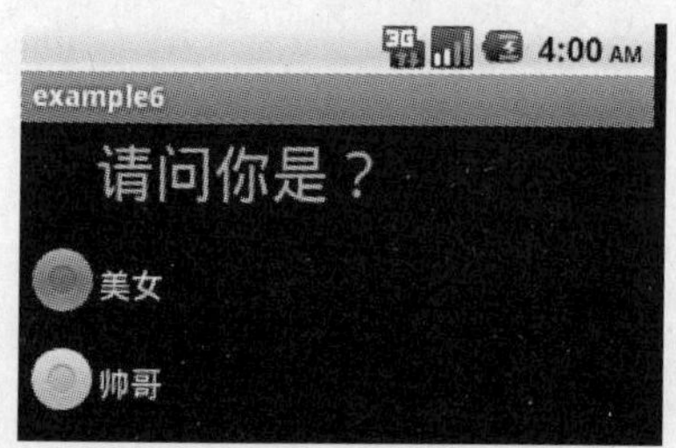

图 4-12　初始效果

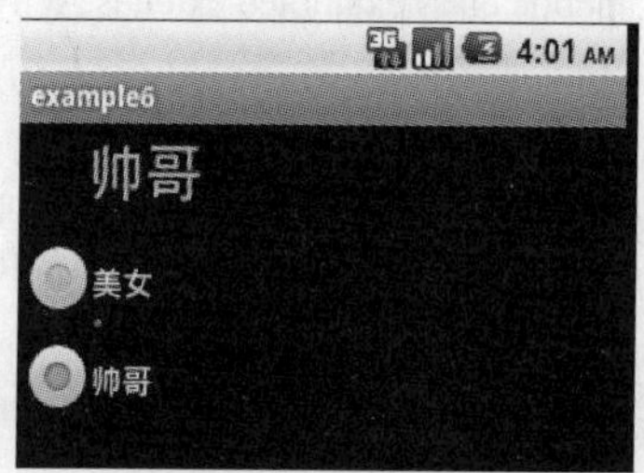

图 4-13　输出提示

4.6　ImageView 实现相框效果

在本节实例中，事先准备了三张图片（两张外框图、一张内框图），将这三张图片放在"res/drawable"文件夹下面，在此使用的图片为 PNG 图形文件，而图案大小调整成了手机屏幕大小，当然也可以依据手机的分辨率，动态调整 ImageView（图片组件）的大小。

然后在布局当中创建了两个 ImageView，且以绝对定位的方式"堆积"在一起，在其下方放上两个按钮，按钮的目的是为了要用来切换图片，创建完成后，要在 Button 事件里处理置换图片的动作。

当单击 Button1 后，ImageView1 会出现 right 的图片；单击 Button2 后，ImageView1 对象会出现 left 的图片，而 ImageView2 对象皆为固定不动（文件名叫 oa）。

本实例实现代码保存在"光盘:\daima\4\example7"，下面开始讲解本实例的具体实现流程。

在主程序文件 example7.java 中，其核心功能是通过 getResources()方法实现的，此方法负责访问 Resource ID，访问资源里的图文件、文字都要用到 getResources()；在此使用 getResources().getDrawable() 来载入"res/drawable"目录里的图文件，并将图片放置在 ImageView 当中。文件 example7.java 的具体实现代码如下。

```
package irdc.example7;

import irdc.example7.R;
```

```
import android.app.Activity;
import android.os.Bundle;
import android.view.View;
import android.widget.Button;
import android.widget.ImageView;

public class example7 extends Activity
{
  /*声明 Button、ImageView 对象*/
  private ImageView mImageView01;
  private ImageView mImageView02;
  private Button mButton01;
  private Button mButton02;

  /** Called when the activity is first created. */
  @Override
  public void onCreate(Bundle savedInstanceState)
  {
    super.onCreate(savedInstanceState);
    setContentView(R.layout.main);

    /*取得 Button、ImageView 对象*/
    mImageView01 = (ImageView)findViewById(R.id.myImageView1);
    mImageView02 = (ImageView)findViewById(R.id.myImageView2);
    mButton01 = (Button) findViewById(R.id.myButton1);
    mButton02 = (Button) findViewById(R.id.myButton2);

    /*设置 ImageView 背景图*/
    mImageView01.setImageDrawable(getResources().
                  getDrawable(R.drawable.right));
    mImageView02.setImageDrawable(getResources().
                  getDrawable(R.drawable.oa));

    /*用 OnClickListener 事件来启动*/
    mButton01.setOnClickListener(new Button.OnClickListener()
    {
      @Override
      public void onClick(View v)
      {
      /*当启动后，ImageView 立刻换背景图*/
      mImageView01.setImageDrawable(getResources().
                    getDrawable(R.drawable.right));
      }
    });
```

```
        mButton02.setOnClickListener(new Button.OnClickListener()
        {
          @Override
          public void onClick(View v)
          {
          mImageView01.setImageDrawable(getResources().
                      getDrawable(R.drawable.left));
          }
        });

    }
}
```

实例执行后的初始效果如图 4-14 所示，当分别单击按钮“pic1”和“pic2”后，会分别显示不同素材的相框，如图 4-15 所示。

图 4-14　初始效果

图 4-15　不同素材相框

对于上述实例，读者可以将两个 ImageButton（图片按钮）堆放在一起，如此一来，不但有背景图，而且还有按钮事件可以触发。代码如下。

```
<ImageButton
  android:id="@+id/myImageButton1"
  android:state_focused="true"
  android:layout_width="320px"
  android:layout_height="280px"
  android:layout_x="0px"
  android:layout_y="36px"
/>
```

ImageButton 的使用方法比较简单，而堆栈的技巧可参考这个范例程序，比较不同的地方就是只要单击图片，即可直接做换图的动作，不需要再单击下面的 Button 做更换，需要注意

的是图片大小要作调整，否则可能会与 ImageButton 不合。

4.7　Spinner 实现选择处理

Spinner 是一个拉菜单，类似于 Swing 中的 ComboBox、HTML 的 <select>。因为手机画面大小有限，所以要在有限的范围选择项目，下拉菜单是唯一、也是较好的选择。

在本节实例中，将自定义下拉菜单里的样式，然后调用 setDropDownViewResource（设置下拉菜单的显示方式）方法，以 XML 的方式定义下拉菜单要显示的模样。实例中除了自定义下拉菜单，还用程序设计了一段动画，当用户以触控的方式点击这个自定义的 Spinner 时，会以一段动画提示用户。

本实例实现代码保存在“光盘:\daima\4\example8”，下面开始讲解本实例的具体实现流程。

在主程序文件 example8.java 中，在新建 ArrayAdapter（数据视图对象时）使用 ArrayAdapter (Context context, int textViewResourceId, T[] objects)这个 Constructor，textViewResourceId 使用 Android 提供的 ResourceID，objects 为必须传递的字符串数组（String Array）。文件 example8.java 的具体实现代码如下。

```
package irdc.example8;

import irdc.example8.R;
import android.app.Activity;
import android.os.Bundle;
import android.view.MotionEvent;
import android.view.View;
import android.view.animation.Animation;
import android.view.animation.AnimationUtils;
import android.widget.AdapterView;
import android.widget.ArrayAdapter;
import android.widget.TextView;
import android.widget.ListView;
import android.widget.Spinner;

public class example8 extends Activity
{
  private static final String[] countriesStr =
  { "美国", "日本", "英国", "法国" };
  private TextView myTextView;
  private Spinner mySpinner;
  private ArrayAdapter<String> adapter;
  Animation myAnimation;

  /** Called when the activity is first created. */
  @Override
  public void onCreate(Bundle savedInstanceState)
  {
```

```
        super.onCreate(savedInstanceState);
        /* 载入 main.xml Layout */
        setContentView(R.layout.main);

        /* 以 findViewById()取得对象 */
        myTextView = (TextView) findViewById(R.id.myTextView);
        mySpinner = (Spinner) findViewById(R.id.mySpinner);

        adapter = new ArrayAdapter<String>(this,
              android.R.layout.simple_spinner_item, countriesStr);
        /* myspinner_dropdown 为自定义下拉菜单模式定义在 res/layout 目录下 */
        adapter.setDropDownViewResource(R.layout.myspinner_dropdown);

        /* 将 ArrayAdapter 添加 Spinner 对象中 */
        mySpinner.setAdapter(adapter);

        /* 将 mySpinner 添加 OnItemSelectedListener */
        mySpinner.setOnItemSelectedListener
         (new Spinner.OnItemSelectedListener()
        {
          @Override
          public void onItemSelected
           (AdapterView<?> arg0, View arg1, int arg2,
              long arg3)
          {
            /* 将所选 mySpinner 的值带入 myTextView 中 */
            myTextView.setText("选择的是" + countriesStr[arg2]);
            /* 显示 mySpinner 对象的内容 */
            arg0.setVisibility(View.VISIBLE);
          }

          @Override
          public void onNothingSelected(AdapterView<?> arg0)
          {
          }
        });

        /* 取得 Animation 定义在 res/anim 目录下 */
        myAnimation = AnimationUtils.loadAnimation(this, R.anim.my_anim);

        /* 将为 mySpinner 添加 OnTouchListener（触摸监听事件） */
        mySpinner.setOnTouchListener(new Spinner.OnTouchListener()
        {

          @Override
          public boolean onTouch(View v, MotionEvent event)
```

```
            {
                /* 将 mySpinner 运行 Animation 效果的动画 */
                v.startAnimation(myAnimation);
                /* 将 mySpinner 对象隐藏 */
                v.setVisibility(View.INVISIBLE);
                return false;
            }

        });

        mySpinner.setOnFocusChangeListener(new Spinner.OnFocusChangeListener()
        {
            @Override
            public void onFocusChange(View v, boolean hasFocus)
            {
                // TODO Auto-generated method stub
            }
        });
    }
}
```

Adapter 的 setDropDownViewResource 可以设置下拉菜单的显示方式，将该 XML 定义在 res/layout 目录下面，可针对下拉菜单中的文本框进行设置，如同本程序里的 R.layout.myspinner_dropdown 即为自定义的下拉菜单文本框样式。除了改变下拉菜单样式外，也对 Spinner 做了一点动态效果，单击下拉菜单时，晃动下拉菜单再出现下拉菜单（myAnimation）。

执行后的初始效果如图 4-16 所示，单击下拉菜单后会显示一个悬浮选项效果，里面显示了 4 个国家供用户选择，如图 4-17 所示。

图 4-16　初始效果

图 4-17　4 个选项

当选择一个选项后，会显示出对应的提示信息，例如选择了“意大利”，效果如图 4-18 所示。

图 4-18　提示信息

4.8 Gallery 实现相簿功能

在上一章的内容中，通过 Gallery（图像库组件）实现了一个相片画廊的效果。在本节实例中，首先将 6 张 PNG 图片导入 Drawable 文件夹当中，并于 onCreate 的同时，载入到 Gallery Widget 中；然后，添加一个 OnItemClick（当单击 Gallery 中不同的图像时触发）的事件，以取得图片的 ID 编号来响应用户点击图片时的状态，完成 Gallery 的高级使用。

本实例的另一个重点是设置 Gallery 图片的宽高，以及设置图片 Layout 的大小，在此编写一个继承自 BaseAdapter（基本容器）的 ImageAdapter（图像构造器）来存放图片，通过 ImageView.setScaleType()方法来改变图片的显示方式，再通过 setLayoutParams()方法来改变 Layout 的宽高。

本实例实现代码保存在“光盘:\daima\4\example9”，下面开始讲解本实例的具体实现流程。

在主程序文件 example9.java 中有 2 个重点需要注意。

1）ImageAdapter 继承 BaseAdapter class 的未实现方法的重写构造。

2）Gallery 的 OnItemClick() 方法与图片及 Layout 宽高设置。

文件 example9.java 的具体实现代码如下。

```
package irdc.example9;

import irdc.example9.R;
import android.app.Activity;
import android.os.Bundle;

/* 本范例需使用到的类 */
import android.content.Context;
import android.content.res.TypedArray;
import android.view.View;
import android.view.ViewGroup;
import android.widget.AdapterView;
import android.widget.BaseAdapter;
import android.widget.Gallery;
import android.widget.ImageView;
import android.widget.Toast;
import android.widget.AdapterView.OnItemClickListener;

public class example9 extends Activity
{
  /** Called when the activity is first created. */
  @Override
  public void onCreate(Bundle savedInstanceState)
  {
    super.onCreate(savedInstanceState);
    setContentView(R.layout.main);
    /*通过 findViewById 取得*/
```

```
    Gallery g = (Gallery) findViewById(R.id.mygallery);
    /* 添加一 ImageAdapter 并设置给 Gallery 对象 */
    g.setAdapter(new ImageAdapter(this));

    /* 设置一个 itemclickListener 并 Toast（提示）被单击图片的位置 */
    g.setOnItemClickListener(new OnItemClickListener()
    {
      public void onItemClick
      (AdapterView<?> parent, View v, int position, long id)
      {
        Toast.makeText
        (example9.this, getString(R.string.my_gallery_text_pre)
        + position+ getString(R.string.my_gallery_text_post),
        Toast.LENGTH_SHORT).show();
      }
    });
  }

  /* 改写 BaseAdapter 自定义一 ImageAdapter class */
  public class ImageAdapter extends BaseAdapter
  {
    /*声明变量*/
    int mGalleryItemBackground;
    private Context mContext;

    /*ImageAdapter 的构造器*/
    public ImageAdapter(Context c)
    {
      mContext = c;

      /* 使用在 res/values/attrs.xml 中的<declare-styleable>定义
      * 的 Gallery 属性.*/
      TypedArray a = obtainStyledAttributes(R.styleable.Gallery);

      /*取得 Gallery 属性的 Index id（图像编号）*/
      mGalleryItemBackground = a.getResourceId
      (R.styleable.Gallery_android_galleryItemBackground, 0);

      /*让对象的 styleable 属性能够反复使用*/
      a.recycle();
    }

    /* 覆盖的方法 getCount,返回图片数目 */
    public int getCount()
    {
      return myImageIds.length;
```

```
        }

        /* 覆盖的方法 getItemId,返回图像的数组 id */

        public Object getItem(int position)
        {
          return position;
        }
        public long getItemId(int position)
        {
          return position;
        }

        /* 覆盖的方法 getView,返回一 View 对象 */
        public View getView
        (int position, View convertView, ViewGroup parent)
        {
          /*产生 ImageView 对象*/
          ImageView i = new ImageView(mContext);
          /*设置图片给 imageView 对象*/
          i.setImageResource(myImageIds[position]);
          /*重新设置图片的宽高*/
          i.setScaleType(ImageView.ScaleType.FIT_XY);
          /*重新设置 Layout 的宽高*/
          i.setLayoutParams(new Gallery.LayoutParams(136, 88));
          /*设置 Gallery 背景图*/
          i.setBackgroundResource(mGalleryItemBackground);
          /*返回 imageView 对象*/
          return i;
        }

        /*建构一 Integer array 并取得预加载 Drawable 的图片 id*/
        private Integer[] myImageIds =
        {
          R.drawable.photo1,
          R.drawable.photo2,
          R.drawable.photo3,
          R.drawable.photo4,
          R.drawable.photo5,
          R.drawable.photo6,
        };
      }
    }
```

这样，整个实例介绍完毕。执行后的效果如图 4-19 所示，通过滑动鼠标可以浏览每一幅图片。

图 4-19　执行效果

4.9　java.io.File 实现文件搜索

通过文件搜索功能，可以快速的帮我们找到需要的文件。同样，在手机中也可以实现文件搜索的功能。在 Java I/O 的 API 中，提供了 java.io.File 对象，只要利用 File 对象的方法，再搭配 Android 的 EditText、TextView 等组件，就可以轻松实现手机的文件搜索引擎。

在本节实例中，使用 EditText、Button 与 TextView 三种组件来实现此功能，用户将要搜索的文件名称或关键字输入 EditText 中，单击 Button 后，程序会在根目录中寻找符合的文件，并将搜索结果显示于 TextView 中；如果找不到符合的文件，则显示找不到文件。

本实例实现代码保存在“光盘:\daima\4\example10”，下面开始讲解本实例的具体实现流程。

在主程序文件 example10.java 中，以 java.io.File 对象来取得根目录下的文件，经过比较后，将符合条件的文件路径写入 TextView 中，若要在 TextView 中换行，需使用“\n”作为换行符号。文件 example10.java 的具体实现代码如下：

```
package irdc.example10;

/* 加载相关类 */
import irdc.example10.R;

import java.io.File;
import android.app.Activity;
import android.os.Bundle;
import android.view.View;
import android.widget.Button;
import android.widget.EditText;
import android.widget.TextView;

public class example10 extends Activity
{
  /*声明对象变量*/
  private Button mButton;
  private EditText mKeyword;
```

```
    private TextView mResult;

    /** Called when the activity is first created. */
    @Override
    public void onCreate(Bundle savedInstanceState)
    {
      super.onCreate(savedInstanceState);
      /* 载入布局文件 main.xml */
      setContentView(R.layout.main);

      /* 初始化对象 */
      mKeyword=(EditText)findViewById(R.id.mKeyword);
      mButton=(Button)findViewById(R.id.mButton);
      mResult=(TextView) findViewById(R.id.mResult);

      /* 将 mButton 对象添加 onClickListener 事件 */
      mButton.setOnClickListener(new Button.OnClickListener()
      {
        public void onClick(View v)
        {
        /*取得输入的关键字*/
          String keyword = mKeyword.getText().toString();
          if(keyword.equals(""))
          {
            mResult.setText("关键字不能为空!!");
          }
          else
          {
            mResult.setText(searchFile(keyword));
          }
        }
      });
    }

    /* 搜索文件的方法 */
    private String searchFile(String keyword)
    {
      String result="";
      File[] files=new File("/").listFiles();
      for( File f : files )
      {
        if(f.getName().indexOf(keyword)>=0)
        {
          result+=f.getPath()+"\n";
        }
      }
```

```
        if(result.equals("")) result="找不到文件!!";
        return result;
    }
}
```

在上述代码中，searchFile(String keyword)方法的功用是为搜索根目录下符合关键字的文件，在搜索文件的过程中，只搜索根目录中的文件，并没有再对子目录下的文件做进一步的比较，如果要再强化这个文件搜索的功能，让它也能搜索包含子目录下的所有文件，可以在程序中利用 File.isDirectory() 这个方法来判断其是否为目录。如果是的话，就继续往下一层寻找；不是的话，就终止向下寻找的动作。这个做法在后面的范例中会有详细的示范，要注意的是手机硬件环境是否能负荷程序做大规模的文件搜索，毕竟手机的硬件配备（处理器、内存）是比不上一般计算机的。

执行后的效果如图 4-20 所示，输入搜索关键字，并单击“搜索处理”按钮后会显示出对应的搜索结果，如图 4-21 所示。

图 4-20　执行效果

图 4-21　搜索效果

4.10　ImageButton 实现按钮置换

Android 的默认按钮通常都是方方正正的，在本实例中可以实现两个按钮之间的交互，即当单击 A 按钮，恢复 B 按钮图片；当单击 B 按钮，恢复 A 按钮的图片。使用的方法为单击的瞬间置换图片，置换的图片方式与先前介绍的相同（ImageButton.setImageDrawable）。

1．实现原理

在实例中使用了 3 张图片，分别是 p1.png、p2.png 以及 p3.png，而在 onCreate()时，画面 Layout 上的两个 ImageButton（图像按钮）各自显示 p1.png 以及 p2.png，当单击任一按钮，则改变自己的图片为 p3.png，并还原另一按钮为默认的图文件，以按钮事件及图文件置换的方式来提醒用户现在所“选择”的图像按钮是什么。

本实例实现代码保存在“光盘:\daima\4\ practice11”，下面开始讲解本实例的具体实现流程。

2．主程序文件

本实例的主程序文件是 practice11.java，它通过 setImageDrawable()方法来改变按钮的图片，图片是放在 res/drawable 目录下面。而 ImageButton.setOnClickListener()的角色则是本程序的关键，通过同时更换两个按钮的图片来表示被选择的图片，是一种类似被选择的假象手法，常应用于切换开关之用。文件 example11.java 的具体实现代码如下。

```
package irdc. practice11;

import irdc. practice11.R;
import android.app.Activity;
import android.os.Bundle;
import android.view.View;
import android.widget.Button;
import android.widget.ImageButton;
import android.widget.TextView;

public class example11 extends Activity
{
  TextView myTextView;
  ImageButton myImageButton_1;
  ImageButton myImageButton_2;

  /** Called when the activity is first created. */
  @Override
  public void onCreate(Bundle savedInstanceState)
  {
    super.onCreate(savedInstanceState);
    /*  载入 main.xml Layout */
    setContentView(R.layout.main);

    /*  以 findViewById()取得 TextView 及 ImageButton 对象  */
    myTextView = (TextView) findViewById(R.id.myTextView);
    myImageButton_1=(ImageButton)findViewById(R.id.myImageButton_1);
    myImageButton_2=(ImageButton)findViewById(R.id.myImageButton_2);

    /*  为 myImageButton_1 按钮添加 OnClickListener 事件  */
    myImageButton_1.setOnClickListener(new Button.OnClickListener()
    {
      public void onClick(View v)
      {
        myTextView.setText("你点击的是 myImageButton_1");
        /*  单击 myImageButton_1 按钮时将 myImageButton_1 图片置换成 p3 图片  */
        myImageButton_1.setImageDrawable(getResources().getDrawable(
              R.drawable.p3));
```

```
                /*  单击 myImageButton_1 按钮时将 myImageButton_2 图片置换成 p2 图片  */
                myImageButton_2.setImageDrawable(getResources().getDrawable(
                    R.drawable.p2));
            }
        });

        /*  为 myImageButton_2 按钮添加 OnClickListener 事件  */
        myImageButton_2.setOnClickListener(new Button.OnClickListener()
        {
            public void onClick(View v)
            {
                myTextView.setText("你点击的是 myImageButton_2");
                /*  单击 myImageButton_2 按钮时将 myImageButton_1 图片置换成 p1 图片  */
                myImageButton_1.setImageDrawable(getResources().getDrawable(
                    R.drawable.p1));
                /*  单击 myImageButton_2 按钮时将 myImageButton_2 图片置换成 p3 图片  */
                myImageButton_2.setImageDrawable(getResources().getDrawable(
                    R.drawable.p3));
            }
        });
    }
}
```

在上述代码中，除了在 res/drawable 放置图片方式外，也可以用系统 Android 操作系统默认的图片。如打电话、简短提示的图标等，只需修改 main.xml 里 ImageButton 的属性。

```
android:src="@drawable/p1"
```

修改为：

```
android:src="@android:drawable/sym_action_call"
```

在上述代码中，通过标识“@android:”就表示引用是 android 提供的，而不是自行导入的。实例执行后的效果如图 4-22 所示，当单击按钮后会显示另一幅图片。

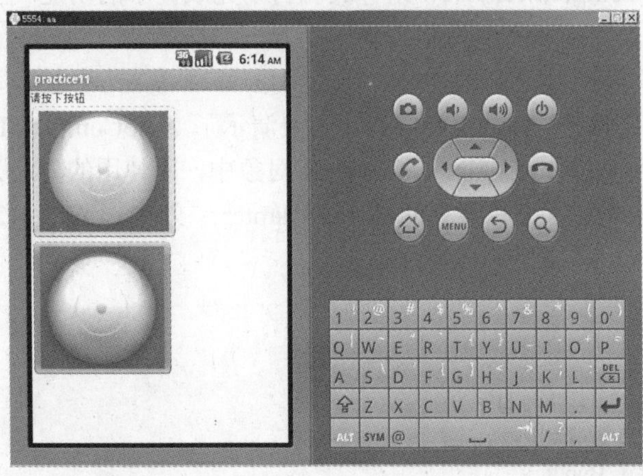

图 4-22　执行效果

4.11 AutoCompleteTextView 实现输入提示

在使用百度搜索时，只需输入几个文字，就会显示可能的关键字供用户选择。如图 4-23 所示。

图 4-23　百度的搜索提示功能

上述功能在 Android 中是非常容易实现的，通过 Android 的 AutoCompleteTextView（自动提示输入来源组件），只要搭配 ArrayAdapter（数据视图）就能设计出类似百度搜索提示的效果。

1．实例介绍

本范例先在 Layout 当中布局一个 AutoCompleteTextView Widget，然后通过预先设置好的字符串数组，将此字符串数组放入 ArrayAdapter，最后利用 AutoCompleteTextView.setAdapter 方法，就可以让 AutoCompleteTextView 组件具有自动完成提示的功能。例如，只要输入“ab”，就会自动带出包含“ab”的所有字符串列表。

2．主程序文件

本实例的主程序文件是 example12.java，主要演示了 AutoCompleteTextView 的用法。此外，将 ArrayAdapter 添加 AutoCompleteTextView 对象中，所使用的方法为 setAdapter()，当中传输唯一的参数类型即为字符串类型的 ArrayAdapter。文件 example12.java 的具体实现代码如下。

```
package irdc.example12;

import irdc.example12.R;
import android.app.Activity;
import android.os.Bundle;
import android.widget.ArrayAdapter;
```

```
import android.widget.AutoCompleteTextView;

public class example12 extends Activity
{
  private static final String[] autoStr = new String[]
  { "a", "abc", "abcd", "abcde" };

  /** Called when the activity is first created. */
  @Override
  public void onCreate(Bundle savedInstanceState)
  {
    super.onCreate(savedInstanceState);
    /* 载入布局文件 main.xml */
    setContentView(R.layout.main);

    /* 新建 ArrayAdapter 对象，并将 autoStr 字符串数组传入这个对象 */
    ArrayAdapter<String> adapter = new ArrayAdapter<String>(this,
        android.R.layout.simple_dropdown_item_1line, autoStr);

    /* 以 findViewById()方法取得 AutoCompleteTextView 对象的内容 */
    AutoCompleteTextView myAutoCompleteTextView =
    (AutoCompleteTextView) findViewById(R.id.myAutoCompleteTextView);

    /* 将 ArrayAdapter 添加 AutoCompleteTextView 对象中 */
    myAutoCompleteTextView.setAdapter(adapter);
  }
}
```

执行后会显示一个输入表单，当输入一个字符时会显示对应的提示词组。如图 4-24 所示。

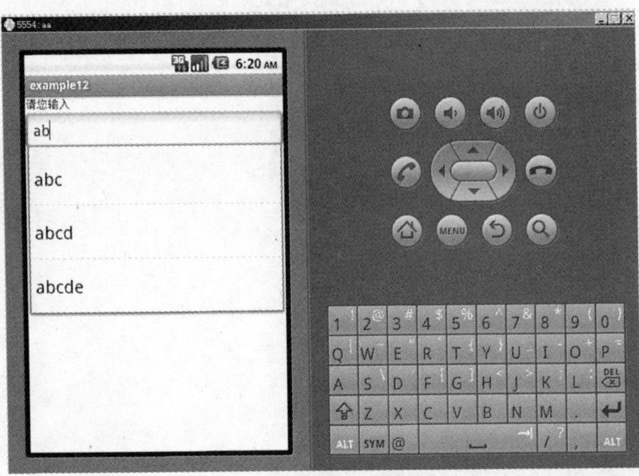

图 4-24　输入提示

另外，在 Android 中有个类似 AutoCompleteTextView 的对象 MultiAutoCompleteTextView，

它继承了 AutoCompleteTextView，差别在于它可以在输入框一直增加新的选择值，其编写方式也有些不同，一定要 setTokenizer，否则会出现错误。

4.12 AnalogClock 实现时钟效果

Android 中的 AnalogClock（时钟组件）是一个时钟对象，本节实例将实现一个时钟效果，并在其下放置一个 TextView，为了做对照，上面放置的为模拟时钟，下面的 TextView 用于模拟电子时钟，将 AnalogClock 的时间以数字钟形式显示。

1．实现原理

在具体实现上，是 android.os.Handler、java.lang.Thread 以及 android.os.Message 三对象的整合应用的结果，通过产生 Thread（线程）对象，在进程内同步调用 System.currentTimeMillis() 取得系统时间，并通过 Message（通知）对象来通知 Handler（操作）对象，Handler 则扮演联系 Activity 与 Thread 之间的桥梁，在收到 Message 对象后，将时间变量的值，显示于 TextView 当中，产生数字时钟的外观与功能。

本实例实现代码保存在“光盘:\daima\4\example13”，下面开始讲解本实例的具体实现流程。

2．主程序介绍

本实例的主程序文件是 example13.java，它需要另外加载 Java 的 Calendar 与 Thread 对象，并在 onCreate()中构造 Handler 与 Thread 两对象，并实现 handelMessage()和 run()两个方法，文件 example13.java 的具体实现代码如下。

```
package irdc.example13;

import android.app.Activity;
import android.os.Bundle;
/*这里我们需要使用 Handler 类与 Message 类来处理运行线程*/
import android.os.Handler;
import android.os.Message;
import android.widget.AnalogClock;
import android.widget.TextView;
/*需要使用 Java 的日历与线程类来取得系统时间*/
import irdc.example13.R;

import java.util.Calendar;
import java.lang.Thread;

public class example13 extends Activity
{
  /*声明一常数作为判别信息用*/
  protected static final int GUINOTIFIER = 0x1234;

  /*声明两个 widget 对象变量*/
  private TextView mTextView;
```

```
public AnalogClock mAnalogClock;

/*声明与时间相关的变量*/
public Calendar mCalendar;
public int mMinutes;
public int mHour;

/*声明关键 Handler 与 Thread 变量*/
public Handler mHandler;
private Thread mClockThread;

/** Called when the activity is first created. */
public void onCreate(Bundle savedInstanceState)
{
  super.onCreate(savedInstanceState);
  setContentView(R.layout.main);

  /*通过 findViewById 取得两个 widget 对象*/
  mTextView=(TextView)findViewById(R.id.myTextView);
  mAnalogClock=(AnalogClock)findViewById(R.id.myAnalogClock);

  /*通过 Handler 来接收运行线程所传递的信息并更新 TextView*/
  mHandler = new Handler()
  {
    public void handleMessage(Message msg)
    {
      /*这里是处理信息的方法*/
      switch (msg.what)
      {

        case example13.GUINOTIFIER:
        /*  在这处理要 TextView 对象 Show 时间的事件  */
          mTextView.setText(mHour+" : "+mMinutes);
          break;
      }
      super.handleMessage(msg);
    }
  };

  /*通过运行线程来持续取得系统时间*/
   mClockThread=new LooperThread();
   mClockThread.start();
}

/*改写一个线程类用来持续取得系统时间*/
class LooperThread extends Thread
{
```

```
            public void run()
            {
               super.run();
               try
               {
                  do
                  {
                     /*取得系统时间*/
                     long time = System.currentTimeMillis();
                     /*通过日历对象来取得小时与分钟*/
                     final Calendar mCalendar = Calendar.getInstance();
                     mCalendar.setTimeInMillis(time);
                     mHour = mCalendar.get(Calendar.HOUR);
                     mMinutes = mCalendar.get(Calendar.MINUTE);

                     /*让运行线程休息一秒*/
                     Thread.sleep(1000);
                      /*重要关键程序:取得时间后发出信息给 Handler*/
                      Message m = new Message();
                      m.what = example13.GUINOTIFIER;
                      example13.this.mHandler.sendMessage(m);
                  }while(example13.LooperThread.interrupted()==false);
                  /*当系统发出中断信息时停止本循环*/
               }
               catch(Exception e)
               {
                  e.printStackTrace();
               }
            }
         }
      }
```

在上述代码中，要达到本范例效果的代码应该只有 2 行，即对 TextView 和线程 Thread 的处理，改为使用 widget.DigitalClock 的方式，具体写法如下。

```
import android.widget.AnalogClock
mDigitalClock =(DigitalClock)findViewById(R.id.myDigitalClock);
```

本实例中使用了 TextView 来模拟 DigitalClock（数字时钟）的做法，实际上，也是参考 AnalogClock（模拟时钟）与 DigitalClock 这两个 Widget 的程序代码所做的练习，对于将来实现 Timer 相关的小对象，会有所帮助。

执行后会显示一个数字时钟效果。如图 4-25 所示。

另外，Android 还提供了 System.currentTimeMillis()、uptimeMillis()和 elapsedRealtime()这三种不同特性的 System Clock 给开发者使用。本范例使用 System.currentTimeMillis() 就是标准的 Clock 用法，需要搭配真实的日期与时间使用，另外两者则是适用于 interval 与 elapse time 来控制程序和 UI。

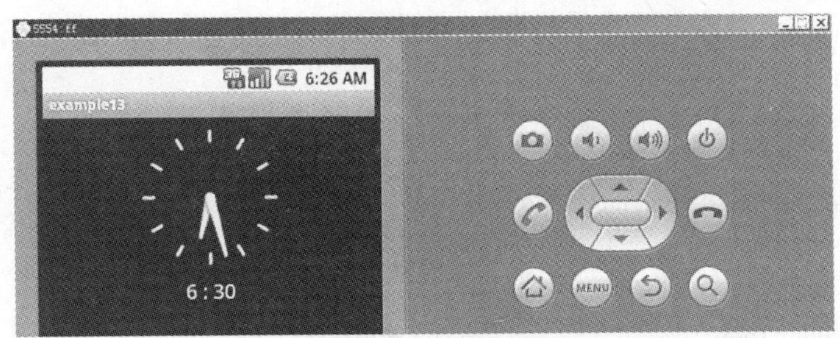

图 4-25　执行效果

4.13　DatePicker 和 TimePicker 实现时间选择

在现实中的许多应用中，经常需要用户输入日期与时间格式的数据，很多站点给用户提供了直接在万年历上来选择日期与时间，选择完毕后会自动将日期与时间带入需要填写的字段中。

1．实例介绍

在 Android 中也提供了类似的组件用于实现上述功能。在本实例中，将演示使用 Android API 中提供的 DatePicker（日期选择组件）与 TimePicker（时间选择组件）两种对象来实现动态输入日期与时间的功能。实例中使用了 DatePicker、TimePicker 与 TextView 三种对象，以 TextView 来显示日期与时间，默认带入目前系统的日期与时间，DatePicker 与 TimePicker 可让用户动态调整日期与时间，当用户调整了 DatePicker 的日期或 TimePicker 时间时，则 TextView 中所显示的日期与时间亦会跟着改变。

本实例实现代码保存在“光盘:\daima\4\example14”，下面开始讲解本实例的具体实现流程。

2．主程序介绍

本实例的主程序文件是 example14.java，以 updateDisplay()这个方法来设置 TextView 中所显示的日期时间，以 java.util.Calendar 对象来取得目前的系统时间，并预先带入 TextView 中。当用户更改了 DatePicker 里的年、月、日时，将触发 DatePicker 的 onDateChange() 事件，运行 updateDisplay()来重新设置 TextView 中显示的日期；同样的原理，当用户更改了 TimePicker 里的时间，会触发 TimePicker 的 onTimeChange() 事件，运行 updateDisplay() 来重新设置 TextView 中显示的时间。

文件 example14.java 的具体实现代码如下。

```
package irdc.example14;

/* 加载相关类 */
import irdc.example14.R;

import java.util.Calendar;
import android.app.Activity;
import android.os.Bundle;
```

```
import android.widget.TextView;
import android.widget.DatePicker;
import android.widget.TimePicker;

public class example14 extends Activity
{
  /*声明日期及时间变量*/
  private int mYear;
  private int mMonth;
  private int mDay;
  private int mHour;
  private int mMinute;
  /*声明对象变量*/
  TextView tv;
  TimePicker tp;
  DatePicker dp;

  /** Called when the activity is first created. */
  @Override
  public void onCreate(Bundle savedInstanceState)
  {
    /*取得目前日期与时间*/
    Calendar c=Calendar.getInstance();
    mYear=c.get(Calendar.YEAR);
    mMonth=c.get(Calendar.MONTH);
    mDay=c.get(Calendar.DAY_OF_MONTH);
    mHour=c.get(Calendar.HOUR_OF_DAY);
    mMinute=c.get(Calendar.MINUTE);

    super.onCreate(savedInstanceState);
    /* 载入 main.xml Layout */
    setContentView(R.layout.main);

    /*取得 TextView 对象，并调用更新显示方法 updateDisplay()
      来设置显示的初始日期时间*/
    tv= (TextView) findViewById(R.id.showTime);
    updateDisplay();

    /*取得 DatePicker 对象，以 init()
      设置初始值，通过 onDateChangeListener()监听时间变化 */
    dp=(DatePicker)findViewById(R.id.dPicker);
    dp.init(mYear,mMonth,mDay,new DatePicker.OnDateChangedListener()
    {
      @Override
      public void onDateChanged(DatePicker view,int year,
                               int monthOfYear,int dayOfMonth)
```

```
    {
      mYear=year;
      mMonth= monthOfYear;
      mDay=dayOfMonth;
      /*调用 updateDisplay()来改变显示日期*/
      updateDisplay();
    }
  });
  /*取得 TimePicker 对象，并设置为 24 小时制显示*/
  tp=(TimePicker)findViewById(R.id.tPicker);
  tp.setIs24HourView(true);

  /*setOnTimeChangedListener，并覆盖 onTimeChanged event*/
  tp.setOnTimeChangedListener(new TimePicker.OnTimeChangedListener()
  {
    @Override
    public void onTimeChanged(TimePicker view,
                              int hourOfDay,
                              int minute)
    {
      mHour=hourOfDay;
      mMinute=minute;
      /*调用 updateDisplay()来改变显示时间*/
      updateDisplay();
    }
  });
}

/*设置显示日期时间的方法*/
private void updateDisplay()
{
  tv.setText(
    new StringBuilder().append(mYear).append("/")
                       .append(format(mMonth + 1)).append("/")
                       .append(format(mDay)).append("   ")
                       .append(format(mHour)).append(":")
                       .append(format(mMinute))
  );
}

/*日期时间显示两位数的方法*/
private String format(int x)
{
  String s=""+x;
  if(s.length()==1) s="0"+s;
```

```
            return s;
        }
    }
```

执行后会显示一个数字时钟效果。用户可以分别单击 + 和 − 来自动选择日期和时间。具体效果如图 4-26 所示。

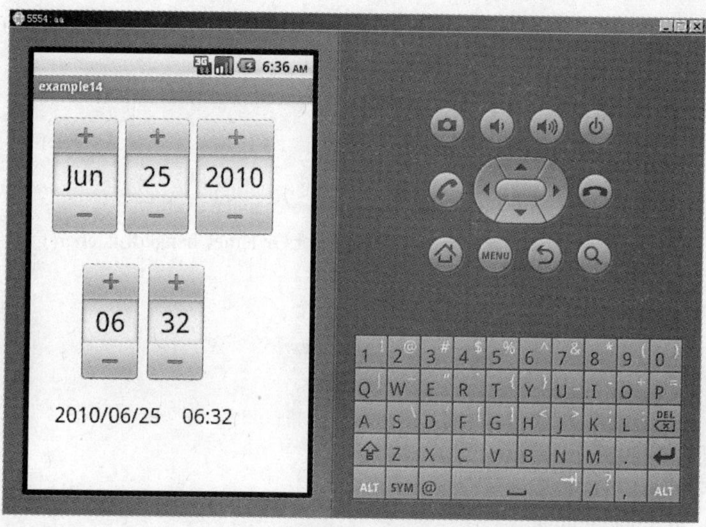

图 4-26 执行效果

4.14 ProgressBar 和 Handler 实现进度条提示

ProgressBar 是一个进度条组件，能够实现网络中的进度条效果。如图 4-27 所示。

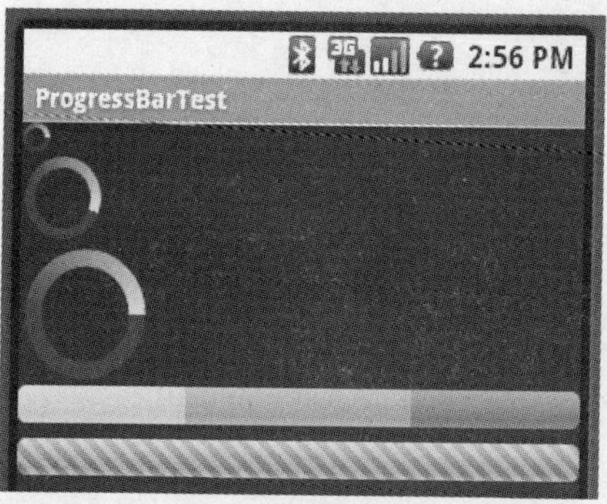

图 4-27 进度条

进度条分不确定和确定 2 种。默认值是不确定（indeterminate=true）。在图 4-27 中的第 1-3 个是不确定进度条，第 4 个是确定进度条。

进度条有 4 种风格可以使用。

- ❑ 默认值是 progressBarStyle，图 4-27 中的的第 2 个
- ❑ 设置成 progressBarStyleSmall 后，图标变小。图 4-27 中的第 1 个。

- 设置成 progressBarStyleLarge 后，图标变大。图 4-27 中的第 3 个。
- 设置成 progressBarStyleHorizontal 后，变成横向长方形。图 4-27 中的第 4 个。

在本节实例中，使用 ProgressBar（进度条组件）和 Handler 实现了后台进度提示效果。不但使用 ProgressBar 实现了进度条，而且利用 Handler 实现了新进程对 Activity 中 Widget 的访问，并将运行状态显示出来。通过 Handler 对象和 Message 对象，将进程里的状态往外传递，最后由 Activity 的 Handler 事件来取得运行状态。

本实例实现代码保存在“光盘:\daima\4\example15”，主程序文件是 example15.java，具体实现代码如下。

```
package irdc.example15;

import irdc.example15.R;
import android.app.Activity;
import android.os.Bundle;
import android.os.Handler;
import android.os.Message;
import android.view.View;
import android.widget.Button;
import android.widget.ProgressBar;
import android.widget.TextView;

public class example15 extends Activity
{
  private TextView mTextView01;
  private Button mButton01;
  private ProgressBar mProgressBar01;
  public int intCounter=0;

  /* 自定义 Handler 信息代码，用以作为识别事件处理 */
  protected static final int GUI_STOP_NOTIFIER = 0x108;
  protected static final int GUI_THREADING_NOTIFIER = 0x109;

  /** Called when the activity is first created. */
  @Override
  public void onCreate(Bundle savedInstanceState)
  {
    super.onCreate(savedInstanceState);
    setContentView(R.layout.main);

    mButton01 = (Button)findViewById(R.id.myButton1);
    mTextView01 = (TextView)findViewById(R.id.myTextView1);

    /* 设置进度条对象 */
    mProgressBar01 = (ProgressBar)findViewById(R.id.myProgressBar1);
```

```
/* 调用 setIndeterminate 方法赋值 indeterminate 模式为 false */
mProgressBar01.setIndeterminate(false);

/* 当单击按钮后，开始运行线程工作 */
mButton01.setOnClickListener(new Button.OnClickListener()
{
  @Override
  public void onClick(View v)
  {

    /* 单击按钮让进度条显示 */
    mTextView01.setText(R.string.str_progress_start);

    /* 将隐藏的进度条显示出来 */
    mProgressBar01.setVisibility(View.VISIBLE);

    /* 指定 mProgressBar01 对象为最多 100 */
    mProgressBar01.setMax(100);

    /* 初始化 mProgressBar01 对象为 0 */
    mProgressBar01.setProgress(0);

    /* 起始一个运行线程 */
    new Thread(new Runnable()
    {
      public void run()
      {
        /* 默认 0 至 9，共运行 10 次的循环叙述 */
        for (int i=0;i<10;i++)
        {
          try
          {
            /* 成员变量，用以识别加载进度 */
            intCounter = (i+1)*20;
            /* 每运行一次循环，即暂停 1 秒 */
            Thread.sleep(1000);

            /* 当 Thread 运行 5 秒后显示运行结束 */
            if(i==4)
            {
              /* 以 Message 对象，传递参数给 Handler */
              Message m = new Message();

              /* 以 what 属性指定 User 自定义 */
              m.what = example15.GUI_STOP_NOTIFIER;
              example15.this.myMessageHandler.sendMessage(m);
```

```
                break;
              }
              else
              {
                Message m = new Message();
                m.what = example15.GUI_THREADING_NOTIFIER;
                example15.this.myMessageHandler.sendMessage(m);
              }
            }
            catch(Exception e)
            {
              e.printStackTrace();
            }
          }
        }
      }).start();
    }
  });
}

/* Handler 建构之后，会聆听传来的信息 */
Handler myMessageHandler = new Handler()
{
  /* @Override */
  public void handleMessage(Message msg)
  {
    switch (msg.what)
    {
      /* 当取得识别为 离开运行线程时所取得的信息 */
      case example15.GUI_STOP_NOTIFIER:

        /* 显示运行结束 */
        mTextView01.setText(R.string.str_progress_done);

        /* 设置 ProgressBar 组件为隐藏状态 */
        mProgressBar01.setVisibility(View.GONE);
        Thread.currentThread().interrupt();
        break;

      /* 当取得识别为 持续在运行线程当中时所取得的信息 */
      case example15.GUI_THREADING_NOTIFIER:
        if(!Thread.currentThread().isInterrupted())
        {
          mProgressBar01.setProgress(intCounter);
          /* 将显示进度显示在文本框当中 */
```

```
                    mTextView01.setText
                    (
                      getResources().getText(R.string.str_progress_start)+
                      "("+Integer.toString(intCounter)+"%)\n"+
                      "Progress:"+
                      Integer.toString(mProgressBar01.getProgress())+
                      "\n"+"Indeterminate:"+
                      Boolean.toString(mProgressBar01.isIndeterminate())
                    );
                  }
                  break;
              }
              super.handleMessage(msg);
          }
      };
  }
```

在本实例中，设计了一个按钮，此按钮将部署在 main.xml 中的进度条控件显示出来。在默认的 main.xml 中没有指定它的 indeterm-inate 属性，所以即使在程序中调用了 ProgressBar 的 setindeterminate()方法，无法改变 ProgressBar.getProgress 的值，此值永远是 0.所以在此使用了循环动画来当作运行过程中的显示素材，并使用了一个 counter 整数来递增，表示运行的百分比。

执行后会显示一个按钮界面，用户单击“按下后开始运行”按钮后会显示进度条提示效果。具体效果如图 4-28 所示。

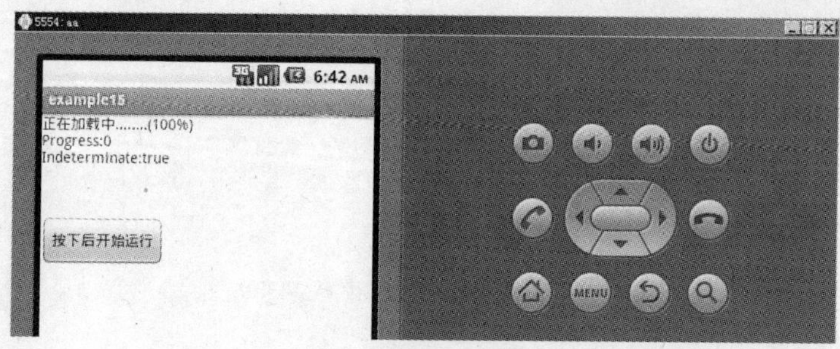

图 4-28　执行效果

4.15　网格视图控件和 ArrayAdapter 实现动态排版

众所周知，GridView 是一个网格化的二维排版配置视图。在本节实例中，将演示如何使用网格视图控件来实现动态排版处理。在具体实现上，插入了 2 个按钮，作为动态放入网格视图组件的开关。一个按钮设置 GridView 为两列显示样式，放入 4 个 Item；另一个按钮指定为 3 列显示样式，放入 9 个 Item（条目），这样就实现了对文字的动态排版处理。

本实例实现代码保存在“光盘:\daima\4\example16”，主程序文件是 example16.java，具体

实现代码如下。

```
package irdc.example16;

import irdc.example16.R;
import android.app.Activity;
import android.os.Bundle;
import android.view.View;
import android.widget.AdapterView;
import android.widget.ArrayAdapter;
import android.widget.Button;
import android.widget.GridView;
import android.widget.TextView;

public class example16 extends Activity
{
  private TextView mTextView01;
  private Button mButton01,mButton02;
  private GridView mGridView01;
  private String[] mGames1,mGames2;
  private ArrayAdapter<String> aryAdapter1;

  /** Called when the activity is first created. */
  @Override
  public void onCreate(Bundle savedInstanceState)
  {
    super.onCreate(savedInstanceState);
    setContentView(R.layout.main);

    /* 4 个字符串数组 */
    mGames1 = new String[]
                  {
                      getResources().getString(R.string.str_list1),
                      getResources().getString(R.string.str_list2),
                      getResources().getString(R.string.str_list3),
                      getResources().getString(R.string.str_list4)
                  };

    /* 9 个字符串数组 */
    mGames2 = new String[]
                  {
                      getResources().getString(R.string.str_list1),
                      getResources().getString(R.string.str_list2),
                      getResources().getString(R.string.str_list3),
                      getResources().getString(R.string.str_list4),
                      getResources().getString(R.string.str_list1),
                      getResources().getString(R.string.str_list2),
```

```
                    getResources().getString(R.string.str_list3),
                    getResources().getString(R.string.str_list4),
                    getResources().getString(R.string.str_list1)
                  };

mButton01 = (Button)findViewById(R.id.myButton1);
mButton02 = (Button)findViewById(R.id.myButton2);
mGridView01 = (GridView)findViewById(R.id.myGridView1);

mTextView01 = (TextView)findViewById(R.id.myTextView1);

mButton01.setOnClickListener(new Button.OnClickListener()
{
  @Override
  public void onClick(View v)
  {

    /* 4 个元素，以 2 列方式呈现(2×2) */
    mGridView01.setNumColumns(2);

    aryAdapter1 = new ArrayAdapter<String>
    (example16.this, R.layout.simple_list_item_1_small, mGames1);

    mGridView01.setAdapter(aryAdapter1);
    mGridView01.setScrollBarStyle(DEFAULT_KEYS_DIALER);
    mGridView01.setSelection(2);
    mGridView01.refreshDrawableState();
  }
});

mButton02.setOnClickListener(new Button.OnClickListener()
{
  @Override
  public void onClick(View v)
  {

    /* 9 个元素，以 3 列方式呈现(3×3) */
    mGridView01.setNumColumns(3);

    aryAdapter1 = new ArrayAdapter<String>
    (example16.this, R.layout.simple_list_item_1_small, mGames2);

    mGridView01.setAdapter(aryAdapter1);
  }
});

mGridView01.setOnItemClickListener
```

```
                (new GridView.OnItemClickListener()
        {
          @Override
          public void onItemClick(AdapterView<?> parent,
                              View v, int position, long arg3)
          {

            /* 判断 Adapter 里的元素个数，判断被点击的是第几个元素名称 */
            switch(aryAdapter1.getCount())
            {
              case 4:
                /* position 为 GridView 里的元素索引值 */
                mTextView01.setText(mGames1[position]);
                break;
              case 9:
                mTextView01.setText(mGames2[position]);
                break;
            }
          }
        });
      }
    }
```

在实例中，设置了 mButton01 和 mButton02 两个按钮，设置各自 OnClickListener（单击事件），在按钮事件中处理要配置 GridView 对象的方法。GridView 的 setNumColumns 方法为设置其字段数量，为了方便配置 GridView 对象。通过使用 ArrarAdapter，Android.widget.ArrarAdapter 默认构造需要 3 个参数，并将初始化的 ArrarAdapter 对象 ArrarAdapter1 以调用 GridView.setAdapter()的方式，将 String 类型的 Item 放入到 GridView 对象中去。

执行后会显示 2 个按钮的界面，当单击“显示 4 个”按钮后会显 2×2 排列的样式，效果如图 4-39 所示；当单击“显示 9 个”按钮后会显 3×3 排列的样式，效果如图 4-40 所示。

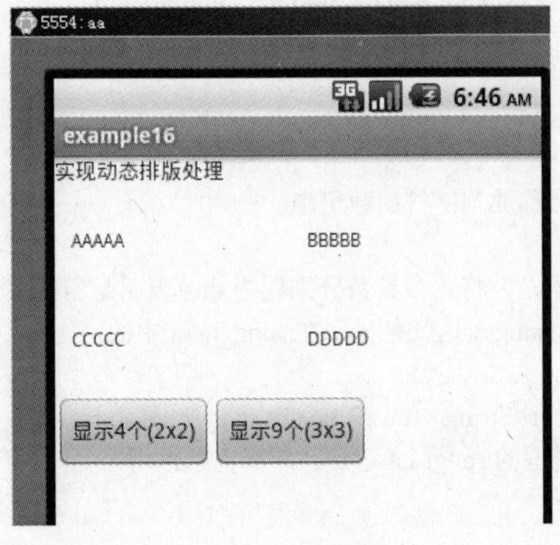

图 4-29　2×2 排列

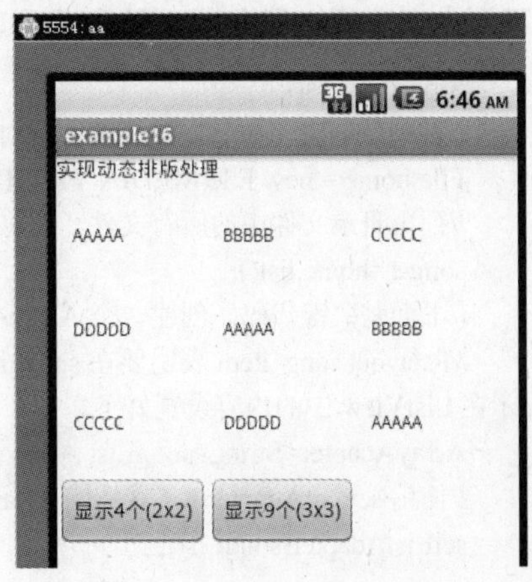

图 4-30　3×3 排列

4.16 使用 ListActivity

ListActivity（列表视图组件）也是一个独立的类，它和 Activity 是同一级别的类。ListActivity 也能够实现布局处理，能用于显示菜单列表、列表明细项目。在本节的内容中，将详细介绍 ListActivity 的基本知识。

4.16.1 ListActivity 介绍

在 android.app 包里包含了几个重要的类，ListActivity 就是其中之一。ListActivity 类其实就是一个含有一个 ListView 组件的 Activity 类。也就是说，如果直接在一个普通的 Activity 中自己加一个 ListView，也是完全可以取代这个 ListActivity 的，只是它更方便而已。

ListActivity 像数组或者是光标一样，将绑定的数据资源以陈列选项的方式显示内容。当读者选择这些选项时，将会触发一个事件。ListActivity 主持操作着一个列表视图对象，这个列表视图能绑定不同的数据资源，典型的就是一个持有查询结果的数组或者是光标。

1．屏幕布局(Screen Layout)

ListActivity 有一个默认的布局，这个布局由单一的、全屏列表构成。通过在 onCreate()方法中使用 setContentView()设置自己的视图布局来定制屏幕布局。如果要完成这些，我们自己的视图必须包含一个 id 为“@android:id/list”的 ListView 对象。如果列表为空时，可以包含另外一个视图对象，这个空的列表必须有一个“android:empty”值的 id，注意到当有一个空的视图显示时，这个列表视图将会在没有任何数据时被隐藏。

2．排布局(Row Layout)

可以在列表中确定单一的排，它是通过在活动所操作的 ListAdapter 对象中设定的一个资源布局来实现的。一个 ListAdapter 持有一个的参数来为每一行确定了所使用到的布局的资源。同时，它还有另外两个参数，这两个参数让你明确与哪个对象相互关联，通常是两个平行的数组。Android 提供了一些标准的行布局资源。这些都是在 R.layout 类中所定义的，名称像 simple_list_item_1，simple_list_item_2，two_line_list_item 之类一样。

下面先来分析一段代码，这段代码就是使用了这个方面的知识。

```
String songss[];
//首先是创建一个 File 类型的实例对象 home
File home = new File(MEDIA_PATH);
//列出目录文件中的所有文件，并将这些文件名称放到字符串数组中
songss=home.list();
//下面是关键代码，创建一个 ArrayAdapter 对象，并将三个参数分别置为 this(表明是当前上下文),
//R.layout.song_item 表明要在 song_item 中显示,songss(这就是显示在 song_item 中的内容)，这样就
将要显示在 ListView 中的内容设置好了。
ArrayAdapter<String>songList = new ArrayAdapter<String> (this,R.layout.song_item,songss);
//使用 setListAdapter()，就是使 ArrayAdapter 类型的 songList 运行，即真正起作用
setListAdapter(songList);
```

通过上述代码介绍了知识点 ListView 中 ListAdapter 的使用。

同时，Android 不仅仅提供了数组，还提供了另外两个标准的列表适配器：处理 static 数据类型的 SimpleAdapter 和处理光标结果查询的 SimpleCursorAdapter。例如，下面是一个将名称和公司信息绑定到一个两行排布局的活动列表视图（所谓的两行排指的是两行数值作一个排布局，也可以是三行作为一个排布局或者是一行作为一个行布局）。

```
public class MyListAdapter extends ListActivity {
    @Override
    protected void onCreate(Bundle icicle){
        super.onCreate(icicle);
        setContentView(R.layout.custom_list_activity_view);
        mCursor = People.query(this.getContentResolver(), null);
        startManagingCursor(mCursor);
        ListAdapter adapter = new SimpleCursorAdapter(
                this,
                android.R.layout.two_line_list_item,
                mCursor,
                new String[] {People.NAME, People.COMPANY},
                new int[]);
        setListAdapter(adapter);
    }
}
```

3．可实现的方法

如果让程序继承 ListActivity，则可以实现下面的方法。

- getListAdapter()：获取列表项目的 Adapter。
- getListView：获取列表的 View。
- getSelectItemId()：获取当前 Keypad 所选择的 Item ID。
- onContentChanged()：ListActivity 列表内容更新事件。
- onListItemClick(ListView,View,int,long)：User 在列表项目中单击触发事件。
- onRestoreInstancsState(Bundle)：还原至此实例状态事件。
- setListAdapter(ListAdapter)：设置 ListAdapter 的列表项目。
- setSelection(int)：设置所选的项目。

在使用 ListActivity 时，并不用像使用 Activity 那样必须使用 setContentView 方法来设置版型 Layout 才能显示页面。ListActivity 在不需要重写 protected void onCreate(Bundle savedInstanceState)的情况下，直接将列表加载到 ListActivity 中，这样使用起来十分方便，经常用于投票选择和多项目列表条中去。

4.16.2　Listactivity 应用方法

在本节的内容中，将通过一个具体实例的实现，来讲解使用 ListActivity 的具体流程。本实例实现代码保存在“光盘:\daima\4\example17”，是通过 Listactivity 和 Menu 联合实现的。本实例的具体实现流程如下。

第一步：打开 eclipse，依次单击“File”→“New”→“Android Project”，新建一个名为“example17”的工程文件。

第二步：编写布局文件 main.xml 实现整体布局。具体代码如下。

```
<?xml version="1.0" encoding="utf-8"?>
<TextView
    xmlns:android="http://schemas.android.com/apk/res/android"
    android:id="@+id/myTextView1"
    android:layout_width="fill_parent"
    android:layout_height="wrap_content"
    android:text="@string/hello"
/>
```

第三步：在 string.xml 添加程序中要使用的字符串，具体代码如下。

```
<?xml version="1.0" encoding="utf-8"?>
<resources>
    <string name="hello">你好雨夜</string>
    <string name="app_name">欢迎你</string>
    <string name="str_list1">精绝古城</string>
    <string name="str_list2">龙岭迷窟</string>
    <string name="str_list3">云南虫谷  </string>
    <string name="str_list4">昆仑神宫</string>
    <string name="str_list5">C++语言</string>
    <string name="str_list6">Java 语言</string>
    <string name="str_list7">PHP 语言</string>
    <string name="str_list8">Basic 语言</string>
    <string name="str_menu_list1">显示列表 1</string>
    <string name="str_menu_list2">显示列表 2</string>
</resources>
```

第四步：编写 example17.java，实现显示功能，其具体实现流程如下。

1）声明对象变量。

2）载入 main.xml Layout。

3）初始化对象，并将 mButton 添加到 onClickListener（监听单击事件）。

4）取得输入的关键字。

5）根据搜索文件的 method 和关键字搜索。

文件 shiyongListActivity.java 的具体实现代码如下：

```
package irdc.example17;

import irdc.example17.R;
import android.app.ListActivity;
import android.os.Bundle;
```

```
import android.view.Menu;
import android.view.MenuItem;
import android.view.View;
import android.widget.ArrayAdapter;
import android.widget.ListView;
import android.widget.Toast;

public class example17 extends ListActivity
{
  private int selectedItem = -1;
  private String[] mString;
  static final private int MENU_LIST1 = Menu.FIRST;
  static final private int MENU_LIST2 = Menu.FIRST+1;
  private ArrayAdapter<String> mla;

  @Override
  protected void onCreate(Bundle savedInstanceState)
  {
    /* TODO 自动引起的方法残余部分 */
    super.onCreate(savedInstanceState);
  }

  @Override
  protected void onListItemClick(ListView l, View v, int position, long id)
  {
    /* TODO 自动引起的方法残余部分 */

    selectedItem = position;
    Toast.makeText(shiyongListActivity.this, mString[selectedItem], Toast.LENGTH_SHORT).show();
    super.onListItemClick(l, v, position, id);
  }

  @Override
  public boolean onCreateOptionsMenu(Menu menu)
  {
    /* menu 组 ID */
    int idGroup1 = 0;

    /*项目的顺序位置*/
    int orderMenuItem1 = Menu.NONE;
    int orderMenuItem2 = Menu.NONE+1;

    menu.add(idGroup1, MENU_LIST1, orderMenuItem1, R.string.str_menu_list1);
```

```
        menu.add(idGroup1, MENU_LIST2, orderMenuItem2, R.string.str_menu_list2);

        return super.onCreateOptionsMenu(menu);
    }

    @Override
    public boolean onOptionsItemSelected(MenuItem item)
    {
        switch(item.getItemId())
        {
            case (MENU_LIST1):
                mString = new String[]
                                    {
                                        getResources().getString(R.string.str_list1),
                                        getResources().getString(R.string.str_list2),
                                        getResources().getString(R.string.str_list3),
                                        getResources().getString(R.string.str_list4)
                                    };

                mla = new ArrayAdapter<String>(shiyongListActivity.this, R.layout.main, mString);
                shiyongListActivity.this.setListAdapter(mla);
                break;
            case (MENU_LIST2):
                mString = new String[]
                                    {
                                        getResources().getString(R.string.str_list5),
                                        getResources().getString(R.string.str_list6),
                                        getResources().getString(R.string.str_list7),
                                        getResources().getString(R.string.str_list8)
                                    };
                mla = new ArrayAdapter<String>(shiyongListActivity.this, R.layout.main, mString);
                shiyongListActivity.this.setListAdapter(mla);
                break;
        }
        return super.onOptionsItemSelected(item);
    }

}
```

执行后将在底部显示 2 个布局块，并分别显示设置的文本，具体效果如图 4-31 所示；当单击第一个布局块后会显示弹出对应的信息，如图 4-32 所示；当单击第二个布局块后会显示弹出对应的信息，如图 4-33 所示。

图 4-31　初始效果

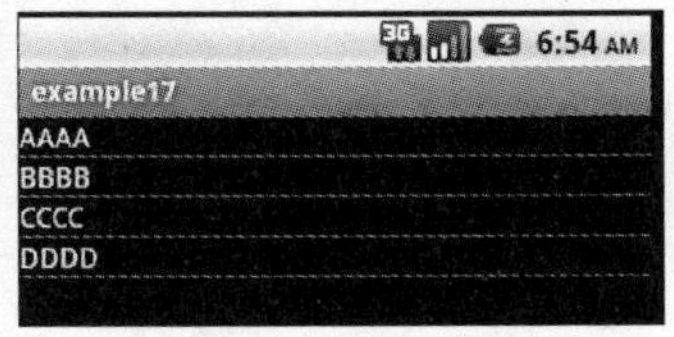

图 4-32　对应的信息

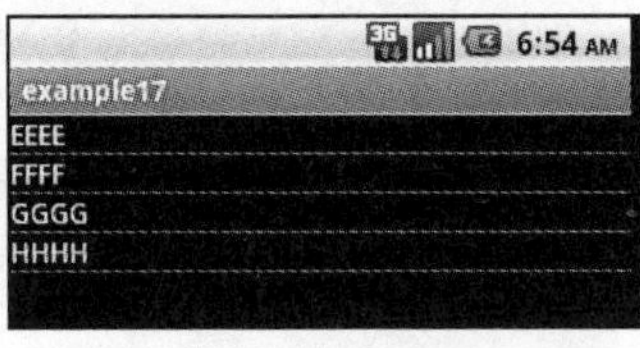

图 4-33　对应的信息

4.17　Matrix 实现图片缩放

在本节的内容中，将通过一个具体实例的实现，来讲解使用 Matrix（颜色矩阵对象）实现图片缩放的具体流程。本实例保存在“光盘:\daima\4\example18”中，其具体实现流程如下。

第一步：打开 eclipse，依次单击“File”→“New”→“Android Project”，新建一个名为“example18”的工程文件。

第二步：编写布局文件 main.xml 实现整体布局。

通过上述步骤，在页面中插入了一幅图片，名称为 suofang.bng，2 个 Button 按钮。

第三步：在 string.xml 添加程序中要使用的字符串，具体代码如下。

```
<?xml version="1.0" encoding="utf-8"?>
  <resources>
    <string name="hello">Hello</string>
    <string name="app_name">雨夜</string>
    <string name="str_button1">缩小处理</string>
    <string name="str_button2">放大处理</string>
  </resources>
```

第四步：编写处理文件 example18.java，其具体实现流程如下。

1）载入布局文件 main.xml Layout。

2）取得屏幕分辨率大小，初始化相关变量。

3）实现缩小按钮 onClickListener 处理 mButton01.setOnClickListener。

4）实现放大按钮 onClickListener 处理 mButton02.setOnClickListener。

5）定义图片缩小方法 small()。

6）定义图片放大方法 big()。

文件 suofang.java 的具体实现代码如下：

```
package irdc.example18;

/*加载相关类 */
import irdc.example18.R;
import android.app.Activity;
import android.graphics.Bitmap;
import android.graphics.BitmapFactory;
import android.graphics.Matrix;
import android.os.Bundle;
import android.util.DisplayMetrics;
import android.view.View;
import android.widget.AbsoluteLayout;
import android.widget.Button;
import android.widget.ImageView;

public class example18 extends Activity
{
  /* 相关变量声明 */
  private ImageView mImageView;
  private Button mButton01;
  private Button mButton02;
  private AbsoluteLayout layout1;
  private Bitmap bmp;
  private int id=0;
  private int displayWidth;
  private int displayHeight;
  private float scaleWidth=1;
  private float scaleHeight=1;

  /** Called when the activity is first created. */
  @Override
  public void onCreate(Bundle savedInstanceState)
  {
    super.onCreate(savedInstanceState);
    /* 载入 main.xml Layout */
    setContentView(R.layout.main);

    /* 取得屏幕分辨率大小 */
    DisplayMetrics dm=new DisplayMetrics();
    getWindowManager().getDefaultDisplay().getMetrics(dm);
    displayWidth=dm.widthPixels;
    /* 屏幕高度须扣除下方 Button 高度 */
    displayHeight=dm.heightPixels-80;
    /* 初始化相关变量 */
    bmp=BitmapFactory.decodeResource(getResources(),
                                     R.drawable.suofang);
```

```
    mImageView = (ImageView)findViewById(R.id.myImageView);
    layout1 = (AbsoluteLayout)findViewById(R.id.layout1);
    mButton01 = (Button)findViewById(R.id.myButton1);
    mButton02 = (Button)findViewById(R.id.myButton2);

    /* 缩小按钮 onClickListener */
    mButton01.setOnClickListener(new Button.OnClickListener()
    {
      @Override
      public void onClick(View v)
      {
        small();
      }
    });

    /* 放大按钮 onClickListener */
    mButton02.setOnClickListener(new Button.OnClickListener()
    {
      @Override
      public void onClick(View v)
      {
        big();
      }
    });
  }

  /* 图片缩小的 method */
  private void small()
  {
    int bmpWidth=bmp.getWidth();
    int bmpHeight=bmp.getHeight();
    /* 设置图片缩小的比例 */
    double scale=0.8;
    /* 计算出这次要缩小的比例 */
    scaleWidth=(float) (scaleWidth*scale);
    scaleHeight=(float) (scaleHeight*scale);

    /* 产生 reSize 后的 Bitmap 对象 */
    Matrix matrix = new Matrix();
    matrix.postScale(scaleWidth, scaleHeight);
    Bitmap resizeBmp = Bitmap.createBitmap(bmp,0,0,bmpWidth,
                                           bmpHeight,matrix,true);

    if(id==0)
    {
      /* 如果是第一次按，就删除原来默认的 ImageView */
```

```
        layout1.removeView(mImageView);
      }
      else
      {
        /* 如果不是第一次按，就删除上次放大缩小所产生的 ImageView */
        layout1.removeView((ImageView)findViewById(id));
      }
      /* 产生新的 ImageView，放入新大小的 Bitmap 对象，再放入界面中 */
      id++;
      ImageView imageView = new ImageView(suofang.this);
      imageView.setId(id);
      imageView.setImageBitmap(resizeBmp);
      layout1.addView(imageView);
      setContentView(layout1);

      /* 因为图片放到最大时放大按钮会 disable，所以在缩小时把他重设为 enable */
      mButton02.setEnabled(true);
    }

    /* 图片放大的方法 */
    private void big()
    {
      int bmpWidth=bmp.getWidth();
      int bmpHeight=bmp.getHeight();
      /* 设置图片放大的比例 */
      double scale=1.25;
      /* 计算这次要放大的比例 */
      scaleWidth=(float)(scaleWidth*scale);
      scaleHeight=(float)(scaleHeight*scale);

      /* 产生新大小后的 Bitmap 对象 */
      Matrix matrix = new Matrix();
      matrix.postScale(scaleWidth, scaleHeight);
      Bitmap resizeBmp = Bitmap.createBitmap(bmp,0,0,bmpWidth,
                                              bmpHeight,matrix,true);

      if(id==0)
      {
        /* 如果是第一次按，就删除原来设置的图像 */
        layout1.removeView(mImageView);
      }
      else
      {
        /* 如果不是第一次按，就删除上次放大缩小所产生的 ImageView */
        layout1.removeView((ImageView)findViewById(id));
      }
```

```
/* 产生新的 ImageView，放入新大小的 Bitmap 对象，再放入 Layout 中 */
id++;
ImageView imageView = new ImageView(suofang.this);
imageView.setId(id);
imageView.setImageBitmap(resizeBmp);
layout1.addView(imageView);
setContentView(layout1);

/* 如果再放大会超过屏幕大小，就把 Button disable */
if(scaleWidth*scale*bmpWidth>displayWidth||
   scaleHeight*scale*bmpHeight>displayHeight)
{
   mButton02.setEnabled(false);
}
   }
}
```

执行后将显示一幅图片和两个按钮，如图 4-34 所示；分别单击“缩小处理”和“放大处理”按钮后，会实现对图片的缩小、放大处理，如图 4-35 所示。

图 4-34　初始效果

图 4-35　缩小后效果

4.18　Bitmap 和 Matrix 实现图片旋转

在本节的内容中，将通过一个具体实例的实现过程，来讲解使用 Bitmap 对象和 Matrix 对象实现图片旋转的具体流程。本实例保存在“光盘:\daima\4\example19”中，其具体实现流程如下。

第一步：打开 eclipse，依次单击“File”→“New”→“Android Project”，新建一个名为“example19”的工程文件。

第二步：编写布局文件 main.xml 实现整体布局。

通过上述步骤，在页面中插入了一幅图片，名称为 hippo.png，2 个 Button 按钮，1 个 TextView 控件。

第三步：在 string.xml 添加程序中要使用的字符串，具体代码如下。

```
<?xml version="1.0" encoding="utf-8"?>
<resources>
  <string name="hello">example19</string>
  <string name="app_name">example19</string>
  <string name="str_button1">向左旋转</string>
  <string name="str_button2">向右旋转</string>
  <string name="str_done">完成</string>
</resources>
```

第四步：编写处理文件 xuanzhuan.java，其具体实现流程如下。

1）加载默认的 Drawable。

2）实现向左旋转按钮处理 mButton1.setOnClickListener 事件。

3）实现向右旋转按钮处理 mButton2.setOnClickListener 事件。

文件 xuanzhuan.java 的具体实现代码如下：

```
package irdc.xuanzhuan;

import irdc.xuanzhuan.R;
import android.app.Activity;
import android.graphics.Bitmap;
import android.graphics.BitmapFactory;
import android.graphics.Matrix;
import android.graphics.drawable.BitmapDrawable;
import android.os.Bundle;
import android.view.View;
import android.widget.Button;
import android.widget.ImageView;
import android.widget.TextView;

public class xuanzhuan extends Activity
{
  private Button mButton1;
  private Button mButton2;
  private TextView mTextView1;
  private ImageView mImageView1;
  private int ScaleTimes;
  private int ScaleAngle;

  /** Called when the activity is first created. */
  @Override
  public void onCreate(Bundle savedInstanceState)
```

```
{
  super.onCreate(savedInstanceState);
  setContentView(R.layout.main);


  mButton1 =(Button) findViewById(R.id.myButton1);
  mButton2 =(Button) findViewById(R.id.myButton2);
  mTextView1 = (TextView) findViewById(R.id.myTextView1);
  mImageView1 = (ImageView) findViewById(R.id.myImageView1);
  ScaleTimes = 1;
  ScaleAngle = 1;

  final Bitmap mySourceBmp =
  BitmapFactory.decodeResource(getResources(), R.drawable.hippo);

  final int widthOrig = mySourceBmp.getWidth();
  final int heightOrig = mySourceBmp.getHeight();

  /* 程序刚运行，加载默认的 Drawable */
  mImageView1.setImageBitmap(mySourceBmp);

  /* 向左旋转按钮 */
  mButton1.setOnClickListener(new Button.OnClickListener()
  {
    @Override
    public void onClick(View v)
    {
      ScaleAngle--;
      if(ScaleAngle<-5)
      {
        ScaleAngle = -5;
      }

      /* ScaleTimes=1，维持 1:1 的宽高比例*/
      int newWidth = widthOrig * ScaleTimes;
      int newHeight = heightOrig * ScaleTimes;

      float scaleWidth = ((float) newWidth) / widthOrig;
      float scaleHeight = ((float) newHeight) / heightOrig;

      Matrix matrix = new Matrix();
      /* 使用 Matrix.postScale 设置维度 */
      matrix.postScale(scaleWidth, scaleHeight);

      /* 使用 Matrix.postRotate 方法旋转 Bitmap*/
      //matrix.postRotate(5*ScaleAngle);
      matrix.setRotate(5*ScaleAngle);
```

```
      /* 创建新的 Bitmap 对象 */
      Bitmap resizedBitmap =
      Bitmap.createBitmap
      (mySourceBmp, 0, 0, widthOrig, heightOrig, matrix, true);

      /**/
      BitmapDrawable myNewBitmapDrawable =
      new BitmapDrawable(resizedBitmap);

      mImageView1.setImageDrawable(myNewBitmapDrawable);
      mTextView1.setText(Integer.toString(5*ScaleAngle));
    }
});

/* 向右旋转按钮 */
mButton2.setOnClickListener(new Button.OnClickListener()
{
  @Override
  public void onClick(View v)
  {
    ScaleAngle++;
    if(ScaleAngle>5)
    {
      ScaleAngle = 5;
    }

    /* ScaleTimes=1，维持 1:1 的宽高比例*/
    int newWidth = widthOrig * ScaleTimes;
    int newHeight = heightOrig * ScaleTimes;

    /* 计算旋转的 Matrix 比例 */
    float scaleWidth = ((float) newWidth) / widthOrig;
    float scaleHeight = ((float) newHeight) / heightOrig;

    Matrix matrix = new Matrix();
    /* 使用 Matrix.postScale 设置维度 */
    matrix.postScale(scaleWidth, scaleHeight);

    /* 使用 Matrix.postRotate 方法旋转 Bitmap*/
    //matrix.postRotate(5*ScaleAngle);
    matrix.setRotate(5*ScaleAngle);

    /* 创建新的 Bitmap 对象 */
    Bitmap resizedBitmap =
    Bitmap.createBitmap
```

```
            (mySourceBmp, 0, 0, widthOrig, heightOrig, matrix, true);

            /**/
            BitmapDrawable myNewBitmapDrawable =
            new BitmapDrawable(resizedBitmap);

            mImageView1.setImageDrawable(myNewBitmapDrawable);
            mTextView1.setText(Integer.toString(5*ScaleAngle));
        }
    });
  }
}
```

执行后将显示一幅图片和两个按钮，分别单击“左旋转”和“右旋转”按钮后，会实现对图片旋转处理，分别如图 4-36 和图 4-37 所示。

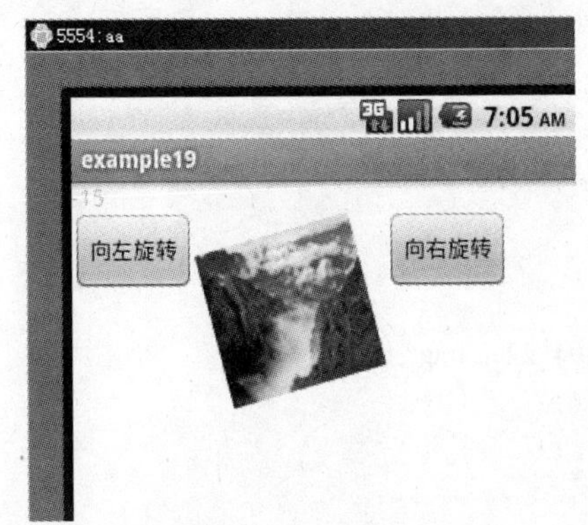

图 4-36　左旋转效果

图 4-37　右旋转后的效果

4.19　decodeFile 加载手机磁盘文件

在 Android 中，如果将图片保存在“res\drawable”目录下，则可以通过 R.drawable.id 来获取图片的 id，即可以任意调用图片素材。但如果没有将图片素材保存在这个目录下，例如要想从存储卡中获取一幅图片时，就需要 Android API 中的 Bitmap 对象来实现。

在本节的内容中，将通过一个具体实例的实现，来讲解使用 decodeFile 方法来加载手机磁盘中图片文件的方法。本实例保存在“光盘:\daima\4\example20”中，其具体实现流程如下。

第一步：打开 eclipse，依次单击“File”→“New”→“Android Project”，新建一个名为“example20”的工程文件。

第二步：编写布局文件 main.xml 实现整体布局，分别插入 1 个 TextView 组件，1 个 ImageView 组件，1 个 Button 组件。

第三步：编写处理文件 example20.java，使用方法 BitmapFactory.decodeFile(fileName)获取 Bitmap 对象，然后用 setImageBitmap()方法设置将获取的信息作为在 ImageView 组件中显示的图像。文件 xuanzhuan.java 的具体实现代码如下。

```
package irdc.example20;

/* 加载相关类 */
import irdc.example20.R;

import java.io.File;
import android.app.Activity;
import android.graphics.Bitmap;
import android.graphics.BitmapFactory;
import android.os.Bundle;
import android.view.View;
import android.widget.Button;
import android.widget.ImageView;
import android.widget.TextView;

public class example20 extends Activity
{
  /*声明对象变量*/
  private ImageView mImageView;
  private Button mButton;
  private TextView mTextView;
  private String fileName="/data/data/irdc.ex04_22/ex04_22_2.png";

  /** Called when the activity is first created. */
  @Override
  public void onCreate(Bundle savedInstanceState)
  {
    super.onCreate(savedInstanceState);
    /* 载入布局文件 main.xml */
    setContentView(R.layout.main);

    /* 取得 Button 对象，并添加 onClickListener */
    mButton = (Button)findViewById(R.id.mButton);
    mButton.setOnClickListener(new Button.OnClickListener()
    {
      public void onClick(View v)
      {
      /* 取得对象 */
        mImageView = (ImageView)findViewById(R.id.mImageView);
        mTextView=(TextView)findViewById(R.id.mTextView);
        /* 检查文件是否存在 */
        File f=new File(fileName);
        if(f.exists())
        {
```

```
            /* 产生 Bitmap 对象，并放入 mImageView 对象中 */
            Bitmap bm = BitmapFactory.decodeFile(fileName);
            mImageView.setImageBitmap(bm);
            mTextView.setText(fileName);
          }
          else
          {
            mTextView.setText("文件不存在");
          }

        }
      });
    }
  }
```

执行后将显示加载的图片，如图 4-38 所示。

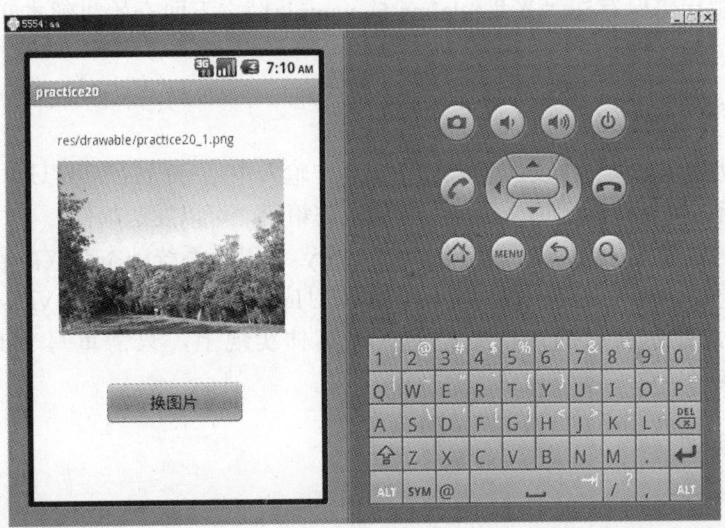

图 4-38　执行效果

第 5 章　手机交互应用服务

作为一个移动手机设备，当然需要具备交互通信的功能。手机中的交互通信方式有多种，例如常见的通话、短信、邮件和蓝牙等。在本章的内容中，将通过具体的实例来详细讲解 Android 中交互式通信应用的具体实现流程。希望读者能够举一反三，为学习本书后面知识打下基础。

5.1　TextView 小试牛刀

本实例的源代码保存在“光盘:\daima\5\example1”，下面开始讲解本实例的具体实现流程。

5.1.1　功能介绍

本实例的功能是提供一个文本输入框，当用户输入电话号码后，可以进行拨号处理；当输入一个网址后，可以登录到这个地址；当输入邮箱后，可以发送邮件。

上述功能是通过 Linkify 对象实现的，Linkify 可以让系统动态获取内容，并作出一个判断，判断是电话、邮箱还是网址，并随之作出对应的处理。通过 TextView 和 EditText 交互，就可以在此显示自己输入的数据值。在具体实现上，只需重写 EditText.setOnkey Listener()即可实现。

5.1.2　具体实现

编写主程序文件 example1.java，具体代码如下。

```
package irdc.example1;


import irdc.example1.R;
import android.app.Activity;
import android.os.Bundle;
import android.text.util.Linkify;
import android.view.KeyEvent;
import android.view.View;
import android.widget.EditText;
import android.widget.TextView;



public class example1 extends Activity
{
```

```
        private TextView mTextView01;
        private EditText mEditText01;

          /** Called when the activity is first created. */
          @Override
        public void onCreate(Bundle savedInstanceState)
          {
          super.onCreate(savedInstanceState);
          setContentView(R.layout.main);

          mTextView01 = (TextView)findViewById(R.id.myTextView1);
          mEditText01 = (EditText)findViewById(R.id.myEditText1);

          mEditText01.setOnKeyListener(new EditText.OnKeyListener()
          {
            @Override
            public boolean onKey(View arg0, int arg1, KeyEvent arg2)
            {
              mTextView01.setText(mEditText01.getText());
              /*判断输入是何种类型，并与系统连接*/
             Linkify.addLinks(mTextView01,Linkify.WEB_URLS|Linkify.
                  EMAIL_ADDRESSES|Linkify.PHONE_NUMBERS);
              return false;
            }
          });
          }
}
```

在上述代码中，设置了 Linkify 的处理事件，Linkify 是在创建 TextView 之后设置的，否则设置的 RE 不会起效果。并且设置了 mTextView01 对象拥有自动链接功能，这样就能转到指定的页面。

执行后的效果如图 5-1 所示，用户可以输入电话号码、邮箱地址或网址，输入后可显示对应的操作。例如，输入 www.163.com 后，会在下面自动显示输入的网址，如图 5-2 所示；当单击网址后会来到对应的页面，如图 5-3 所示。

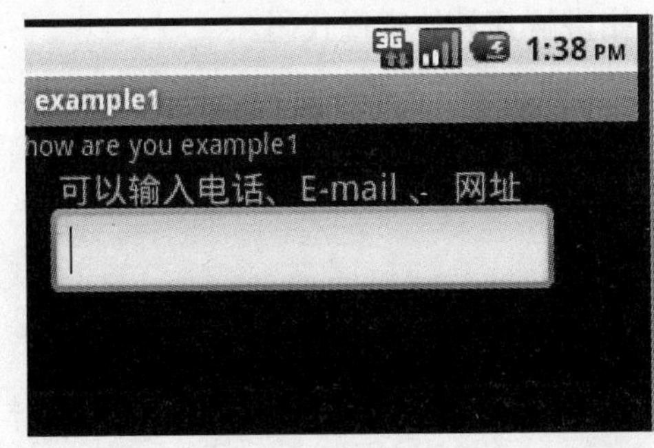

图 5-1　执行效果

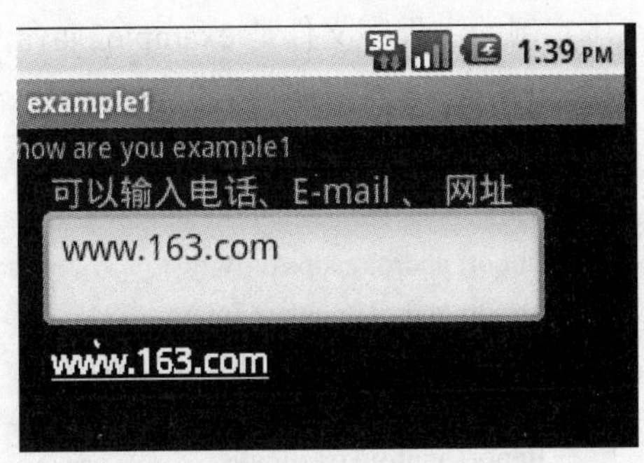

图 5-2　自动显示输入数据

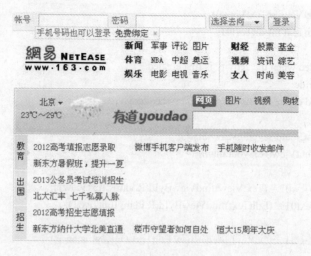

图 5-3　163 主页

5.2　拨打电话

本实例的源代码保存在“光盘:\daima\5\example2”，下面开始讲解本实例的具体实现流程。

5.2.1　功能介绍

本实例实现了一个基本的手机拨打电话的过程，首先使用了一个 EditText 用于获取输入的电话号码，当单击 Button 后执行拨打电话程序，并且通过自定义的 isPhoneNumberValid（String phoneNumber）来确保用户输入的是合法的电话号码。

在具体拨打电话时，首先要在文件 AndiroidManifest.xml 中添加 uses-permission 权限，这样就实现了对拨打电话的声明；然后通过自定义的 Intent 对象，通过“ACTION_CALL”键和 Uri.parse()方法将用户输入的电话号码写入；最后，通过 startActivity()方法即可完成程序拨打电话的功能。

5.2.2　具体实现

本实例的主程序文件是 example2.java，下面开始讲解其具体实现代码。

1）引用类和对象，具体代码如下。

```
package irdc.example2;

import android.app.Activity;
import android.content.Intent;
/*引用 Uri 类才能使用 Uri.parse()*/
import android.net.Uri;
import android.os.Bundle;
import android.view.View;
```

```
/*引用 widget.Button 才能声明使用 Button 对象*/
import android.widget.Button;
import android.widget.Toast;
/*引用 widget.EditText 才能声明使用 EditText 对象*/
import android.widget.EditText;
/*引用 java.util.regex 才能使用 Regular 表达式*/
import irdc.example2.R;

import java.util.regex.Matcher;
import java.util.regex.Pattern;

public class example2 extends Activity
{
 /*声明 Button 与 EditText 对象名称*/
  private Button mButton1;
  private EditText mEditText1;

  /** Called when the activity is first created. */
  @Override
  public void onCreate(Bundle savedInstanceState)
  {
    super.onCreate(savedInstanceState);
    setContentView(R.layout.main);

    /*通过 findViewById 构造器来构造 EditText 与 Button 对象*/
    mEditText1 = (EditText) findViewById(R.id.myEditText1);
    mButton1 = (Button) findViewById(R.id.myButton1);
```

2）设置 Button 按钮的 OnClickListener 来监听 OnClick 事件，具体代码如下。

```
    mButton1.setOnClickListener(new Button.OnClickListener()
    {
      @Override
      public void onClick(View v)
      {
        try
        {
          /*取得 EditText 中用户输入的字符串*/
          String strInput = mEditText1.getText().toString();
          if (isPhoneNumberValid(strInput)==true)
          {
            /*建构一个新的 Intent
            运行 action.CALL 的常数与通过 Uri 将字符串带入*/
            Intent myIntentDial = new
            Intent
            (
              "android.intent.action.CALL",
```

```
                Uri.parse("tel:"+strInput)
            );
            /*在 startActivity()方法中
            带入自定义的 Intent 对象以运行拨打电话的工作 */
            startActivity(myIntentDial);
            mEditText1.setText("");
          }
          else
          {
            mEditText1.setText("");
            Toast.makeText(
            example2.this, "输入的电话格式不符",
            Toast.LENGTH_LONG).show();
          }
        }
        catch(Exception e)
        {
          e.printStackTrace();
        }
      }
    });
  }
```

3）定义 isPhoneNumberValid(String phoneNumber)方法，检查字符串是否为电话号码的方法，并返回 true 或者 false 的判断值。具体代码如下。

```
public static boolean isPhoneNumberValid(String phoneNumber)
{
  boolean isValid = false;
  /* 可接受的电话格式有:
   * ^\\(? : 可以使用 "(" 作为开头
   * (\\d{3}): 紧接着三个数字
   * \\)? : 可以使用")"接续
   * [- ]? : 在上述格式后可以使用具选择性的 "-".
   * (\\d{4}) : 再紧接着四个数字
   * [- ]? : 可以使用具选择性的 "-" 接续.
   * (\\d{4})$: 以四个数字结束.
   * 可以比较下列数字格式:
   * (123)456-78900, 123-4560-7890, 12345678900, (123)-4560-7890
  */
  String expression = "^\\(?(\\d{3})\\)?[- ]?(\\d{3})[- ]?(\\d{5})$";
  String expression2 ="^\\(?(\\d{3})\\)?[- ]?(\\d{4})[- ]?(\\d{4})$";
  CharSequence inputStr = phoneNumber;
  /*创建 Pattern 对象*/
  Pattern pattern = Pattern.compile(expression);
  /*将 Pattern 以参数传入 Matcher 作 Regular 表达式*/
  Matcher matcher = pattern.matcher(inputStr);
```

```
        /*创建 Pattern2 对象*/
        Pattern pattern2 =Pattern.compile(expression2);
        /*将 Pattern2 以参数传入 Matcher2 作 Regular expression*/
        Matcher matcher2= pattern2.matcher(inputStr);
        if(matcher.matches()||matcher2.matches())
        {
          isValid = true;
        }
        return isValid;
      }
    }
```

程序执行后的效果如图 5-4 所示；如果输入的电话号码不规范则输出对应的提示，如图 5-5 所示；当输入规范的号码并单击“开始拨打”按钮后，会实现拨号处理，显示拨打界面，如图 5-6 所示。

图 5-4　执行效果

输入的电话格式不符

图 5-5　自动显示输入数据

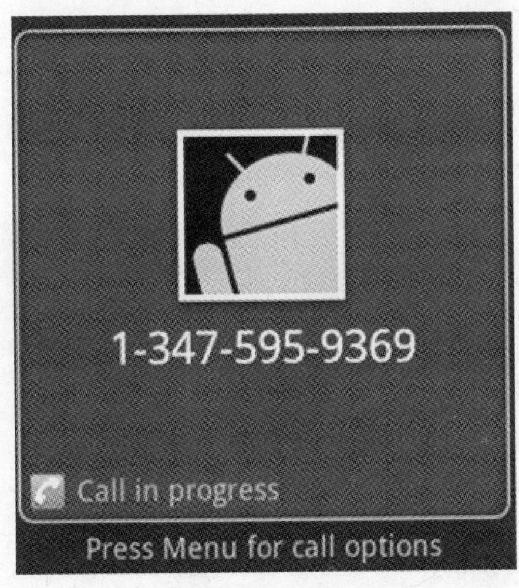

图 5-6　拨打界面

5.3　发送短信

除了拨打电话外，发送短信也是另外一个极为重要的功能。在本节的内容中，将通过一个具体实例的实现，介绍 Android 系统实现短信发送功能的方法。本实例的源代码保存在

“光盘:\daima\5\example3”，下面开始讲解本实例的具体实现流程。

5.3.1 功能介绍

在本实例中，定义了两个 EditText 控件，用于分别获取收信人电话和短信正文，并设置了判断手机号码规范化的方法，并且设置了短信的字数不超过 70 个字符。

在具体实现上，是通过 SmsManage（短信管理）对象的 sendTextMessage()方法来完成的。在 sendTextMessage()方法中要传入 5 个值，分别是：收件人地址 String、发送地址 String、正文 String、发送服务 PendingIntent 和送达服务服务 PendingIntent。

5.3.2 具体实现

本实例的主程序文件是 example3.java，下面开始讲解其具体实现代码。

1）声明一个 Button 变量和两个 EditText 变量，EditText 供获取输入收信人电话号码和短信内容，Button 按钮用于激活发信处理程序。具体代码如下。

```
public class example3 extends Activity
{
  /*声明一个 Button 变量与两个 EditText 变量*/
  private Button mButton1;
  private EditText mEditText1;
  private EditText mEditText2;

  /** Called when the activity is first created. */
  @Override
  public void onCreate(Bundle savedInstanceState)
  {
    super.onCreate(savedInstanceState);
    setContentView(R.layout.main);

    /*
    * 通过 findViewById 构造器来建构
    * EditText1，EditText2 与 Button 对象
    */
    mEditText1=(EditText) findViewById(R.id.myEditText1);
    mEditText2=(EditText) findViewById(R.id.myEditText2);
    mButton1=(Button) findViewById(R.id.myButton1);

    /*将默认文字加载 EditText 中*/
    mEditText1.setText("请输入号码");
    mEditText2.setText("请输入内容!!");

    /*设置 onClickListener 方法，用于当用户单击 EditText 时做出反应*/
    mEditText1.setOnClickListener(new EditText.OnClickListener()
    {
      public void onClick(View v)
      {
```

```
            /*单击 EditText 时清空正文*/
            mEditText1.setText("");
        }
    }
);
```

2）设置 onClickListener()方法，用于当用户单击 EditText 时做出反应。具体代码如下。

```
/*设置 onClickListener 方法，用于当用户单击 EditText 时做出反应*/
mEditText2.setOnClickListener(new EditText.OnClickListener()
{
  public void onClick(View v)
  {
    /*单击 EditText 时清空正文*/
    mEditText2.setText("");
  }
}
);
```

3）设置 onClickListener 方法，用于用户单击 Button 时做出反应，具体代码如下。

```
/*设置 onClickListener 方法，当用户单击 Button 时做出反应*/
mButton1.setOnClickListener(new Button.OnClickListener()
{
  @Override
  public void onClick(View v)
  {
    /*由 EditText1 取得短信收件人电话*/
    String strDestAddress = mEditText1.getText().toString();
    /*由 EditText2 取得短信文字内容*/
    String strMessage = mEditText2.getText().toString();
    /*建构一取得 default instance 的 SmsManager 对象 */
    SmsManager smsManager = SmsManager.getDefault();
```

先检查收件人电话格式与短信字数是否超过 70 字符，然后通过 smsManager.SendTextMessage 实现短信发送，具体实现代码如下。

```
    /*检查收件人电话格式与短信字数是否超过 70 字符*/
    if(isPhoneNumberValid(strDestAddress)==true &&
       iswithin70(strMessage)==true)
    {
      try
      {
        /*
         * 两个条件都检查通过的情况下，发送短信
```

```
     * 先建构一 PendingIntent 对象并使用 getBroadcast()方法广播
     * 将 PendingIntent，电话，短信文字等参数
     * 传入 sendTextMessage()方法发送短信
     */
     PendingIntent mPI = PendingIntent.getBroadcast
     (example3.this, 0, new Intent(), 0);
     smsManager.sendTextMessage
     (strDestAddress, null, strMessage, mPI, null);
   }
   catch(Exception e)
   {
     e.printStackTrace();
   }
   Toast.makeText
   (
     example3.this,"送出成功!!" ,
     Toast.LENGTH_SHORT
   ).show();
   mEditText1.setText("");
   mEditText2.setText("");
 }
 else
 {
   /* 电话格式与短信文字不符合条件时，以 Toast 提醒 */
   if (isPhoneNumberValid(strDestAddress)==false)
   { /*且字数超过 70 字符*/
     if(iswithin70(strMessage)==false)
     {
       Toast.makeText
       (
         example3.this,
         "电话号码格式错误+短信内容超过 70 字，请检查!!",
         Toast.LENGTH_SHORT
       ).show();
     }
     else
     {
       Toast.makeText
       (
         example3.this,
         "电话号码格式错误，请检查!!",
```

```
                Toast.LENGTH_SHORT
              ).show();
            }
          }
          /*字数超过 70 字符*/
          else if (iswithin70(strMessage)==false)
          {
            Toast.makeText
            (
              example3.this,
              "短信内容超过 70 字，请删除部分内容!!",
              Toast.LENGTH_SHORT
            ).show();
          }
        }
      }
    });

}
/*检查字符串是否为电话号码可接受的格式，并返回 true 或 false 的判断值*/
public static boolean isPhoneNumberValid(String phoneNumber)
{
  boolean isValid = false;
  /*  可接受的电话格式有:
   * ^\\(? : 可以使用 "(" 作为开头
   * (\\d{3}): 紧接着三个数字
   * \\)? : 可以使用")"接续
   * [- ]? : 在上述格式后可以使用具选择性的 "-"
   * (\\d{3}) : 再紧接着三个数字
   * [- ]? : 可以使用具选择性的 "-" 接续
   * (\\d{5})$: 以五个数字结束
   *  可以比较下列数字格式:
   * (123) 456-7890, 123-456-7890, 1234567890, (123)-456-7890
  */
  String expression =
  "^\\(?(\\d{3})\\)?[- ]?(\\d{3})[- ]?(\\d{5})$";

  /*  可接受的电话格式有:
   * ^\\(? : 可以使用 "(" 作为开头
   * (\\d{3}): 紧接着三个数字
   * \\)? : 可以使用")"接续
```

```
            * [- ]? : 在上述格式后可以使用具选择性的 "-"
            * (\\d{4}) : 再紧接着四个数字
            * [- ]? : 可以使用具选择性的 "-" 接续
            * (\\d{4})$: 以四个数字结束
            *  可以比较下列数字格式:
            * (02)3456-7890, 02-3456-7890, 0234567890, (02)-3456-7890
          */
          String expression2=
          "^\\(?(\\d{3})\\)?[- ]?(\\d{4})[- ]?(\\d{4})$";

          CharSequence inputStr = phoneNumber;
          /*创建 Pattern*/
          Pattern pattern = Pattern.compile(expression);
          /*将 Pattern  以参数传入 Matcher 作 Regular expression*/
          Matcher matcher = pattern.matcher(inputStr);
          /*创建 Pattern2*/
          Pattern pattern2 =Pattern.compile(expression2);
          /*将 Pattern2  以参数传入 Matcher2 作 Regular expression*/
          Matcher matcher2= pattern2.matcher(inputStr);
          if(matcher.matches()||matcher2.matches())
          {
            isValid = true;
          }
          return isValid;
        }

        public static boolean iswithin70(String text)
        {
          if (text.length()<= 70)
          {
            return true;
          }
          else
          {
            return false;
          }
        }
      }
```

在上述代码中，通过 PendingIntent.getBroadcast()方法自定义了 PendingIntent 类，此类能

够处理即将发生的事情。然后通过 Broadcast 进行广播处理，然后使用 SmsManager 对象中 SmsManager.getDefault()方法来构建预处理的数据，使用 sendTextMessage()方法将有关的数据以参数形式带入，这样即可完成发短信的任务。

执行后的效果如图 5-7 所示，输入手机号码，编写短信内容后，单击“发送”按钮后即可完成短信发送功能，系统会提示成功信息。如图 5-8 所示。

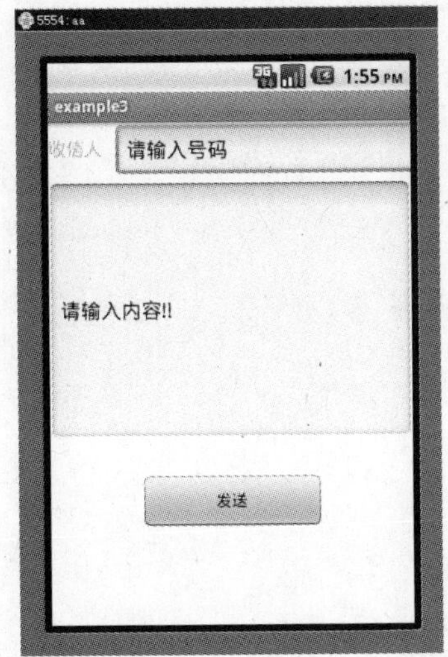

图 5-7　初始效果

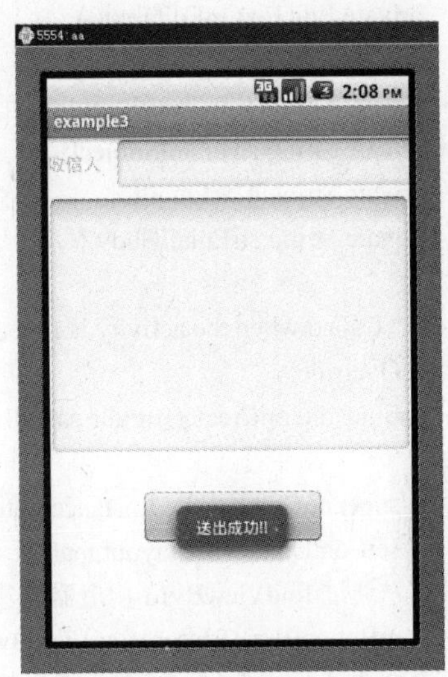

图 5-8　发送成功

如果短信内容和收信人号码格式不规范，会输出对应的错误提示。

5.4　自制发送 Email 程序

除了拨打电话和发送短信外，发送邮件也是另外一个极为重要的功能。在本节的内容中，将通过一个具体实例的实现过程，介绍 Android 系统实现邮件发送功能的方法。本实例的源代码保存在“光盘:\daima\5\example4”，下面开始讲解本实例的具体实现流程。

5.4.1　功能介绍

在本实例中，自定义了 Intent 类，并使用 Android.content.Intent.ACTION_SEND 的参数设置通过手机寄发 Email 的服务，整个过程还算简单。

在具体实现上，邮件的收发过程是通过 Android 内置的 Gmail 程序实现的，并不是使用 SMTP 的 Protocol 实现的。为了确保邮件能够发出，必须在收件人字段上输入标准的邮件地址格式，如果格式不规范，则发送按钮处于不可用状态。

5.4.2　具体实现

本实例的主程序文件是 example4.java，下面开始讲解其具体实现代码。

1）分别声明四个 EditText、一个 Button 以及四个 String 变量，用于输入邮箱地址、邮件、主题和副本，具体代码如下。

```
public class example4 extends Activity
{
  /*声明四个 EditText、一个 Button 以及四个 String 变量*/
  private EditText mEditText01;
  private EditText mEditText02;
  private EditText mEditText03;
  private EditText mEditText04;
  private Button mButton01;
  private String[] strEmailReciver;
  private String strEmailSubject;
  private String[] strEmailCc;
  private String strEmailBody ;

  /** Called when the activity is first created. */
  @Override
  public void onCreate(Bundle savedInstanceState)
  {
    super.onCreate(savedInstanceState);
    setContentView(R.layout.main);
    /*通过 findViewById 构造器来建构 Button 对象*/
    mButton01 = (Button)findViewById(R.id.myButton1);
    /*通过 findViewById 构造器来构造所有 EditText 对象*/
    mButton01.setEnabled(false);
    /*设置 OnKeyListener，当 key 事件发生时进行反应*/
    mEditText01 = (EditText)findViewById(R.id.myEditText1);
    mEditText02 = (EditText)findViewById(R.id.myEditText2);
    mEditText03 = (EditText)findViewById(R.id.myEditText3);
    mEditText04 = (EditText)findViewById(R.id.myEditText4);
```

2）定义 setOnKeyListener 方法，如果用户键入为正规 Email 文字，则按钮可用，反之则按钮不可用，具体代码如下。

```
/*若用户键入为正规 Email 文字，则设置按钮 enable（可用）反之则 disable */
mEditText01.setOnKeyListener(new EditText.OnKeyListener()
{
  @Override
  public boolean onKey(View v, int keyCode, KeyEvent event)
  {
    /*如果是邮件地址格式，则按钮可按下*/
    if(isEmail(mEditText01.getText().toString()))
    {
      mButton01.setEnabled(true);
    }
    else
    {
      mButton01.setEnabled(false);
    }
```

```
        return false;
    }
});
```

3）定义 onClickListener 响应按钮单击事件，当单击按钮后实现邮件发送处理。具体代码如下。

```
/*定义 onClickListener 响应按钮*/
mButton01.setOnClickListener(new Button.OnClickListener()
{
  @Override
  public void onClick(View v)
  {
    Intent mEmailIntent = new Intent(android.content.Intent.ACTION_SEND);
    mEmailIntent.setType("plain/text");

    strEmailReciver = new String[]{mEditText01.getText().toString()};
    strEmailCc = new String[]{mEditText02.getText().toString()};
    strEmailSubject = mEditText03.getText().toString();
    strEmailBody = mEditText04.getText().toString();

    mEmailIntent.putExtra(android.content.Intent.EXTRA_EMAIL, strEmailReciver);
    mEmailIntent.putExtra(android.content.Intent.EXTRA_CC, strEmailCc);
    mEmailIntent.putExtra(android.content.Intent.EXTRA_SUBJECT, strEmailSubject);
    mEmailIntent.putExtra(android.content.Intent.EXTRA_TEXT, strEmailBody);
    startActivity(Intent.createChooser(mEmailIntent, getResources().getString(R.string.str_message)));
  }
});
}
```

4）定义 isEmail(String strEmail)方法，检查是否是规范的邮件地址格式。具体代码如下。

```
public static boolean isEmail(String strEmail)
{
  String strPattern = "^[a-zA-Z][\\w\\.-]*[a-zA-Z0-9]@[a-zA-Z0-9][\\w\\.-]*[a-zA-Z0-9]\\.[a-zA-Z][a-
zA-Z\\.]*[a-zA-Z]$";
  Pattern p = Pattern.compile(strPattern);
  Matcher m = p.matcher(strEmail);
  return m.matches();
}
}
```

执行后的效果如图 5-9 所示，输入手机号码，编写短信内容后，单击“发送”按钮后即可完成短信发送功能，系统会提示成功信息。如图 5-10 所示。

注意：Android 模拟器中不会内置 Gmail Client 端程序，所以当使用本实例发送邮件后，会显示“No Application can perform this action”的提示。但是在现实手机设备上，如果运行本实例程序，会调用 Gmail 程序，成功实现邮件发送。

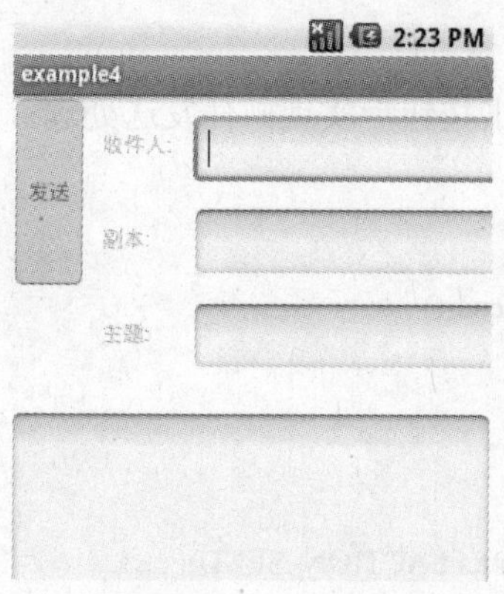

图 5-9　初始效果

图 5-10　发送成功

5.5　手机震动效果

除了基本的通话和短信外，震动也是另外一个极为重要的功能。通过手机震动，能够帮助我们及时感知打来的电话或发来的短信，在本节的内容中，将通过一个具体实例的实现过程，介绍 Android 系统实现手机震动功能的方法。本实例的源代码保存在“光盘:\daima\5\example5”，下面开始讲解本实例的具体实现流程。

5.5.1　实现原理

在本实例中，读者将会了解到触发手机震动事件的方法。Android 中的震动事件是 Vibration，需要设置震动的时间长短和周期，设置单位是毫秒。如果要建立手机震动，则必须建立 Vibrator 对象，并通过调用 Vibrate()方法来实现震动目的。在 Vibrator 构造器中有 4 个参数，前三个用于设置震动大小，最后那个用于设置震动持续时间。

在本范例中的震动方式是不同的，分为一直持续和只震动一轮两种。

5.5.2　具体实现

本实例的主程序文件是 example5.java，下面开始讲解其具体实现代码。

1）设置 ToggleButton（开关）的对象，检测 ToggleButton 是否被启动，如果单击“ON”按钮则启动震动模式，如果单击“OFF”按钮则关闭震动模式。具体代码如下。

```
public class example5 extends Activity
{
  private Vibrator mVibrator01;

  /** Called when the activity is first created. */
  @Override
  public void onCreate(Bundle savedInstanceState)
```

```
{
   super.onCreate(savedInstanceState);
   setContentView(R.layout.main);

   /*设置 ToggleButton 对象*/
   mVibrator01 = ( Vibrator )getApplication().getSystemService
   (Service.VIBRATOR_SERVICE);

   final ToggleButton mtogglebutton1 =
   (ToggleButton) findViewById(R.id.myTogglebutton1);

   final ToggleButton mtogglebutton2 =
   (ToggleButton) findViewById(R.id.myTogglebutton2);

   final ToggleButton mtogglebutton3 =
   (ToggleButton) findViewById(R.id.myTogglebutton3);
```

2）设置短震动模式，通过 mVibrator01.vibrate(new long[]{100,10,100,1000},-1)设置了震动周期，具体代码如下。

```
/* 短震动 */
mtogglebutton1.setOnClickListener(new OnClickListener()
{
   public void onClick(View v)
   {
      if (mtogglebutton1.isChecked())
      {
         /* 设置震动的周期 */
         mVibrator01.vibrate( new long[]{100,10,100,1000},-1);
         /*用 Toast 显示震动启动*/
         Toast.makeText
         (
            example5.this,
            getString(R.string.str_ok),
            Toast.LENGTH_SHORT
         ).show();
      }
      else
      {
         /* 取消震动 */
         mVibrator01.cancel();
         /*用 Toast 显示震动已被取消*/
         Toast.makeText
         (
            example5.this,
            getString(R.string.str_end),
```

```
            Toast.LENGTH_SHORT
        ).show();
      }
    }
});
```

3）设置长震动模式，通过 mVibrator01.vibrate(new long[]{100,100,100,1000},0);设置了震动周期，具体代码如下。

```
/* 长震动 */
mtogglebutton2.setOnClickListener(new OnClickListener()
{
  public void onClick(View v)
  {
    if (mtogglebutton2.isChecked())
    {
      /*设置震动的周期*/
      mVibrator01.vibrate(new long[]{100,100,100,1000},0);

      /*用 Toast 显示震动启动*/
      Toast.makeText
      (
        example5.this,
        getString(R.string.str_ok),
        Toast.LENGTH_SHORT
      ).show();
    }
    else
    {
      /* 取消震动 */
      mVibrator01.cancel();

      /* 用 Toast 显示震动取消 */
      Toast.makeText
      (
        example5.this,
        getString(R.string.str_end),
        Toast.LENGTH_SHORT
      ).show();
    }
  }
});
```

4）设置节奏震动模式，通过 mVibrator01.vibrate(new long[]{1000,50,1000,50,1000},0);设置了震动周期，具体代码如下。

```
/* 节奏震动 */
```

```
            mtogglebutton3.setOnClickListener(new OnClickListener()
            {
              public void onClick(View v)
              {
                if (mtogglebutton3.isChecked())
                {
                  /* 设置震动的周期 */
                  mVibrator01.vibrate( new long[]{1000,50,1000,50,1000},0);

                  /*用 Toast 显示震动启动*/
                  Toast.makeText
                  (
                    example5.this, getString(R.string.str_ok),
                    Toast.LENGTH_SHORT
                  ).show();
                }
                else
                {
                  /* 取消震动 */
                  mVibrator01.cancel();
                  /* 用 Toast 显示震动取消 */
                  Toast.makeText
                  (
                    example5.this,
                    getString(R.string.str_end),
                    Toast.LENGTH_SHORT
                  ).show();
                }
              }
            });
          }
        }
```

另外，还需要在文件 AndroidManifest.xml 中设置权限，即必须允许 Android.permission. VIBRATE 的权限，具体代码如下。

```
<manifest
  xmlns:android="http://schemas.android.com/apk/res/android"
  android:versionCode="1"
  android:versionName="1.0.0" package="irdc.example5">
  <application
    android:icon="@drawable/icon"
    android:label="@string/app_name">
    <activity
      android:label="@string/app_name"
      android:name=".example5">
      <intent-filter>
```

```
            <action android:name="android.intent.action.MAIN" />
            <category android:name="android.intent.category.LAUNCHER" />
        </intent-filter>
    </activity>
</application>
<uses-permission android:name="android.permission.VIBRATE" />
</manifest>
```

执行后的效果如图 5-11 所示，当选择一种模式，并单击按钮后会启动对应的震动模式，如图 5-12 所示的是启动了“短时间”震动模式。

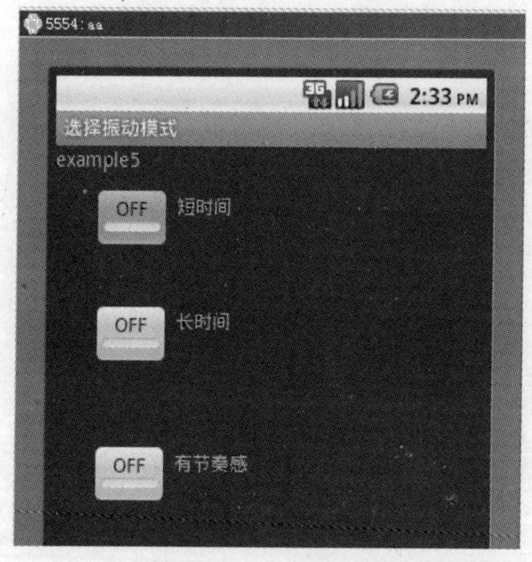

图 5-11　初始效果

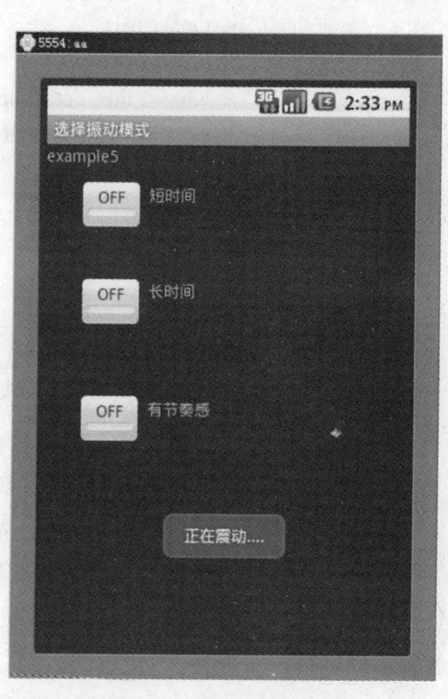

图 5-12　短时间震动模型

5.6　图文提醒

在本书前面的内容中，已经讲解过 Toast 提醒的基本知识。Toast 能够在手机中实现一个醒目的提示效果，提示用户即将需要完成的操作。在本节的内容中，将通过一个具体实例的实现，介绍 Android 系统实现图文提醒功能的方法。本实例的源代码保存在“光盘:\daima\5\example6”，下面开始讲解本实例的具体实现流程。

5.6.1　实现原理

在本实例中，在 Toast 中放置一幅图片，并且辅助了一行文字，这样即可实现图文提醒功能。图文提醒和单独的文字提醒相比，具有更好的视觉效果。

在具体实现上，在 Toast 中放置一个 Layout（布局界面），这个 Layout 中包含了图片和文字，这些图片和文字就是提醒的内容。

5.6.2　具体实现

本实例的主程序文件是 example6.java，下面开始讲解其具体实现代码。文件 example6.java 的具体实现流程如下。

1）设置 Button 用 OnClickListener 启动事件 setOnClickListener(new Button.OnClickListener()。

2）分别创建 ImageView 对象和 LinearLayout（线性布局）对象。

3）设置 mTextView 去抓取 string 值。

4）判断 mTextView 的内容为何，并与系统实现连接。

5）用 Toast 方式将内容显示出来。

文件 example6.java 的具体代码如下。

```
public class example6 extends Activity
{
  private Button mButton01;

  /** Called when the activity is first created. */
  @Override
  public void onCreate(Bundle savedInstanceState)
  {
    super.onCreate(savedInstanceState);
    setContentView(R.layout.main);
    mButton01 = (Button)findViewById(R.id.myButton1);

    /*设置 Button 用 OnClickListener 启动事件*/
    mButton01.setOnClickListener(new Button.OnClickListener()
    {

      @Override
      public void onClick(View v)
      {
        // TODO Auto-generated method stub

        /* 创建 ImageView 对象 */
        ImageView mView01 = new ImageView(example6.this);
        TextView mTextView = new TextView(example6.this);

        /* 创建 LinearLayout 对象 */
        LinearLayout lay = new LinearLayout(example6.this);

        /* 设置 mTextView 去抓取 string 值 */
        mTextView.setText(R.string.app_url);
```

```
            /* 判断 mTextView 的内容为何，并与系统做连接 */
            Linkify.addLinks
            (
              mTextView,Linkify.WEB_URLS|
              Linkify.EMAIL_ADDRESSES|
              Linkify.PHONE_NUMBERS
            );

            /*用 Toast 方式显示信息*/
            Toast toast = Toast.makeText
                        (
                          example6.this,
                          mTextView.getText(),
                          Toast.LENGTH_LONG
                        );

            /* 自定义 View 对象 */
            View textView = toast.getView();

            /* 以水平方式排列 */
            lay.setOrientation(LinearLayout.HORIZONTAL);

            /* 在 ImageView Widget 里指定显示的图片 */
            mView01.setImageResource(R.drawable.icon);

            /* 在 Layout 里添加刚创建的 View */
            lay.addView(mView01);

            /* 在 Toast 里显示提示文字 */
            lay.addView(textView);

            /* 以 Toast，setView 方法将界面传入 */
            toast.setView(lay);

            /* 显示 Toast */
            toast.show();
          }
        });
      }
    }
```

执行后的效果如图 5-13 所示，当单击“有图片的提醒”按钮后，会弹出图文效果的提醒，如图 5-14 所示。

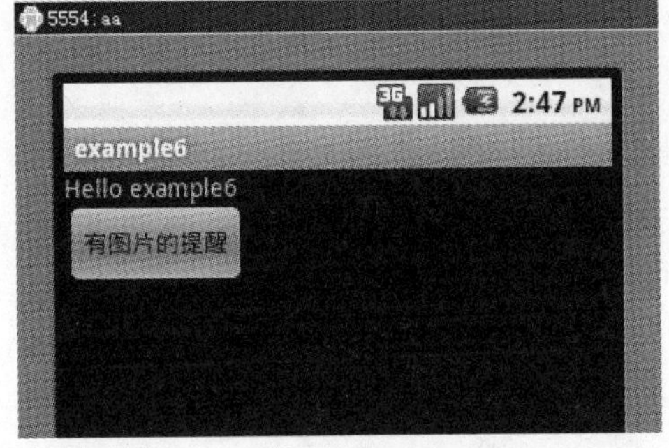

图 5-13　初始效果

图 5-14　图文提醒

5.7　状态栏提醒

在手机的最顶端，常用于显示时间、信号强度和电池用量，这就是状态栏。在实现提醒处理时，也可以在状态栏中显示一幅图片，并结合图片和文字来实现更加美观的提示。在本节的内容中，将通过一个具体实例的实现，介绍在 Android 状态栏中实现图文提醒功能的方法。本实例的源代码保存在“光盘:\daima\5\example7”，下面开始讲解本实例的具体实现流程。

5.7.1　实现原理

在本实例中，在状态栏中放置一幅提醒图标。在具体实现上，Android API 为了管理提示信息，定义了 NotificationManage（通知管理对象），只需将 Notification（通知信息）添加到 NotificationManage 即可将信息显示在状态栏。

5.7.2　具体实现

本实例的主程序文件是 example7.java 和 example7_1.java，下面先讲解文件 example7.java 的具体实现代码。

1）先声明对象变量，分别设置在线、离开、忙碌、一会回来和离线 5 种状态。具体代码如下。

```
public class example7 extends Activity
{
```

```
/*声明对象变量*/
private NotificationManager myNotiManager;
private Spinner mySpinner;
private ArrayAdapter<String> myAdapter;
private static final String[] status =
{ "在线","离开","忙碌","一会回来","离线" };

@Override
protected void onCreate(Bundle savedInstanceState)
{
  super.onCreate(savedInstanceState);
  /* 载入 main.xml Layout */
  setContentView(R.layout.main);
```

2）初始化 NotificationManager 对象，然后使用 myspinner_dropdown 自定义下拉菜单模式，并将 ArrayAdapter 添加 Spinner 对象中，根据具体状态显示对应的提示图标。具体代码如下。

```
/* 初始化对象 */
myNotiManager=
  (NotificationManager)getSystemService(NOTIFICATION_SERVICE);
mySpinner=(Spinner)findViewById(R.id.mySpinner);
myAdapter=new ArrayAdapter<String>(this,
          android.R.layout.simple_spinner_item,status);
/* 应用 myspinner_dropdown 自定义下拉菜单模式 */
myAdapter.setDropDownViewResource(R.layout.myspinner_dropdown);
/* 将 ArrayAdapter 添加 Spinner 对象中 */
mySpinner.setAdapter(myAdapter);

/* 将 mySpinner 添加 OnItemSelectedListener */
mySpinner.setOnItemSelectedListener(
  new Spinner.OnItemSelectedListener()
{
  @Override
  public void onItemSelected(AdapterView<?> arg0,View arg1,
                             int arg2,long arg3)
  {
    /* 依照选择的 item 来判断要发哪一个提醒*/
    if(status[arg2].equals("在线"))
    {
      setNotiType(R.drawable.msn,"在线");
    }
    else if(status[arg2].equals("离开"))
    {
      setNotiType(R.drawable.away,"离开");
    }
```

```
        else if(status[arg2].equals("忙碌中"))
        {
          setNotiType(R.drawable.busy,"忙碌中");
        }
        else if(status[arg2].equals("马上回来"))
        {
          setNotiType(R.drawable.min,"马上回来");
        }
        else
        {
          setNotiType(R.drawable.offine,"离线");
        }
      }
```

3）定义 onNothingSelected(AdapterView<?> arg0)，供用户选择状态。具体代码如下。

```
        @Override
        public void onNothingSelected(AdapterView<?> arg0)
        {
        }
      });
    }

    /* 发出 Notification 的方法*/
    private void setNotiType(int iconId, String text)
    {
      /* 创建新的 Intent，作为单击 Notification 留言条时，
       * 会运行的 Activity */
      Intent notifyIntent=new Intent(this,example7_1.class);
      notifyIntent.setFlags( Intent.FLAG_ACTIVITY_NEW_TASK);
      /* 创建 PendingIntent 作为设置递延运行的 Activity */
      PendingIntent appIntent=PendingIntent.getActivity(example7.this,0,notifyIntent,0);
```

4）创建 Notication（通知）对象，并设置相关对应的参数。具体代码如下。

```
      Notification myNoti=new Notification();
      /* 设置 statusbar 显示的 icon */
      myNoti.icon=iconId;
      /* 设置 statusbar 显示的文字信息 */
      myNoti.tickerText=text;
      /* 设置 notification 发生时同时发出默认声音 */
      myNoti.defaults=Notification.DEFAULT_SOUND;
      /* 设置提醒留言条的参数 */
      myNoti.setLatestEventInfo(example7.this,"MSN 登录状态",
                      text,appIntent);
```

```
        /* 送出 Notification */
        myNotiManager.notify(0,myNoti);
      }
    }
```

然后介绍文件 example7_1.java，其功能是引入 example7_1 extends Activity，并发出 Toast 提示，实现模拟 QQ 和 MSN 的登录效果。具体代码如下。

```
package irdc.example7;

/* import 相关类 */
import android.app.Activity;
import android.os.Bundle;
import android.widget.Toast;

/* 当用户单击提醒留言条时，会运行的 Activity */
public class example7_1 extends Activity
{
  @Override
  protected void onCreate(Bundle savedInstanceState)
  {
    super.onCreate(savedInstanceState);

    /* 发出 Toast */
    Toast.makeText(example7_1.this,
                  "模拟 MSN 切换登录状态",
                  Toast.LENGTH_SHORT
                 ).show();
    finish();
  }
}
```

执行后将首先显示默认的“在线”状态，并在状态栏中显示 QQ 的在线图标，如图 5-15 所示。可以在下拉框中继续选择一种状态，对应的状态栏也会显示对应的图标，如图 5-16 所示。例如，当选择“离开”按钮后，会在状态栏中显示对应的提示图标，如图 5-17 所示。这是一个 MSN 的图标。

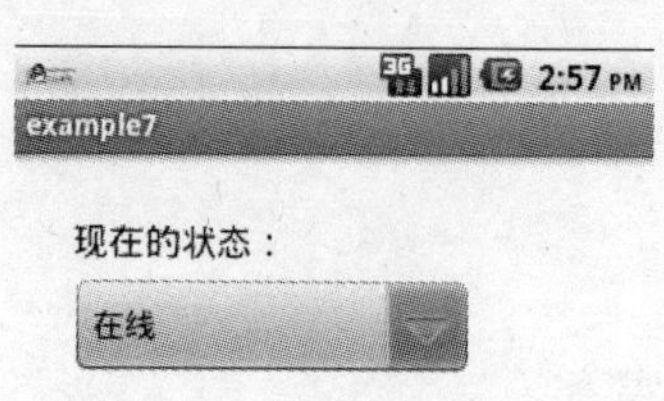

图 5-15　初始效果

在线
离开
忙碌
一会回来
离线

图 5-16　图文提醒

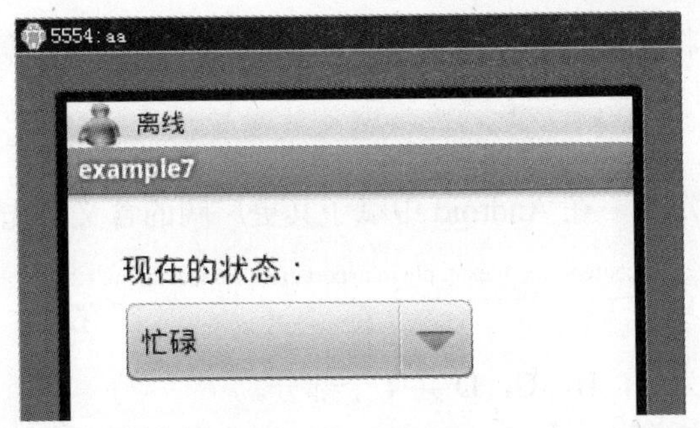

图 5-17　离开图标

5.8　ContentResolver 检索手机通讯录

在手机应用中，联系人检索是一个十分常见的功能之一，用户可以通过手机的查询系统，迅速找到通讯录中的联系人电话。在本节的内容中，将通过一个具体实例的实现，介绍通过 ContentResolver 实现检索手机通讯录的方法。本实例的源代码保存在“光盘:\daima\5\example8”，下面开始讲解本实例的具体实现流程。

5.8.1　实现原理

在百度主页中，只需输入要检索关键字的首字符，系统就能够提示显示出很多对应的关键字。同样，在手机联系人检索应用中，也可以实现类似的功能效果。即输入要查询的联系人信息，系统能够根据输入的关键字首字符，自动显示出通讯录中对应的联系人信息。

5.8.2　ContentResolver 介绍

Android 是如何实现应用程序之间数据共享的？一个应用程序可以将自己的数据完全暴露出去，外界根本看不到，也不用看到这个应用程序暴露的数据是如何存储的，或者是使用数据库还是使用文件，还是通过网上获得，这些一切都不重要，重要的是外界可以通过这一套标准及统一的接口和这个程序里的数据打交道，例如，添加（insert）、删除（delete）、查询（query）或修改（update），当然需要一定的权限才可以。

如何将应用程序的数据暴露出去？Android 提供了 Content Provider，一个程序可以通过实现一个 Content Provider 的抽象接口将自己的数据完全暴露出去，而且 Content providers 是以类似数据库中表的方式将数据暴露。Content Providers 存储和检索数据，通过它可以让所有的应用程序访问数据，这也是应用程序之间唯一共享数据的方法。要想使应用程序的数据公开化，可通过两种方法：创建一个属于你自己的 Content Provider 或者将你的数据添加到一个已经存在的 Content Provider 中。前提是我的数据有相同数据类型并且有写入 Content Provider 的权限。

如何通过一套标准及统一的接口获取其他应用程序暴露的数据？Android 提供了 ContentResolver，外界的程序可以通过 ContentResolver 接口访问 Content Provider 提供的数据。

在下面的内容中，将着重讲解如何获取其他应用程序共享的数据，比如获取 Android 手机电话簿中的信息。

1．什么是 URI

URI 是网络资源的定义，在 Android 中赋予其更广阔的含义，先看个例子，如下，

```
content://com.example.transportationprovider/trains/122
   A                    B                       C    D
```

在上例中，将其分为 A，B，C，D 共 4 个部分。

A：标准前缀，用来说明一个 Content Provider 控制这些数据，无法改变的。

B：URI 的标识，它定义了是哪个 Content Provider 提供这些数据。对于第三方应用程序，为了保证 URI 标识的唯一性，它必须是一个完整的、小写的类名。这个标识在 <provider> 元素的 authorities 属性中说明。

```
<provider name=".TransportationProvider"
authorities="com.example.transportationprovider">
```

C：路径，Content Provider 使用这些路径来确定当前需要什么类型的数据，URI 中可能不包括路径，也可能包括多个；

D：如果 URI 中包含，表示需要获取的记录的 ID；如果没有 ID，就表示返回全部；

由于 URI 通常比较长，而且有时候容易出错，且难以理解。所以，在 Android 当中定义了一些辅助类，并且定义了一些常量来代替这些长字符串，例如，People.CONTENT_URI。

2．ContentResolver 简介

ContentResolver 是通过 URI 来查询 Content Provider 中提供的数据。除了 URI 外，还必须知道需要获取的数据段的名称，以及此数据段的数据类型。如果你需要获取一个特定的记录，就必须知道当前记录的 ID，也就是 URI 中 D 部分。

前面也提到了 Content Providers 是以类似数据库中表的方式将数据暴露出去，那么 ContentResolver 也将采用类似数据库的操作来从 Content Providers 中获取数据。现在简要介绍 ContentResolver 的主要接口，具体如表 5-1 所示。

表 5-1　ContentResolver 返回值

返 回 值	函 数 声 明
final Uri	insert(Uri url, ContentValues values)Inserts a row into a table at the given URL.
final int	delete(Uri url, String where, String[] selectionArgs)Deletes row(s) specified by a content URI.
final Cursor	query(Uri uri, String[] projection, String selection, String[] selectionArgs, String sortOrder)Query the given URI, returning a Cursor over the result set.
final int	update(Uri uri, ContentValues values, String where, String[] selectionArgs)Update row(s) in a content URI.

3．ContentResolver 简单实例

通过上面内容的学习，了解了如何获取、使用 ContentResolver 的方法。下面启动 Eclipse，制作一个 ContentResolver 简单实例。

1）打开 Eclipse，创建一个名为“ContentResolver”的项目，如图 5-18 所示。

2）编写主程序代码，具体如下。

Project name: ContentProvider

Contents

◉ Create new project in workspace

○ Create project from existing source

☑ Use default location

Location: D:/Java/ContentProvider　Browse...

Build Target

	Target Name	Vendor	Platform	AP...
☐	Android 1.1	Android Open Source Project	1.1	2
☑	Android 1.5	Android Open Source Project	1.5	3
☐	Google APIs	Google Inc.	1.5	3

Android + Google APIs

Properties

Application name: moandroid.showcontact

Package name: moandroid.showcontact

☑ Create Activity: showcontact

Min SDK Version: 3

图 5-18　新建项目

```
package moandroid.showcontact;
import android.app.ListActivity;
import android.database.Cursor;
import android.os.Bundle;

import android.provider.Contacts.Phones;
import android.widget.ListAdapter;
import android.widget.SimpleCursorAdapter;
public class showcontact extends ListActivity {
protected void onCreate(Bundle savedInstanceState) {

super.onCreate(savedInstanceState);
Cursor c = getContentResolver().query(Phones.CONTENT_URI, null, null, null, null);
startManagingCursor(c);
ListAdapter adapter = new SimpleCursorAdapter(this,
android.R.layout.simple_list_item_2, c,
new String[] { Phones.NAME, Phones.NUMBER },
new int[] { android.R.id.text1, android.R.id.text2 });
setListAdapter(adapter);
}
}
```

3）在 AndroidManifest.XML 中<application>元素前增加如下许可，代码如下。

```
<uses-permission android:name="android.permission.READ_CONTACTS" />
```

最后运行程序，在模拟器启动后，单击 Menu 返回到 Home 界面，打开 Contacts 选择

Contacts 标签页，添加两个联系人信息。返回到 Home，选择 moandroid.showcontact 运行，刚添加的两个联系人信息将显示在界面上。

5.8.3 具体实现

经过前面知识的学习，读者对 ContentResolver 的知识又有了更进一步的认识，下面开始介绍本节范例的具体实现过程。本实例的主程序文件是 example8.java 和 Contacts Adapter.java。

1. 主程序文件 example8.java

下面先讲解文件 example8.java 的具体实现代码。

1）通过 String[]获取通讯录中的字段，具体代码如下。

```
public class example8 extends Activity
{
  private AutoCompleteTextView myAutoCompleteTextView;
  private TextView myTextView1;
  private Cursor contactCursor;
  private ContactsAdapter myContactsAdapter;
  /* 要获取通讯录的字段 */
  public static final String[] PEOPLE_PROJECTION = new String[]
  { Contacts.People._ID, Contacts.People.PRIMARY_PHONE_ID,
      Contacts.People.TYPE, Contacts.People.NUMBER, Contacts.People.LABEL,
      Contacts.People.NAME };
```

2）通过 ContentResolver content = getContentResolver();和 contactCursor = content.query 来获取通讯录里的数据，具体代码如下。

```
/** Called when the activity is first created. */
@Override
public void onCreate(Bundle savedInstanceState)
{
  super.onCreate(savedInstanceState);
  setContentView(R.layout.main);

  myAutoCompleteTextView = (AutoCompleteTextView)
  findViewById(R.id.myAutoCompleteTextView);
  myTextView1 = (TextView) findViewById(R.id.myTextView1);

  /* 定义 ContentResolver 对象 */
  ContentResolver content = getContentResolver();

  /* 取得通讯录的焦点 */
  contactCursor = content.query
  (
    Contacts.People.CONTENT_URI,
    PEOPLE_PROJECTION, null, null,
```

```
        Contacts.People.DEFAULT_SORT_ORDER
    );

    /* 将 Cursor 传入自己实现的 ContactsAdapter 容器 */
    myContactsAdapter = new ContactsAdapter(this, contactCursor);

    myAutoCompleteTextView.setAdapter(myContactsAdapter);
```

在上述代码中，是以 Cursor 来获取通讯录中的存储内容的，例如电话号码和姓名等。

3）定义 myAutoCompleteTextView.setOnItemClickListener 事件，即当用户单击联系人姓名后的拦截事件处理，并通过 ContactsAdapter.getCursor()来获取联系人的电话号码。具体代码如下。

```
        myAutoCompleteTextView.setOnItemClickListener
        (new AdapterView.OnItemClickListener()
        {
          @Override
          public void onItemClick
          (AdapterView<?> arg0, View arg1, int arg2, long arg3)
          {
            /* 取得焦点 */
            Cursor c = myContactsAdapter.getCursor();
            /* 移到所单击的位置 */
            c.moveToPosition(arg2);
            String number = c.getString
            (c.getColumnIndexOrThrow(Contacts.People.NUMBER));
            /* 当找不到电话时显示无输入电话 */
            number = number == null ? "无输入电话" : number;
            myTextView1.setText(c.getString
            (c.getColumnIndexOrThrow(Contacts.People.NAME))
            + "的电话是" + number);
          }
        }); }
    }
```

2．主程序文件 ContactsAdapter.java

接下来看文件 ContactsAdapter.java，在此定义了继承于 CursorAdapt 的类 Contacts-Adapter，并以下拉数据对象 curse 作为下拉菜单数据的类，并且重写了 runQueryOn BackgroundThread(CharSequence constraint)，设置当输入*时，显示出所有的联系人电话信息。具体代码如下。

```
public class ContactsAdapter extends CursorAdapter
{
  private ContentResolver mContent;

  public ContactsAdapter(Context context, Cursor c)
```

```
{
    super(context, c);

    mContent = context.getContentResolver();
}

@Override
public void bindView(View view, Context context, Cursor cursor)
{
    /* 取得通讯录人员的名字 */
    ((TextView) view).setText(cursor.getString(cursor
            .getColumnIndexOrThrow(Contacts.People.NAME)));
}

@Override
public View newView(Context context, Cursor cursor, ViewGroup parent)
{
    final LayoutInflater inflater = LayoutInflater.from(context);
    final TextView view = (TextView) inflater.inflate(
            android.R.layout.simple_dropdown_item_1line, parent, false);
    view.setText(cursor.getString(cursor
            .getColumnIndexOrThrow(Contacts.People.NAME)));
    return view;
}

@Override
public String convertToString(Cursor cursor)
{
    return cursor.getString(cursor.getColumnIndexOrThrow(Contacts.People.NAME));
}

@Override
public Cursor runQueryOnBackgroundThread(CharSequence constraint)
{
    if (getFilterQueryProvider() != null)
    {
        return getFilterQueryProvider().runQuery(constraint);
    }

    StringBuilder buffer = null;
    String[] args = null;
    if (constraint != null)
    {
```

```
            buffer = new StringBuilder();
            buffer.append("UPPER(");
            buffer.append(Contacts.ContactMethods.NAME);
            buffer.append(") GLOB ?");
            args = new String[]
            { constraint.toString().toUpperCase() + "*" };
        }
        return mContent.query(Contacts.People.CONTENT_URI,
            example8.PEOPLE_PROJECTION, buffer == null ? null : buffer.toString(),
            args, Contacts.People.DEFAULT_SORT_ORDER);
    }
}
```

3. 打开权限

在文件 AndroidManifest.xml 中，需要建打开访问通讯录的权限，具体代码如下。

```
<manifest xmlns:android="http://schemas.android.com/apk/res/android"
      package="irdc.example8"
      android:versionCode="1"
      android:versionName="1.0.0">
    <application android:icon="@drawable/icon" android:label="@string/app_name">
        <activity android:name=".example8"
                  android:label="@string/app_name">
            <intent-filter>
                <action android:name="android.intent.action.MAIN" />
                <category android:name="android.intent.category.LAUNCHER" />
            </intent-filter>
        </activity>
    </application>
<uses-permission android:name="android.permission.READ_CONTACTS"></uses-permission>
</manifest>
```

执行后的初始效果如图 5-19 所示。当输入一个联系人字符后，系统能够根据联系人自动提示显示对应的信息，如图 5-20 所示。当输入“*”后会显示通讯录中所有联系人的信息，如图 5-21 所示。

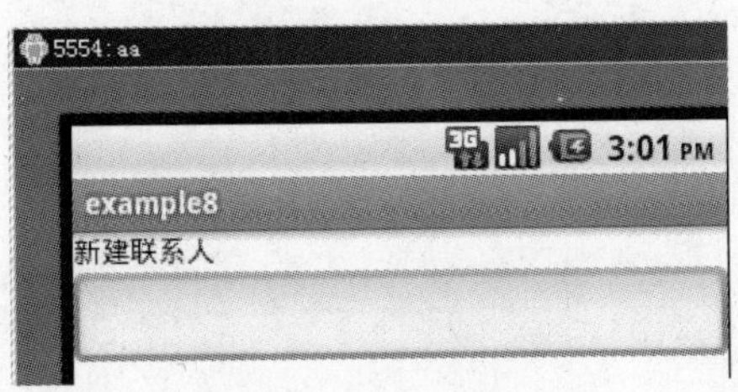

图 5-19　初始效果

图 5-20　提示的信息

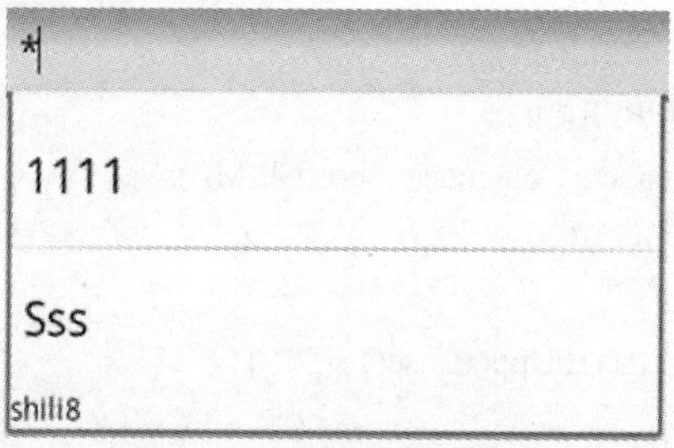

*

1111

Sss

shili8

图 5-21　所有联系人信息

5.9　手机文件管理器

在手机应用中，经常利用手机来存储各种各样的文件信息，这样就需要一个很好的工具来实现对各个文件的管理。在本节的内容中，将通过一个具体实例的实现，介绍 Android 实现文件管理的基本流程。本实例的源代码保存在“光盘:\daima\5\example9”，下面开始讲解本实例的具体实现流程。

5.9.1　实现原理

在本实例中，通过 ListActivity（列表视图组件）和 Java I/O 来查找根目录下的所有文件，并通过 setListAdapter( )方法将存放文件信息的 ArrayAdapter 对象传递给 ListView（列表视图）。而 Android API 提供的 ArrayAdapter 对象只允许存入 String 数组或 List 对象，为此在显示文件列表时只能以字符串的形式来显示文件名。如果要同时显示文件名、图标和文件夹名，则需要定义一个实现 Adapter 接口的对象。

在 Android API 中提供了 BaseAdapter 对象，只要继承了此对象，就可以实现属于自己的 Adapter，在本实例中，事先准备了素材图片作为图标，用于代表不同的文件或文件夹类型。

5.9.2　具体实现

本实例的主程序文件是 example9.java 和 MyAdapter.java。

1．主程序文件 example8.java

下面先讲解文件 example9.java 的具体实现代码。

1）声明需要的变量，其中 items 表示存放显示的名称，paths 表示存放文件路径，rootPath 表示起始目录。具体代码如下。

```
public class example9 extends ListActivity

{
  /* 变量声明
     items：存放显示的名称
     paths：存放文件路径
     rootPath：起始目录           */
  private List<String> items=null;
  private List<String> paths=null;
  private String rootPath="/";
```

```
private TextView mPath;
```

2）载入主布局文件 main.xml，然后初始化 mPath 用于显示目前路径。具体代码如下。

```
@Override
protected void onCreate(Bundle icicle)
{
  super.onCreate(icicle);

  /* 载入 main.xml Layout */
  setContentView(R.layout.main);
  /* 初始化 mPath，用以显示目前路径 */
  mPath=(TextView)findViewById(R.id.mPath);
  getFileDir(rootPath);
}
```

3）定义方法 getFileDir(String filePath)，用于获取文件的具体架构。具体代码如下。

```
/* 取得文件架构的方法 */
private void getFileDir(String filePath)
{
  /* 设置目前所在路径 */
  mPath.setText(filePath);
  items=new ArrayList<String>();
  paths=new ArrayList<String>();
  File f=new File(filePath);
  File[] files=f.listFiles();

  if(!filePath.equals(rootPath))
  {
    /* 第一笔设置为[回到根目录] */
    items.add("b1");
    paths.add(rootPath);
    /* 第二笔设置为[回到上一层] */
    items.add("b2");
    paths.add(f.getParent());
  }
  /* 将所有文件添加 ArrayList 中 */
  for(int i=0;i<files.length;i++)
  {
    File file=files[i];
    items.add(file.getName());
    paths.add(file.getPath());
  }

  /* 使用自定义的 MyAdapter 来将数据传入 ListActivity */
  setListAdapter(new MyAdapter(this,items,paths));
}
```

4）定义按钮触发事件处理方法 onListItemClick，如果是文件夹就再运行 getFileDir()，如果是文件就运行 openFile()。具体代码如下。

```
/* 设置 ListItem 被单击时要做的动作 */
@Override
protected void onListItemClick(ListView l,View v,int position,long id)
{
  File file=new File(paths.get(position));
  if (file.isDirectory())
  {
    /* 如果是文件夹就再运行 getFileDir() */
    getFileDir(paths.get(position));
  }
  else
  {
    /* 如果是文件就运行 openFile() */
    openFile(file);
  }
}
```

5）定义方法 openFile(File f)，用于在手机上打开指定的文件。具体代码如下。

```
/* 在手机上打开文件的方法 */
private void openFile(File f)
{
  Intent intent = new Intent();
  intent.addFlags(Intent.FLAG_ACTIVITY_NEW_TASK);
  intent.setAction(android.content.Intent.ACTION_VIEW);

  /* 调用 getMIMEType()来取得 MimeType */
  String type = getMIMEType(f);
  /* 设置 intent 的 file 与 MimeType */
  intent.setDataAndType(Uri.fromFile(f),type);
  startActivity(intent);
}
```

6）定义 getMIMEType(File f)方法，用于判断文件的类型。其中“end”是结尾的扩展名。只需获取“end”即可。具体代码如下。

```
/* 判断文件 MimeType 的方法 */
private String getMIMEType(File f)
{
  String type="";
  String fName=f.getName();
  /* 取得扩展名 */
  String end=fName.substring(fName.lastIndexOf(".")+1,
              fName.length()).toLowerCase();
```

```
      /* 依附档名的类型决定 MimeType */
      if(end.equals("m4a")||end.equals("mp3")||end.equals("mid")
          ||end.equals("xmf")||end.equals("ogg")||end.equals("wav"))
      {
        type = "audio";
      }
      else if(end.equals("3gp")||end.equals("mp4"))
      {
        type = "video";
      }
      else if(end.equals("jpg")||end.equals("gif")||end.equals("png")
            ||end.equals("jpeg")||end.equals("bmp"))
      {
        type = "image";
      }
      else
      {
        /* 如果无法直接打开，就跳出软件列表给用户选择 */
        type="*";
      }
      type += "/*";
      return type;
    }
  }
```

2. 文件 MyAdapter.java

接下来讲解文件 MyAdapter.java 的具体实现代码。

1）分别声明如下 4 个变量。

- mIcon1：回到根目录的图文件。
- mIcon2：回到上一层的图档。
- mIcon3：文件夹的图文件。
- mIcon4：文件的图档。

```
package irdc.example9;


/* 加载相关类 */
import irdc.example9.R;
/* 自定义的 Adapter，继承 android.widget.BaseAdapter */
public class MyAdapter extends BaseAdapter
{
  /* 变量声明 */
  private LayoutInflater mInflater;
  private Bitmap mIcon1;
  private Bitmap mIcon2;
```

```
private Bitmap mIcon3;
private Bitmap mIcon4;
private List<String> items;
private List<String> paths;
```

2）定义 MyAdapter 构造器，并分别传入了 items、paths 和 mInflater 这 3 个参数，最后对前面定义的 4 个变量进行了赋值。具体代码如下。

```
/* MyAdapter 的构造器，传入 3 个参数  */
public MyAdapter(Context context,List<String> it,List<String> pa)
{
  /* 参数初始化 */
  mInflater = LayoutInflater.from(context);
  items = it;
  paths = pa;
  mIcon1 = BitmapFactory.decodeResource(context.getResources(),R.drawable.back01);
  mIcon2 = BitmapFactory.decodeResource(context.getResources(),R.drawable.back02);
  mIcon3 = BitmapFactory.decodeResource(context.getResources(),R.drawable.folder);
  mIcon4 = BitmapFactory.decodeResource(context.getResources(),R.drawable.doc);
}
```

3）分别覆盖 3 个方法：getCount()、getItem(int position)和 getItemId(int position)。具体代码如下。

```
/* 因继承 BaseAdapter，需覆盖以下方法 */
@Override
public int getCount()
{
  return items.size();
}

@Override
public Object getItem(int position)
{
  return items.get(position);
}

@Override
public long getItemId(int position)
{
  return position;
}
```

4）使用自定义的 file_row 作为布局，然后分别设置“回到根目录”的文字与 icon（图标格式）和“回到上一层”的文字与 icon。具体代码如下。

```
@Override
public View getView(int position,View convertView,ViewGroup parent)
```

```
{
  ViewHolder holder;

  if(convertView == null)
  {
    /* 使用自定义的 file_row 作为 Layout */
    convertView = mInflater.inflate(R.layout.file_row, null);
    /* 初始化 holder 的 text 与 icon */
    holder = new ViewHolder();
    holder.text = (TextView) convertView.findViewById(R.id.text);
    holder.icon = (ImageView) convertView.findViewById(R.id.icon);

    convertView.setTag(holder);
  }
  else
  {
    holder = (ViewHolder) convertView.getTag();
  }

  File f=new File(paths.get(position).toString());
  /* 设置[回到根目录]的文字与 icon */
  if(items.get(position).toString().equals("b1"))
  {
    holder.text.setText("Back to /");
    holder.icon.setImageBitmap(mIcon1);
  }
  /* 设置[回到上一层]的文字与 icon */
  else if(items.get(position).toString().equals("b2"))
  {
    holder.text.setText("Back to ..");
    holder.icon.setImageBitmap(mIcon2);
  }
  /* 设置[文件或文件夹]的文字与 icon */
  else
  {
    holder.text.setText(f.getName());
    if(f.isDirectory())
    {
      holder.icon.setImageBitmap(mIcon3);
    }
    else
    {
      holder.icon.setImageBitmap(mIcon4);
    }
  }
  return convertView;
```

```
        }
        /* class ViewHolder */
        private class ViewHolder
        {
            TextView text;
            ImageView icon;
        }
    }
```

执行后的初始效果如图 5-22 所示。当选择一个文件夹后会继续显示此文件夹里面的内容，如图 5-23 所示。

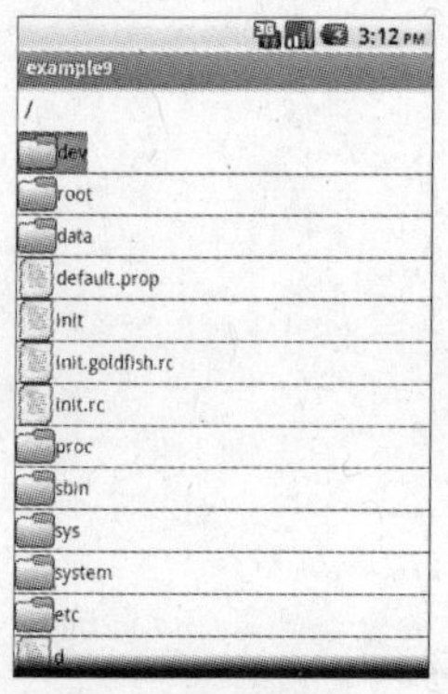

图 5-22　初始效果

图 5-23　文件信息

5.10　清除、还原手机桌面

在 Android 手机系统中，有一个默认的开机主界面，如图 5-24 所示。

图 5-24　默认主界面

实际上，可以通过编程的方法来对默认主界面进行控制。在本节的内容中，将通过一个具体实例的实现，介绍分别清理和还原主界面的方法。本实例的源代码保存在“光盘:\daima\5\example10”，下面开始讲解本实例的具体实现流程。

5.10.1 实现原理

在本实例中，首先在文件 AndroidManifest.xml 中设置操作权限。然后重写了 ContextWrapper 类的 clearWallpaper()方法，这样将轻松的实现了对默认主界面的设置操作。

5.10.2 具体实现

本实例的主程序文件是 example10.java，下面开始讲解解其具体实现流程。

1）设置 Button 按钮用 OnClickListener 来启动事件，具体代码如下。

```
public class example10 extends Activity
{
    private Button mButton1;
    /** Called when the activity is first created. */
  @Override
  public void onCreate(Bundle savedInstanceState)
  {
    super.onCreate(savedInstanceState);
    setContentView(R.layout.main);

    mButton1 =(Button) findViewById(R.id.myButton1);

    /*设置 Button 用 OnClickListener 来启动事件*/
    mButton1.setOnClickListener(new Button.OnClickListener()
    {
      @Override
      public void onClick(View arg0)
      {
        try
        {
          /*清除背景图案  */
          clearWallpaper();
          Toast.makeText(example10.this, getString(R.string.str_done)
              ,Toast.LENGTH_SHORT).show();
        }
        catch (IOException e)
        {
          e.printStackTrace();
        }
      }
    });
  }
```

2）重写 clearWallpaper()方法，清除用户当前的设置桌面，从而还原为原来的默认设置。具体代码如下。

```
        @Override
        public void clearWallpaper() throws IOException
        {
          super.clearWallpaper();
        }
      }
```

接下来需要在文件 AndroidManifest.xml 中设置 permission 权限，具体代码如下。

```
<manifest xmlns:android="http://schemas.android.com/apk/res/android"
      package="irdc.example10"
      android:versionCode="1"
      android:versionName="1.0.0">
    <application android:icon="@drawable/icon" android:label="@string/app_name">
        <activity android:name=".example10"
                  android:label="@string/app_name">
            <intent-filter>
                <action android:name="android.intent.action.MAIN" />
                <category android:name="android.intent.category.LAUNCHER" />
            </intent-filter>
        </activity>
    </application>
    <uses-permission android:name="android.permission.SET_WALLPAPER" />
</manifest>
```

执行后的初始效果如图 5-25 所示。当单击“删除背景”按钮后，会还原到默认的背景主界面。

图 5-25 初始效果

5.11 手机背景图变换处理

在手机应用中，可以对手机的背景图片进行置换处理。在本节的内容中，将通过一个具体实例的实现，介绍 Android 背景图片置换处理的基本流程。本实例的源代码保存在“光

盘:\daima\5\example11”，下面开始讲解本实例的具体实现流程。

5.11.1　实现原理

在本实例中，首先在界面中设置一个 Gallery 组件，然后利用 Gallery 的特性实现了图片的滑动处理，最后将 Gallery 中的 Drawable 图片设置为手机桌面。

在具体设计时，需要将置换的图片必须和手机屏幕的大小比例保持一致，即宽和高的比例一致。

5.11.2　具体实现

本实例的主程序文件是 example11.java，下面开始讲解其具体实现代码。

1）设置设置用于置换的图片素材名，分别如下。

- drawable\google.png。
- drawable\helloking.png。
- drawable\chamberlain.png。
- drawable\king.png。
- drawable\with.png。

具体代码如下所示。

```
public class example11 extends Activity
{
  protected static InputStream is;
  private ImageAdapter mImageAdapter01;

  /** Called when the activity is first created. */
  @Override
  public void onCreate(Bundle savedInstanceState)
  {
    super.onCreate(savedInstanceState);
    setContentView(R.layout.main);

    /*设置图档*/
    Integer[] myImageIds =
    {
      R.drawable.google,
      R.drawable.helloking,
      R.drawable.chamberlain,
      R.drawable.king,
      R.drawable.with,
    };
```

2）定义 ImageAdapter 对象 mImageAdapter01，然后设置图的显示方式是 Gallery，最后分别设置弹出窗口的图式、弹出窗口的信息和确认窗口。具体代码如下。

```
mImageAdapter01 = new ImageAdapter(example11.this, myImageIds);
```

```
/*设置图为 Gallery 的显示方式*/
Gallery g = (Gallery) findViewById(R.id.mygallery);
g.setAdapter(mImageAdapter01);

g.setOnItemClickListener(new Gallery.OnItemClickListener()
{
  @Override
  public void onItemClick
  (AdapterView<?> parent, View v, final int position, long id)
  {
    new AlertDialog.Builder(example11.this)
    .setTitle(R.string.app_about)
    /*设置弹出窗口的图式*/
    .setIcon(mImageAdapter01.myImageIds[position])
    /*设置弹出窗口的信息*/
    .setMessage(R.string.app_about_msg)
    /*确认窗口*/
    .setPositiveButton(R.string.str_ok,
    new DialogInterface.OnClickListener()
    {
      public void onClick(
      DialogInterface dialoginterface, int i)
      {
        Resources resources = getBaseContext().getResources();
        is = resources.openRawResource
        (mImageAdapter01.myImageIds[position]);
        try
        {
```

3）通过 setWallpaper(is)来更换桌面，用 Toast 来显示桌面已更换。具体代码如下。

```
          setWallpaper(is);
          /*用 Toast 来显示桌面已更换*/
          Toast.makeText
          (
            example11.this,
            getString(R.string.my_gallery_text_pre),
            Toast.LENGTH_SHORT
          ).show();
        }
        catch (Exception e)
        {
          e.printStackTrace();
        };
      }
    }).setNegativeButton
```

```
(R.string.str_no, new DialogInterface.OnClickListener()
{
```

4）设置跳出窗口的返回事件处理方法 onClick(DialogInterface dialoginterface, int i)，然后用 Toast 来显示桌面设置已取消。具体代码如下。

```
                /*设置跳出窗口的返回事件*/
                public void onClick(DialogInterface dialoginterface, int i)
                {
                  /*用 Toast 来显示桌面已取消*/
                  Toast.makeText
                  (
                    example11.this,
                    getString(R.string.my_gallery_text_no),
                    Toast.LENGTH_SHORT
                  ).show();
                }
              }).show();
          }
        });
      }
```

5）定义方法 ImageAdapter(Context c, Integer[] aid)，用于保存素材图片。然后分别用 getCount、getItem 和 getItemId 方法来获取具体是哪幅图片。具体代码如下。

```
public class ImageAdapter extends BaseAdapter
{
  int mGalleryItemBackground;
  private Context mContext;
  private Integer[] myImageIds;

  public ImageAdapter(Context c, Integer[] aid)
  {
    mContext = c;
    myImageIds = aid;
    TypedArray a = obtainStyledAttributes(R.styleable.Gallery);
    mGalleryItemBackground = a.getResourceId
    (
      R.styleable.Gallery_android_galleryItemBackground, 0
    );
    a.recycle();
  }

  @Override
  public int getCount()
  {
    return myImageIds.length;
```

```
    }

    @Override
    public Object getItem(int    position)
    {
      return null;
    }

    @Override
    public long getItemId(int    position)
    {
      return position;
    }
```

6）分别产生 ImageView 对象，设置图片给 imageView 对象，重新设置界面的宽高，重新设置图片的宽高，设置 Gallery 背景图，最后返回 ImageView 对象。具体代码如下。

```
    @Override
    public View getView
    (int position, View convertView, ViewGroup parent)
    {

      /*定义 ImageView 对象*/
      ImageView i = new ImageView(mContext);
      /*设置图片给 imageView 对象*/
      i.setImageResource(myImageIds[position]);
      /*重新设置图片的宽高*/
      i.setScaleType(ImageView.ScaleType.FIT_XY);
      /*重新设置 Layout 的宽高*/
      i.setLayoutParams(new Gallery.LayoutParams(138, 108));
      /*设置 Gallery 背景图*/
      i.setBackgroundResource(mGalleryItemBackground);
      /*返回 ImageView 对象*/
      return i;
    }
  }

  @Override
  public void setWallpaper(InputStream data) throws IOException
  {
    super.setWallpaper(data);
  }
}
```

执行后的初始效果如图 5-26 所示。此时可以选择一幅图片作为手机屏幕背景，首先将弹出提示框供用户确认，如图 5-27 所示。

图 5-26　初始效果

图 5-27　文件信息

当单击图 5-27 中的“确定”按钮后，此幅图片会成为手机屏幕的背景，如图 5-28 所示。

图 5-28　选中的图片作为屏幕背景

5.12　对文件的一些操作——修改和删除

在手机应用中，除了能够查看文件和文件夹外，还需要能够对文件进行一些常见的操作。在本节的内容中，将通过一个具体实例的实现，介绍对手机内文件进行修改和删除的基本流程。本实例的源代码保存在“光盘:\daima\5\example12”，下面开始讲解本实例的具体实现流程。

5.12.1 实现原理

在本实例基于前面的 5.9 中的实例，即先实现 5.9 中文件浏览功能，然后再添加文件修改和删除功能，实现对指定文件或文件夹的名字修改或删除操作。上述的名字修改和删除操作，是通过 Java I/O 实现的。

5.12.2 Java I/O 基本类库介绍

I/O 类库通常使用“流”这个抽象概念，它代表任何有能力产生数据的数据源对象或者是有能力接收数据的接收端对象。“流”屏蔽了实际的 I/O 设备中处理数据的细节。例如下面的代码，

```
BufferedReader in = new BufferedReader(
new FileReader("IOStreamDemo.java"));
BufferedReader stdin = new BufferedReader(
new InputStreamReader(System.in));// java.io.BufferedReader(java.io.Reader)
//而 System.in 是一个输入流，因此用 InputStreamReader 进行转换（适配器模式）
```

1）要想打开一个文件用于字符输入，可以使用以 String 或 File 对象作为文件名的 FileReader。为了提高速度，希望对那个文件进行缓冲，那么将作为结果的引用传给一个 BufferedReader 构造器。例如下面的代码，

```
BufferedReader in = new BufferedReader(new FileReader("IOStreamDemo.java"));
String s, s2 = new String();
while ((s = in.readLine()) != null)
        s2 += s + "\n";
in.close();
```

注意：BufferedReader 也提供 readLine()方法，所以这是我们的最终对象和进行读取的接口。

另外，也可以用 System.in 读取来自控制台的输入。System.in 是一个 InputStream，而 BufferedReader 需要的是 Reader 参数，因此引入 InputStreamReader 来进行转换（适配器模式）。例如下面的代码，

```
BufferedReader stdin = new BufferedReader(new InputStreamReader( System.in));
System.out.println("Enter a line");
System.out.println(stdin.readLine());
```

2）从内存读入。用字符串创建一个 StringReader。例如下面的代码，

```
// 2. Input from memory
StringReader in2 = new StringReader(s2);
int c;
while((c = in2.read()) != -1)
System.out.print((char)c);
```

3）格式化的内存输入。

如果要读取格式化数据（JDK5.0 Documentation: read primitive Java data types），此时要用到 DataInputStream，它是一个面向字节的 I/O 类。例如下面的代码，

```
DataInputStream in3 = new DataInputStream(new ByteArrayInputStream(s2.getBytes()));
while(true)
System.out.print((char)in3.readByte());
```

4）文件输出

首先创建一个与指定文件连接的 FileWriter。通常会用 BufferedWriter 将其包装起来用以缓冲输出。然后为了格式化把它转换成 PrintWriter。例如下面的代码，

```
BufferedReader in4 = new BufferedReader(new StringReader(s2));
PrintWriter out1 = new PrintWriter(new BufferedWriter(new FileWriter("D:\\IODemo.txt")));
int lineCount = 1;
while(((s = in4.readLine()) != null))
out1.println(lineCount++ + ": " + s);
out1.close();
```

为了序列化一个对象，首先要创建某些 OutputSteam 对象，然后将其封装在一个 ObjectOutputStream 对象内。这时，只需调用 writeObject()即可将对象序列化，并将其发送给 OutputStream。要将一个序列重组为一个对象，需要将一个 InputStream 封装在 ObjectInputStream 内，然后调用 readObject()。

5.12.3 具体实现

本实例的主程序文件是 example12.java 和 MyAdapter.java，下面开始讲解其具体实现代码。

1. 文件 example12.java

下面先讲解文件 example12.java 的具体实现代码。

1）声明需要的三个对象。

- items：存放显示的名称
- paths：存放文件路径
- rootPath：起始目录

具体代码如下。

```
public class example12 extends ListActivity
{
  /* 对象声明
     items：存放显示的名称
     paths：存放文件路径
     rootPath：起始目录
  */
  private List<String> items=null;
  private List<String> paths=null;
```

```
private String rootPath="/";
private TextView mPath;
private View myView;
private EditText myEditText;

@Override
protected void onCreate(Bundle icicle)
{
  super.onCreate(icicle);
  /* 载入 main.xml Layout */

  setContentView(R.layout.main);
  /* 初始化 mPath，用以显示目前路径 */
  mPath=(TextView)findViewById(R.id.mPath);
  getFileDir(rootPath);
}
```

2）定义方法 getFileDir(String filePath)，用于获取手机内的文件架构。具体过程如下。

第一步：设置目前所在路径。

第二步：分别实现“回到根目录”和“回上层”操作。

第三步：使用自定义的 MyAdapter 来将数据传入 ListActivity。

具体代码如下。

```
private void getFileDir(String filePath)
{
  /* 设置目前所在路径 */
  mPath.setText(filePath);
  items=new ArrayList<String>();
  paths=new ArrayList<String>();

  File f=new File(filePath);
  File[] files=f.listFiles();

  if(!filePath.equals(rootPath))
  {
    /* 第一笔设置为[回到根目录] */
    items.add("b1");
    paths.add(rootPath);
    /* 第二笔设置为[回上层] */
    items.add("b2");
    paths.add(f.getParent());
  }
  /* 将所有文件添加 ArrayList 中 */
  for(int i=0;i<files.length;i++)
  {
    File file=files[i];
```

```
            items.add(file.getName());
            paths.add(file.getPath());
        }

        /* 使用自定义的 MyAdapter 来将数据传入 ListActivity */
        setListAdapter(new MyAdapter(this,items,paths));
    }
```

3）设置 ListItem（列表项）被单击时要做的动作处理事件 onListItemClick(ListView l,View v,int position,long id)，如果是文件夹就再运行 getFileDir()方法，如果是文件则调用 fileHandle()方法。具体代码如下。

```
@Override
protected void onListItemClick(ListView l,View v,int position,long id)
{
  File file = new File(paths.get(position));

  if(file.isDirectory())
  {
    /* 如果是文件夹就再运行 getFileDir()方法 */
    getFileDir(paths.get(position));
  }
  else
  {
    /* 如果是文件则调用 fileHandle()方法 */
    fileHandle(file);
  }
}
```

4）定义处理文件的 fileHandle(final File file)方法，具体过程如下。

第一步：获取文件的 OnClickListener 事件监听。

第二步：如果 which==0，则选择要打开的文件；如果 which==1，则修改文件名；如果 which 是其他值，则删除选中文件。

具体代码如下。

```
/* 处理文件的方法 */
private void fileHandle(final File file){
  /* 单击文件时的 OnClickListener */
  OnClickListener listener1=new DialogInterface.OnClickListener()
  {
    public void onClick(DialogInterface dialog,int which)
    {
      if(which==0)
      {
        /* 选择的 item 为打开文件 */
```

```
            openFile(file);
          }
          else if(which==1)
          {
            /* 选择的 item 为更改文件名 */
            LayoutInflater factory=LayoutInflater.from(example12.this);
            /* 初始化 myChoiceView，使用 rename_alert_dialog 为 layout */
            myView=factory.inflate(R.layout.rename_alert_dialog,null);
            myEditText=(EditText)myView.findViewById(R.id.mEdit);
            /* 将原始文件名先放入 EditText 中 */
            myEditText.setText(file.getName());

            /* 新建一个更改文件名的 Dialog 的确定按钮的监听 */
            OnClickListener listener2=
            new DialogInterface.OnClickListener()
            {
              public void onClick(DialogInterface dialog, int which)
              {
                /* 取得修改后的文件路径 */
                String modName=myEditText.getText().toString();
                final String pFile=file.getParentFile().getPath()+"/";
                final String newPath=pFile+modName;

                /* 判断文件名是否已存在 */
                if(new File(newPath).exists())
                {
                  /* 排除修改文件名时没修改直接送出的状况 */
                  if(!modName.equals(file.getName()))
                  {
                    /* 跳出 Alert 警告档名重复，并确认是否修改 */
                    new AlertDialog.Builder(example12.this)
                        .setTitle("注意!")
                        .setMessage("文件名已经存在，是否要覆盖?")
                        .setPositiveButton("确定",
                         new DialogInterface.OnClickListener()
                        {
                          public void onClick(DialogInterface dialog,int which)
                          {
                            /* 档名重复仍然修改会覆盖掉已存在的文件 */
                            file.renameTo(new File(newPath));
                            /* 重新产生文件列表的 ListView */
                            getFileDir(pFile);
                          }
                        })
                        .setNegativeButton("取消",
```

```
                    new DialogInterface.OnClickListener()
                    {
                      public void onClick(DialogInterface dialog,int which)
                      {
                      }
                    }).show();
              }
            }
            else
            {
              /* 文件名不存在，直接做修改动作 */
              file.renameTo(new File(newPath));
              /* 重新产生文件列表的 ListView */
              getFileDir(pFile);
            }
          }
        };

        /* create 更改文件名时跳出的 Dialog */
        AlertDialog renameDialog=
          new AlertDialog.Builder(example12.this).create();
        renameDialog.setView(myView);

        /* 设置更改文件名单击确认后的 Listener */
        renameDialog.setButton("确定",listener2);
        renameDialog.setButton2("取消",
        new DialogInterface.OnClickListener()
        {
          public void onClick(DialogInterface dialog, int which)
          {
          }
        });
        renameDialog.show();
      }
      else
      {
        /* 选择的 item 为删除文件 */
        new AlertDialog.Builder(example12.this).setTitle("注意!")
            .setMessage("确定要删除文件吗?")
            .setPositiveButton("确定",
             new DialogInterface.OnClickListener()
            {
              public void onClick(DialogInterface dialog,
                                  int which)
              {
```

```
                    /* 删除文件 */
                    file.delete();
                    getFileDir(file.getParent());
                  }
                })
                .setNegativeButton("取消",
                 new DialogInterface.OnClickListener()
                {
                  public void onClick(DialogInterface dialog,int which)
                  {
                  }
                }).show();
          }
        }
      };
```

5）当用户选择一个文件时，系统自动弹出一个要如何处理文件的 ListDialog 对话框。具体代码如下。

```
      String[] menu={"打开文件","更改文件名","删除文件"};
      new AlertDialog.Builder(example12.this)
          .setTitle("你要做甚么?")
          .setItems(menu,listener1)
          .setPositiveButton("取消",
           new DialogInterface.OnClickListener()
          {
            public void onClick(DialogInterface dialog, int which)
            {
            }
          })
          .show();
    }
```

6）定义 openFile(File f)方法，用于在手机上打开指定的文件。具体代码如下。

```
    /* 在手机上打开文件的方法 */
    private void openFile(File f)
    {
      Intent intent = new Intent();
      intent.addFlags(Intent.FLAG_ACTIVITY_NEW_TASK);
      intent.setAction(android.content.Intent.ACTION_VIEW);

      /* 调用 getMIMEType()来取得文件类型 */
      String type = getMIMEType(f);
      /* 设置 intent 的 file 与 MimeType */
      intent.setDataAndType(Uri.fromFile(f),type);
```

```
        startActivity(intent);
    }
```

7）定义方法 getMIMEType(File f)，用于判断指定文件的类型。具体代码如下。

```
    /* 判断文件 MimeType 的方法 */
    private String getMIMEType(File f)
    {
      String type="";
      String fName=f.getName();
      /* 取得扩展名 */
      String end=fName.substring(fName.lastIndexOf(".")+1,
                                    fName.length()).toLowerCase();

      /* 依附文件名的类型决定 MimeType */
      if(end.equals("m4a")||end.equals("mp3")||end.equals("mid")
          ||end.equals("xmf")||end.equals("ogg")||end.equals("wav"))
      {
        type = "audio";
      }
      else if(end.equals("3gp")||end.equals("mp4"))
      {
        type = "video";
      }
      else if(end.equals("jpg")||end.equals("gif")||end.equals("png")
            ||end.equals("jpeg")||end.equals("bmp"))
      {
        type = "image";
      }
      else
      {
        /* 如果无法直接打开，就跳出软件列表给用户选择 */
        type="*";
      }
      type += "/*";
      return type;
    }
}
```

2．文件 MyAdapter.java

接下来讲解文件 MyAdapter.java 的具体实现代码。

1）分别声明如下 4 个变量。

- mIcon1：回到根目录的图文件。
- mIcon2：回到上一层的图文件。
- mIcon3：文件夹的图文件。
- mIcon4：文件的图文件。

```
package irdc.example9;

/* import 相关 class */
import irdc.example9.R;

import java.io.File;
import java.util.List;
import android.content.Context;
import android.graphics.Bitmap;
import android.graphics.BitmapFactory;
import android.view.LayoutInflater;
import android.view.View;
import android.view.ViewGroup;
import android.widget.BaseAdapter;
import android.widget.ImageView;
import android.widget.TextView;

/* 自定义的 Adapter，继承于 android.widget.BaseAdapter */
public class MyAdapter extends BaseAdapter
{
  /* 变量声明 */
  private LayoutInflater mInflater;
  private Bitmap mIcon1;
  private Bitmap mIcon2;
  private Bitmap mIcon3;
  private Bitmap mIcon4;
  private List<String> items;
  private List<String> paths;
```

2）定义 MyAdapter 构造器，并分别传入了 items、paths 和 mInflater 这 3 个参数，最后对前面定义的 4 个变量进行了赋值。具体代码如下。

```
/* MyAdapter 的构造器，传入 3 个参数  */
public MyAdapter(Context context,List<String> it,List<String> pa)
{
  /* 参数初始化 */
  mInflater = LayoutInflater.from(context);
  items = it;
  paths = pa;
  mIcon1 = BitmapFactory.decodeResource(context.getResources(),R.drawable.back01);
  mIcon2 = BitmapFactory.decodeResource(context.getResources(),R.drawable.back02);
  mIcon3 = BitmapFactory.decodeResource(context.getResources(),R.drawable.folder);
  mIcon4 = BitmapFactory.decodeResource(context.getResources(),R.drawable.doc);
}
```

3）分别覆盖 getCount()、getItem(int position)和 getItemId(int position)。具体代码如下。

```
/* 因继承 BaseAdapter，所以需要覆盖以下方法 */
@Override
public int getCount()
{
  return items.size();
}

@Override
public Object getItem(int position)
{
  return items.get(position);
}

@Override
public long getItemId(int position)
{
  return position;
}
```

4）使用自定义的 file_row 作为 Layout，然后分别设置“回到根目录”的文字与 icon 和“回到上一层”的文字与 icon。具体代码如下。

```
@Override
public View getView(int position,View convertView,ViewGroup parent)
{
  ViewHolder holder;

  if(convertView == null)
  {
    /* 使用自定义的 file_row 作为 Layout */
    convertView = mInflater.inflate(R.layout.file_row, null);
    /* 初始化 holder 的 text 与 icon */
    holder = new ViewHolder();
    holder.text = (TextView) convertView.findViewById(R.id.text);
    holder.icon = (ImageView) convertView.findViewById(R.id.icon);

    convertView.setTag(holder);
  }
  else
  {
    holder = (ViewHolder) convertView.getTag();
  }

  File f=new File(paths.get(position).toString());
  /* 设置[回到根目录]的文字与 icon */
  if(items.get(position).toString().equals("b1"))
  {
```

```
                holder.text.setText("Back to /");
                holder.icon.setImageBitmap(mIcon1);
            }
            /* 设置[回到上一层]的文字与 icon */
            else if(items.get(position).toString().equals("b2"))
            {
                holder.text.setText("Back to ..");
                holder.icon.setImageBitmap(mIcon2);
            }
            /* 设置[文件或文件夹]的文字与 icon */
            else
            {
                holder.text.setText(f.getName());
                if(f.isDirectory())
                {
                    holder.icon.setImageBitmap(mIcon3);
                }
                else
                {
                    holder.icon.setImageBitmap(mIcon4);
                }
            }
            return convertView;
        }
        /* class ViewHolder */
        private class ViewHolder
        {
            TextView text;
            ImageView icon;
        }
    }
```

执行后的初始效果如图 5-29 所示。当选择一个文件夹后会弹出一个操作对话框，用户可以选择“打开文件”、“更改文件名”和“删除文件”这 3 种操作，如图 5-30 所示。

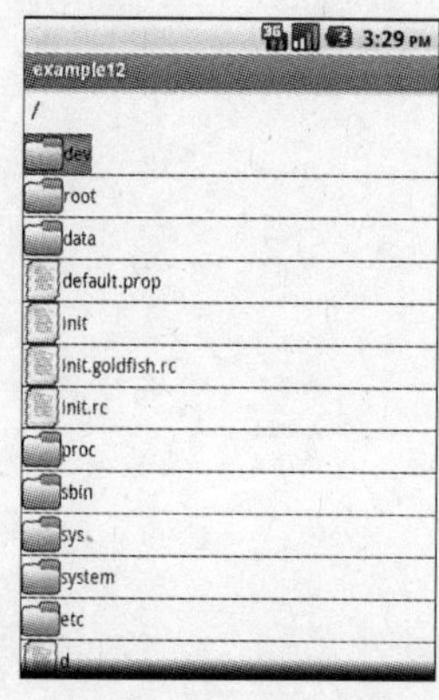

图 5-29　初始效果

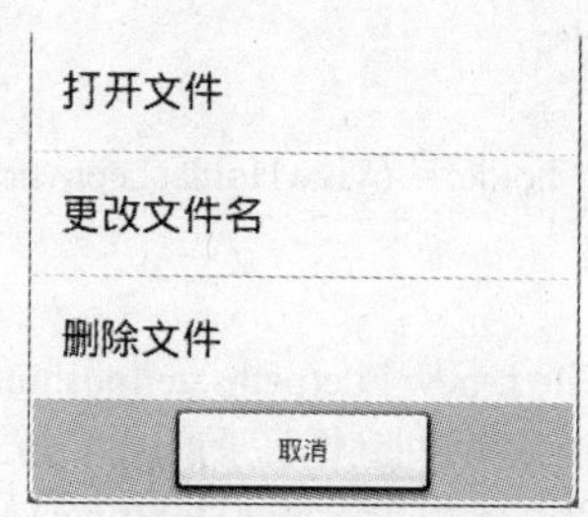

图 5-30　文件信息

当选择图 5-30 中的一种操作后，会执行对应的功能。

5.13　获取 File 和 Cache 的路径

当 Android 程序运行后，可以通过程序来获取正在运行程序的路径，此路径通常位于"/data/data/package name"，此处的 package name 就是程序的 Package 名称。在本节的内容中，将通过一个具体实例的实现，介绍获取 File 和 Cache 的路径的基本流程。本实例的源代码保存在"光盘:\daima\5\example13"，下面开始讲解本实例的具体实现流程。

5.13.1　实现原理

在本实例中，首先设计了两个按钮，分别触发获取 File 和 Cache 路径的事件。当用户选择后，会弹出一个新页面，此新页面以 ListActivity 来显示里面的目录或文件。当打开目录时，还可以看到该目录下的文件。上述功能是通过 getCacheDir 方法和 getFilesDir 方法实现的，两个方法都能获得当前的手机自带的存储空间中的当前包文件的路径。

5.13.2　具体实现

本实例的主程序文件是 example13.java 和 example13_1.java，下面开始讲解其具体实现代码。

1．文件 example13.java

下面先讲解文件 example13.java 的具体实现代码。

1）先定义两个按钮变量 myButton1 和 myButton2，然后定义两个文件变量 cacheDir 和 fileDir。具体代码如下。

```
public class example13 extends Activity
{
  private Button myButton1;
  private Button myButton2;
  private File cacheDir;
  private File fileDir;
```

2）根据单击的按钮，分别获取目前 Cache 目录和目前 File 目录。具体代码如下。

```
/** Called when the activity is first created. */
@Override
public void onCreate(Bundle savedInstanceState)
{
  super.onCreate(savedInstanceState);
  setContentView(R.layout.main);

  myButton1 = (Button) findViewById(R.id.myButton1);
  myButton2 = (Button) findViewById(R.id.myButton2);

  /*  取得目前 Cache 目录  */
```

```
cacheDir = this.getCacheDir();
/*  取得目前 File 目录 */
fileDir = this.getFilesDir();
```

3）分别定义两个按钮的单击处理事件 myButton1.setOnClickListener 和 myButton2.setOnClickListener。具体代码如下。

```
myButton1.setOnClickListener(new Button.OnClickListener()
{

  @Override
  public void onClick(View arg0)
  {

    String path = fileDir.getParent() + java.io.File.separator
        + fileDir.getName();

    showListActivity(path);

  }
});

myButton2.setOnClickListener(new Button.OnClickListener()
{

  @Override
  public void onClick(View arg0)
  {
    String path = cacheDir.getParent() + java.io.File.separator
      + cacheDir.getName();

    showListActivity(path);
  }
});

}
```

4）调用 example13_1，然后将获取路径传入。具体代码如下。

```
private void showListActivity(String path)
{
  Intent intent = new Intent();
  intent.setClass(example13.this, example13_1.class);

  Bundle bundle = new Bundle();
  bundle.putString("path", path);
  intent.putExtras(bundle);
```

```
        startActivity(intent);
    }

}
```

2．文件 example13_1.java

接下来看文件 example13_1.java 的具体实现代码。

1）先定义类 example13_1，并继承了 ListActivity。具体代码如下。

```
public class example13_1 extends ListActivity
{
  private List<String> items = null;
  private String path;

  /** Called when the activity is first created. */
  public void onCreate(Bundle savedInstanceState)
  {
    super.onCreate(savedInstanceState);
    setContentView(R.layout.mylist);

    Bundle bunde = this.getIntent().getExtras();
    /* 取得 example13 所传的路径 */
    path = bunde.getString("path");

    this.setTitle(path);

    java.io.File file = new java.io.File(path);
    /* 列出该路径下的所有文件 */
    fill(file.listFiles());
  }
```

2）定义 onListItemClick 方法，用于显示传入的文件和文件目录。具体代码如下。

```
@Override
protected void onListItemClick
(ListView l, View v, int position, long id)
{
  File file = new File
  (path + java.io.File.separator + items.get(position));

  if (file.isDirectory())
  {
    fill(file.listFiles());
  }
}
```

3）定义 fill(File[] files)方法，将取得的文件名放入 ArrayList。具体代码如下。

```
    private void fill(File[] files)
    {
      items = new ArrayList<String>();
      if (files == null)
      {
        return;
      }

      /* 取得文件名放入 ArrayList */
      for (File file : files)
      {
        items.add(file.getName());
      }

      /* 将 ArrayList 放入 ArrayAdapter */
      ArrayAdapter<String> fileList = new ArrayAdapter<String>
      (this, android.R.layout.simple_list_item_1, items);
      setListAdapter(fileList);
    }
  }
```

3. 设置 Activity

接下来在文件 AndroidManifest.xml 中设置两个 Activity，一个在 LAUNCHER 启动时运行，另一个为取得文件夹信息时所需要显示的 Activity。具体代码如下。

```
<manifest xmlns:android="http://schemas.android.com/apk/res/android"
      package="irdc.example13"
      android:versionCode="1"
      android:versionName="1.0.0">
    <application android:icon="@drawable/icon" android:label="@string/app_name">
        <activity android:name=".example13"
                  android:label="@string/app_name">
            <intent-filter>
                <action android:name="android.intent.action.MAIN" />
                <category android:name="android.intent.category.LAUNCHER" />
            </intent-filter>
        </activity>
        <activity android:name="example13_1"></activity>
    </application>
</manifest>
```

执行后会显示两个按钮，如图 5-31 所示。当单击一个按钮后，会显示对应的目录信息。例如，单击“现在 File”按钮后的效果如图 5-32 所示。

图 5-31　初始效果

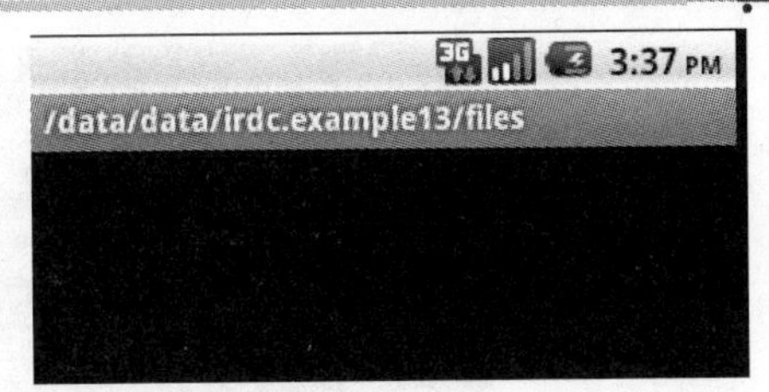

图 5-32　文件信息

5.14　控制 Wi-Fi 服务

在手机应用中，Wi-Fi 也是一个十分重要的通信功能。在本节的内容中，将通过一个具体实例的实现，介绍在 Android 系统中对 Wi-Fi 进行控制基本流程。本实例的源代码保存在“光盘:\daima\5\example14”，下面开始讲解本实例的具体实现流程。

5.14.1　Wi-Fi 简介

Wi-Fi 是一种可以将个人电脑、手持设备（如 PDA、手机）等终端以无线方式互相连接的技术。Wi-Fi 是一个无线网路通信技术的品牌，由 Wi-Fi 联盟（Wi-Fi Alliance）所持有。目的是改善基于 IEEE 802.11 标准的无线网路产品之间的互通性。现时一般人会把 Wi-Fi 及 IEEE 802.11 混为一谈。甚至把 Wi-Fi 等同于无线网际网路。

Wi-Fi 就是一种无线联网的技术，以前通过网线连接电脑，而现在则是通过无线电波来连网；常见的就是一个无线路由器，在这个无线路由器的电波覆盖的有效范围都可以采用 Wi-Fi 连接方式进行联网，如果无线路由器连接了一条 ADSL 线路或者别的上网线路，则又被称为“热点”。

现在市面上常见的无线路由器多为 54M 速度，再上一个等级就是 108M 的速度，当然这个速度并不是你访问互联网的速度，访问互联网的速度主要是取决于 Wi-Fi“热点”的互联网线路。Wi-Fi 是一种帮助用户访问电子邮件、Web 和流式媒体的技术。它为用户提供了无线的宽带互联网访问。同时，它也是在家里、办公室或在旅途中上网的快速、便捷的途径。能够访问 Wi-Fi 网络的地方被称为“热点”。

Wi-Fi 热点是通过在互联网连接上安装访问点来创建的。

5.14.2　实现原理

在 Android 系统中，存在一个无线控制模块，打开方式如下。依次单击“Menu”→“Settings”→“Wireless$networks”→“Wi-Fi settings”，出现图 5-33 所示界面。

图 5-33 所示的是 Wi-Fi 的控制界面，可以控制 Wi-Fi 的打开和关闭，而本实例的目的是以编程的方式实现类似的功能。

在本实例中，一共用到了 WifiManager 的五种状态，具体说明如下。

- WifiManager.WIFI_STATE_ENABLING：表示 Wi-Fi 已经打开。
- WifiManager.WIFI_STATE_DISABLING：表示 Wi-Fi 正在关闭而无法关闭。
- WifiManager.WIFI_STATE_DISABLED：表示 Wi-Fi 已经关闭。

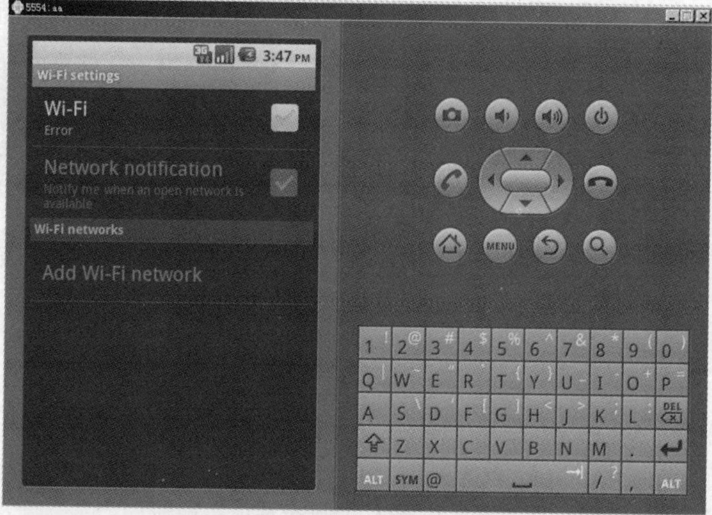

图 5-33　WiFi 控制界面

- WifiManager.WIFI_STATE_ENABLING：表示 Wi-Fi 正在打开而无法关闭。
- WifiManager.WIFI_STATE_ENABLED：表示 Wi-Fi 已经打开无法再打开。
- WifiManager.WIFI_STATE_DISABLING：表示 Wi-Fi 正在关闭。
- WifiManager.WIFI_STATE_DISABLED：表示 Wi-Fi 已经关闭。
- WifiManager.WIFI_STATE_UNKNOWN：表示 Wi-Fi 无法识别。

在具体实现上，会首先定义一个复选框 CheckBox，然后捕捉 CheckBox 的单击事件，根据对应的状态显示对应的提示。例如，可以用下面的代码检测了 Wi-Fi 是否启动。

```
WifiManager wm = (WifiManager) context.getSystemService(Context.WIFI_SERVICE);
if(wm.getWifiState() == WifiManager.WIFI_STATE_ENABLED){
    return true;
}
```

5.14.3　具体实现

本实例的主程序文件是 example14.java，下面开始讲解其具体实现代码。

1）引入 wifi.WifiManager 和 widget.CheckBox，然后创建 WiFiManager 对象 mWiFiManager01。具体代码如下。

```
public class example14 extends Activity
{
  private TextView mTextView01;
  private CheckBox mCheckBox01;

  /* 创建 WiFiManager 对象 */
  private WifiManager mWiFiManager01;
```

2）分别定义两个变量：mTextView01 和 mCheckBox01，分别用于显示提示文本和获取

复选框的选择状态。具体代码如下。

```
/** Called when the activity is first created. */
@Override
public void onCreate(Bundle savedInstanceState)
{
  super.onCreate(savedInstanceState);
  setContentView(R.layout.main);

  mTextView01 = (TextView) findViewById(R.id.myTextView1);
  mCheckBox01 = (CheckBox) findViewById(R.id.myCheckBox1);
```

3）以 getSystemService( )方法取得 WIFI_SERVICE（WIFI 服务），具体代码如下。

```
mWiFiManager01 = (WifiManager)
  this.getSystemService(Context.WIFI_SERVICE);
```

4）通过 if 语句来判断运行程序后的 WiFi 状态是否打开或打开中，这样便可显示对应的提示信息。具体代码如下。

```
/* 判断运行程序后的 WiFi 状态是否打开或打开中 */
if(mWiFiManager01.isWifiEnabled())
{
  /* 判断 WiFi 状态是否“已打开” */
  if(mWiFiManager01.getWifiState()==
    WifiManager.WIFI_STATE_ENABLED)
  {
    /* 若 WiFi 已打开，将选取项打勾 */
    mCheckBox01.setChecked(true);
    /* 更改选取项文字为关闭 WiFi*/
    mCheckBox01.setText(R.string.str_uncheck);
  }
  else
  {
    /* 若 WiFi 未打开，将选取项勾选取消 */
    mCheckBox01.setChecked(false);
    /* 更改选取项文字为打开 WiFi*/
    mCheckBox01.setText(R.string.str_checked);
  }
}
else
{
  mCheckBox01.setChecked(false);
  mCheckBox01.setText(R.string.str_checked);
}
```

5）通过 mCheckBox01.setOnClickListener 来捕捉 CheckBox 的单击事件，用 onClick

(View v)方法获取用户的单击，然后使用 if 语句，根据操作需求来执行对应操作，并根据需要输出对应的提示信息。具体代码如下。

```
mCheckBox01.setOnClickListener(
new CheckBox.OnClickListener()
{
  @Override
  public void onClick(View v)
  {

    /* 当选取项为取消选取状态 */
    if(mCheckBox01.isChecked()==false)
    {
      /* 尝试关闭 WiFi 服务 */
      try
      {
        /* 判断 WiFi 状态是否为已打开 */
        if(mWiFiManager01.isWifiEnabled() )
        {
          /* 关闭 WiFi */
          if(mWiFiManager01.setWifiEnabled(false))
          {
            mTextView01.setText(R.string.str_stop_wifi_done);
          }
          else
          {
            mTextView01.setText(R.string.str_stop_wifi_failed);
          }
        }
        else
        {
          /* WiFi 状态不为已打开状态时 */
          switch(mWiFiManager01.getWifiState())
          {
            /* WiFi 正在打开过程中，导致无法关闭... */
            case WifiManager.WIFI_STATE_ENABLING:
              mTextView01.setText
              (
                getResources().getText
                (R.string.str_stop_wifi_failed)+":"+
                getResources().getText
                (R.string.str_wifi_enabling)
              );
              break;
            /* WiFi 正在关闭过程中，导致无法关闭... */
            case WifiManager.WIFI_STATE_DISABLING:
```

```
                mTextView01.setText
                (
                  getResources().getText
                  (R.string.str_stop_wifi_failed)+":"+
                  getResources().getText
                  (R.string.str_wifi_disabling)
                );
                break;
              /* WiFi 已经关闭 */
              case WifiManager.WIFI_STATE_DISABLED:
                mTextView01.setText

                (
                  getResources().getText
                  (R.string.str_stop_wifi_failed)+":"+
                  getResources().getText
                  (R.string.str_wifi_disabled)
                );
                break;
              /* 无法取得或辨识 WiFi 状态 */
              case WifiManager.WIFI_STATE_UNKNOWN:
              default:
                mTextView01.setText
                (
                  getResources().getText
                  (R.string.str_stop_wifi_failed)+":"+
                  getResources().getText
                  (R.string.str_wifi_unknow)
                );
                break;
            }
            mCheckBox01.setText(R.string.str_checked);
          }
        }
        catch (Exception e)
        {
          Log.i("HIPPO", e.toString());
          e.printStackTrace();
        }
      }
      else if(mCheckBox01.isChecked()==true)
      {
        /* 尝试打开 WiFi 服务 */
        try
        {
          /* 确认 WiFi 服务是关闭且不在打开作业中 */
```

```
if(!mWiFiManager01.isWifiEnabled() &&
 mWiFiManager01.getWifiState()!=
 WifiManager.WIFI_STATE_ENABLING )
{
  if(mWiFiManager01.setWifiEnabled(true))
  {
    switch(mWiFiManager01.getWifiState())
    {
      /* WiFi 正在打开过程中，导致无法打开... */
      case WifiManager.WIFI_STATE_ENABLING:
        mTextView01.setText
        (
          getResources().getText
          (R.string.str_wifi_enabling)
        );
        break;
      /* WiFi 已经为打开，无法再次打开... */
      case WifiManager.WIFI_STATE_ENABLED:
        mTextView01.setText
        (
          getResources().getText
          (R.string.str_start_wifi_done)
        );
        break;
      /* 其他未知的错误 */
      default:
        mTextView01.setText
        (
          getResources().getText
          (R.string.str_start_wifi_failed)+":"+
          getResources().getText
          (R.string.str_wifi_unknow)
        );
        break;
    }
  }
  else
  {
    mTextView01.setText(R.string.str_start_wifi_failed);
  }
}
else
{
  switch(mWiFiManager01.getWifiState())
  {
    /* WiFi 正在打开过程中，导致无法打开... */
```

```
        case WifiManager.WIFI_STATE_ENABLING:
          mTextView01.setText
          (
            getResources().getText
            (R.string.str_start_wifi_failed)+":"+
            getResources().getText
            (R.string.str_wifi_enabling)
          );
          break;
        /* WiFi 正在关闭过程中，导致无法打开... */
        case WifiManager.WIFI_STATE_DISABLING:
          mTextView01.setText
          (
            getResources().getText
            (R.string.str_start_wifi_failed)+":"+
            getResources().getText
            (R.string.str_wifi_disabling)
          );
          break;
        /* WiFi 已经关闭 */
        case WifiManager.WIFI_STATE_DISABLED:
          mTextView01.setText
          (
            getResources().getText
            (R.string.str_start_wifi_failed)+":"+
            getResources().getText
            (R.string.str_wifi_disabled)
          );
          break;
        /* 无法取得或识别 WiFi 状态 */
        case WifiManager.WIFI_STATE_UNKNOWN:
        default:
          mTextView01.setText
          (
            getResources().getText
            (R.string.str_start_wifi_failed)+":"+
            getResources().getText
            (R.string.str_wifi_unknow)
          );
          break;
      }
    }
    mCheckBox01.setText(R.string.str_uncheck);
  }
  catch (Exception e)
  {
```

```
                    Log.i("HIPPO", e.toString());
                    e.printStackTrace();
                }
            }
        }
    });
}
```

6）定义 mMakeTextToast(String str, boolean isLong)，用于根据当前操作显示对应的提示信息。具体代码如下。

```
public void mMakeTextToast(String str, boolean isLong)
{
  if(isLong==true)
  {
    Toast.makeText(example14.this, str, Toast.LENGTH_LONG).show();
  }
  else
  {
    Toast.makeText(example14.this, str, Toast.LENGTH_SHORT).show();
  }
}

@Override
protected void onResume()
{
  // TODO Auto-generated method stub

  /* 在 onResume 重写事件为取得打开程序当下 Wi-Fi 的状态 */
  try
  {
    switch(mWiFiManager01.getWifiState())
    {
      /* WiFi 已经在打开状态... */
      case WifiManager.WIFI_STATE_ENABLED:
        mTextView01.setText
        (
          getResources().getText(R.string.str_wifi_enabling)
        );
        break;
      /* WiFi 正在打开过程中状态... */
      case WifiManager.WIFI_STATE_ENABLING:
        mTextView01.setText
        (
          getResources().getText(R.string.str_wifi_enabling)
        );
        break;
```

```
            /* WiFi 正在关闭过程中... */
            case WifiManager.WIFI_STATE_DISABLING:
              mTextView01.setText
              (
                getResources().getText(R.string.str_wifi_disabling)
              );
              break;
            /* WiFi 已经关闭 */
            case WifiManager.WIFI_STATE_DISABLED:
              mTextView01.setText
              (
                getResources().getText(R.string.str_wifi_disabled)
              );
              break;
            /* 无法取得或识别 Wi-Fi 状态 */
            case WifiManager.WIFI_STATE_UNKNOWN:
            default:
              mTextView01.setText
              (
                getResources().getText(R.string.str_wifi_unknow)
              );
              break;
          }
        }
        catch(Exception e)
        {
          mTextView01.setText(e.toString());
          e.getStackTrace();
        }
        super.onResume();
      }

      @Override
      protected void onPause()
      {
        super.onPause();
      }
    }
```

接下来需要在文件 AndroidManifest.xml 中添加对 Wi-Fi 的访问以及对网络状态的权限。具体代码如下。

```
<manifest xmlns:android="http://schemas.android.com/apk/res/android" package="irdc.example14"
android:versionCode="1" android:versionName="1.0.0">
- <application android:icon="@drawable/icon" android:label="@string/app_name">
- <activity android:name=".example14" android:label="@string/app_name">
- <intent-filter>
```

```
        <action android:name="android.intent.action.MAIN" />
        <category android:name="android.intent.category.LAUNCHER" />
        </intent-filter>
        </activity>
        </application>
    <!--  新增存取 Wi-Fi 以及网络等相关权限
        -->
        <uses-permission android:name="android.permission.CHANGE_NETWORK_STATE" />
        <uses-permission android:name="android.permission.CHANGE_WIFI_STATE" />
        <uses-permission android:name="android.permission.ACCESS_NETWORK_STATE" />
        <uses-permission android:name="android.permission.ACCESS_WIFI_STATE" />
        <uses-permission android:name="android.permission.INTERNET" />
        <uses-permission android:name="android.permission.WAKE_LOCK" />
        </manifest>
```

执行后会显示两个按钮，如图 5-34 所示。当选择复选框后会执行对应的操作处理，并且显示对应的提示信息，如图 5-35 所示。

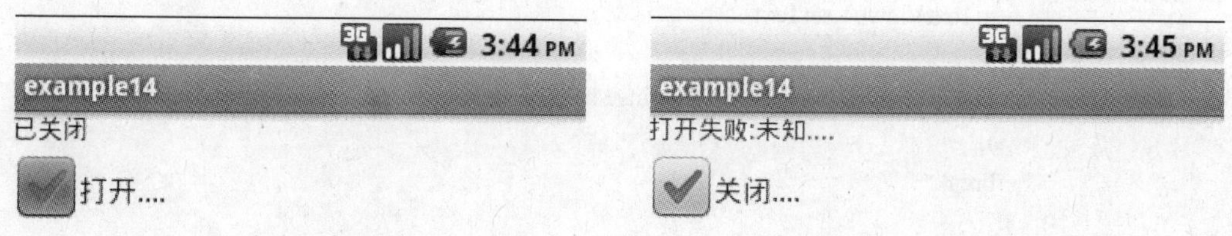

图 5-34　初始效果　　图 5-35　文件信息

5.15　获取 SIM 卡内信息

在手机应用中，通常需要获取 SIM 卡内的信息。在本节的内容中，将通过一个具体实例的实现，介绍在 Android 系统中获取 SIM 卡内信息的基本流程。本实例的源代码保存在"光盘:\daima\5\example15"，下面开始讲解本实例的具体实现流程。

5.15.1　SIM 卡简介

SIM 卡是客户识别模块（Subscriber Identity Module，SIM）的缩写，也称为智能卡、用户身份识别卡，GSM 数字移动电话机必须装上此卡方能使用。它在电脑芯片上存储了数字移动电话客户的信息，加密的密钥等内容，可供 GSM 对网络客户身份进行鉴别，并对客户通话时的语音信息进行加密。SIM 卡的使用，完全防止了并机和通话被窃听行为，并且 SIM 卡的制作是严格按照 GSM 国际标准和规范来完成的，从而可靠的保障了客户的正常通信。

为防止他人擅用您的 SIM 卡， SIM 卡设置了个人识别密码——PIN 码，只要设置了 PIN 码用户在每次打开手机时，屏幕上会显示要求输入 4 位 PIN 码，初始的 PIN 码为 1234 或 0000，连续输入三次错误码的 PIN 码，手机会显示"输入 PUK 码"或"已锁"字样，说明 SIM 卡已被锁上，这时千万不要再按手机键盘，因为如果输入的 PIN 解锁码累计超过 10 次，SIM 卡将自动报废。当 SIM 卡被锁时，需携带 SIM 卡身份证到移动营业厅，由服务人

员为您解锁。

5.15.2　实现原理

在本实例中，通过 Android API 中的 TelephonyManager（即 Android.telephony. Telephony Manager）对象提供的几个方法实现对 SIM 卡信息的读取。另外，TelephonyManager 还能获取手机的号码。看下面的代码。

```
TelephonyManager tm = (TelephonyManager)this.getSystemService(Context.TELEPHONY_SERVICE);
numberText.setText(tm.getLine1Number());
```

其中，上面加粗部分能够获取本机号码，除此之外，TelephonyManager（手机通信管理）类还提供了多种获取手机信息的函数，例如 imei( )、imsi( )等。看下面的代码。

```
package com.pingan.innovation;

import android.app.Activity;
import android.content.Context;
import android.os.Bundle;
import android.telephony.TelephonyManager;
import android.widget.TextView;

public class PhoneInfo extends Activity {
  private TextView numberText;
  private TextView imeiText;
  private TextView onText;
  private TextView snText;
  private TextView imsiText;
  private TextView ssText;
  private TextView ntText;


    @Override
    public void onCreate(Bundle savedInstanceState) {
            super.onCreate(savedInstanceState);
            setContentView(R.layout.main);

            numberText = (TextView) findViewById(R.id.numberText);
            imeiText = (TextView) findViewById(R.id.imeiText);
            onText = (TextView) findViewById(R.id.onText);
            snText = (TextView) findViewById(R.id.snText);
            imsiText = (TextView) findViewById(R.id.imsiText);
            ssText = (TextView) findViewById(R.id.ssText);
            ntText = (TextView) findViewById(R.id.ntText);
            TelephonyManager tm = (TelephonyManager)this.getSystemService(Context.TELEPHONY_
            SERVICE);
            numberText.setText(tm.getLine1Number());
```

```
                imeiText.setText(tm.getDeviceId());
                onText.setText(tm.getNetworkOperatorName());
                snText.setText(tm.getSimSerialNumber());
                imsiText.setText(tm.getSubscriberId());
                ssText.setText(tm.getNetworkCountryIso());
                ntText.setText(tm.getNetworkOperator());
            }
        }
```

通过上述代码中的几个函数，分别获取了手机号码、imei、运营商名称、sim 卡序列号、IMSI、sim 卡所在国家以及运营商编号。运行效果如图 5-36 所示。

手机号码： 15555218135

IMEI： 000000000000000

运营商： Android

sim卡序列号： 89014103211118510720

IMSI： 310260000000000

ISO： us

运营商编号： 310260

图 5-36　运行结果

5.15.3　具体实现

本实例的主程序文件是 example15.java 和 MyAdapter.java，下面开始讲解其具体实现代码。

1．文件 example15.java

下面先讲解文件 example15.java 的具体实现代码。

1）先载入布局文件 main.xml，然后通过 add(getResources().getText(R.string.str_list0).toString())将取得的信息写入 List（列表）中，最后通过 if 语句来设置 SIM 卡的状态。具体代码如下。

```
public class example15 extends ListActivity
{
  private TelephonyManager telMgr;
  private List<String> item=new ArrayList<String>();
  private List<String> value=new ArrayList<String>();

  @SuppressWarnings("static-access")
  @Override
```

```
public void onCreate(Bundle savedInstanceState)
{
  super.onCreate(savedInstanceState);
  /* 载入 main.xml Layout */
  setContentView(R.layout.main);
  telMgr = (TelephonyManager)getSystemService(TELEPHONY_SERVICE);

  /* 将取得的信息写入 List 中 */
  /* 取得 SIM 卡状态 */
  item.add(getResources().getText(R.string.str_list0).toString());
  if(telMgr.getSimState()==telMgr.SIM_STATE_READY)
  {
    value.add("良好");
  }
  else if(telMgr.getSimState()==telMgr.SIM_STATE_ABSENT)
  {
    value.add("无 SIM 卡");
  }
  else
  {
    value.add("SIM 卡被锁定或未知的状态");
  }
```

2）分别获取 SIM 卡的卡号、SIM 卡供货商代码、SIM 卡供货商名称及 SIM 卡国别，然后使用自定义的 MyAdapter 来将数据传入 ListActivity。具体代码如下。

```
  item.add(getResources().getText(R.string.str_list1).toString());
  if(telMgr.getSimSerialNumber()!=null)
  {
    value.add(telMgr.getSimSerialNumber());
  }
  else
  {
    value.add("无法取得");
  }

  /* 取得 SIM 卡供货商代码 */
  item.add(getResources().getText(R.string.str_list2).toString());
  if(telMgr.getSimOperator().equals(""))
  {
    value.add("无法取得");
  }
  else
  {
    value.add(telMgr.getSimOperator());
  }
```

```
    /* 取得 SIM 卡供货商名称 */
    item.add(getResources().getText(R.string.str_list3).toString());
    if(telMgr.getSimOperatorName().equals(""))
    {
      value.add("无法取得");
    }
    else
    {
      value.add(telMgr.getSimOperatorName());
    }

    /* 取得 SIM 卡国别 */
    item.add(getResources().getText(R.string.str_list4).toString());
    if(telMgr.getSimCountryIso().equals(""))
    {
      value.add("无法取得");
    }
    else
    {
      value.add(telMgr.getSimCountryIso());
    }

    /* 使用自定义的 MyAdapter 来将数据传入 ListActivity */
    setListAdapter(new MyAdapter(this,item,value));
  }
}
```

2. 文件 MyAdapter.java

接下来讲解文件 MyAdapter.java 的具体实现代码。

1）先声明 3 个变量 mInflater、items 和 values，然后定义 MyAdapter 构造器，传入三个参数。具体代码如下。

```
/* 自定义的 Adapter，继承 android.widget.BaseAdapter */
public class MyAdapter extends BaseAdapter
{
  /* 变量声明 */
  private LayoutInflater mInflater;
  private List<String> items;
  private List<String> values;
  /* MyAdapter 的构造器，传入三个参数  */
  public MyAdapter(Context context,List<String> item,
                   List<String> value)
  {
    /* 参数初始化 */
    mInflater = LayoutInflater.from(context);
    items = item;
```

```
    values = value;
  }
```

2）分别覆盖方法 getCount()、getItem(int position)、getItemId(int position)及 getView(int position, View convertView，ViewGroup par)，具体代码如下。

```
/* 因继承 BaseAdapter，需覆盖以下方法 */
@Override
public int getCount()
{
  return items.size();
}

@Override
public Object getItem(int position)
{
  return items.get(position);
}

@Override
public long getItemId(int position)
{
  return position;
}

@Override
public View getView(int position,View convertView,ViewGroup par)
{
  ViewHolder holder;

  if(convertView == null)
  {
    /* 使用自定义的 file_row 作为 Layout */
    convertView = mInflater.inflate(R.layout.row_layout,null);
    /* 初始化 holder 的 text 与 icon */
    holder = new ViewHolder();
    holder.text1=(TextView)convertView.findViewById(R.id.myText1);
    holder.text2=(TextView)convertView.findViewById(R.id.myText2);

    convertView.setTag(holder);
  }
  else
  {
    holder = (ViewHolder) convertView.getTag();
  }
  /* 设置要显示的信息 */
  holder.text1.setText(items.get(position).toString());
```

```
        holder.text2.setText(values.get(position).toString());

        return convertView;
    }

    /* class ViewHolder */
    private class ViewHolder
    {
      /* text1：信息名称
       * text2：信息内容 */
      TextView text1;
      TextView text2;
    }
}
```

最后，还需要在文件 AndroidManifest.xml 中设置读取电话状态的权限，具体代码如下。

```
<!-- 设置 READ_PHONE_STATE 权限 -->
<uses-permission android:name="android.permission.READ_PHONE_STATE"></uses-permission>
```

执行后会显示对应的获取信息，如图 5-37 所示。

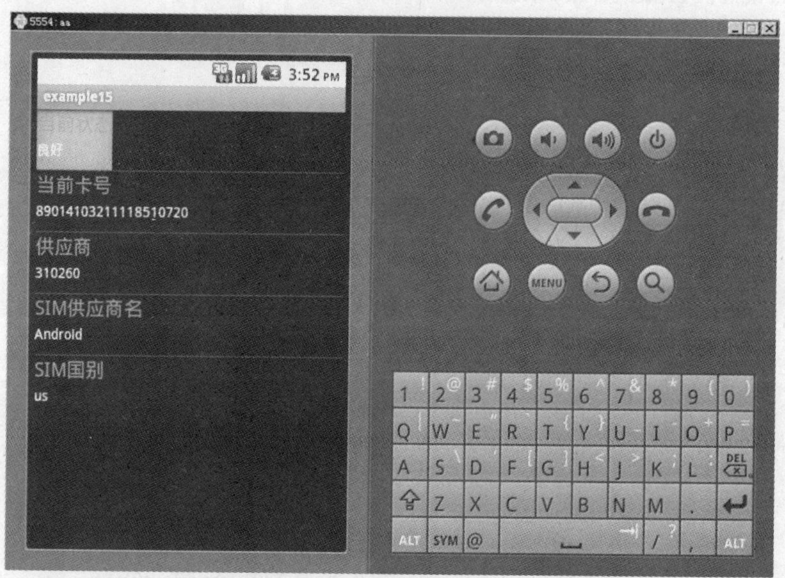

图 5-37　执行效果

5.16　实现触摸拨号按钮

在触摸屏手机应用中，通常需要触摸一个按钮来实现拨号处理。在本节的内容中，将通过一个具体实例的实现，介绍在 Android 中实现类似触摸拨号按钮的基本流程。本实例的源代码保存在“光盘:\daima\5\example16”，下面开始讲解本实例的具体实现流程。

5.16.1 实现原理

在本实例中，通过 Intent 方式将电话号码传递给内置的拨号程序，然后内置拨号程序实现拨号处理操作。利用了 startActivity()方法将程序焦点交给内置的拨号程序，这样原来的 Activity 会成为失去焦点，并且还会发生 onPause()事件，直到关闭拨号程序，焦点也交还给原来的 Activity。

在具体实现上，先插入了一个按钮，当单击按钮后会调用手机内置的默认拨号界面。

5.16.2 具体实现

本实例的主程序文件是 example16.java，主要功能是当用户单击按钮后通过 android.intent.action.CALL_BUTTON 调用默认的拨号界面。具体代码如下。

```
public class example16 extends Activity
{
  private ImageButton myImageButton;

  /** Called when the activity is first created. */
  @Override
  public void onCreate(Bundle savedInstanceState)
  {
    super.onCreate(savedInstanceState);
    setContentView(R.layout.main);

    myImageButton = (ImageButton) findViewById(R.id.myImageButton);

    myImageButton.setOnClickListener(new ImageButton.OnClickListener()
    {

      @Override
      public void onClick(View v)
      {
        /* 调用拨号的画面 */
        Intent myIntentDial = new Intent("android.intent.action.CALL_BUTTON");

        startActivity(myIntentDial);
      }
    });
  }
}
```

执行后会显示对应的按钮，如图 5-38 所示。单击按钮后，会自动来到系统内置的默认拨号界面，如图 5-39 所示。

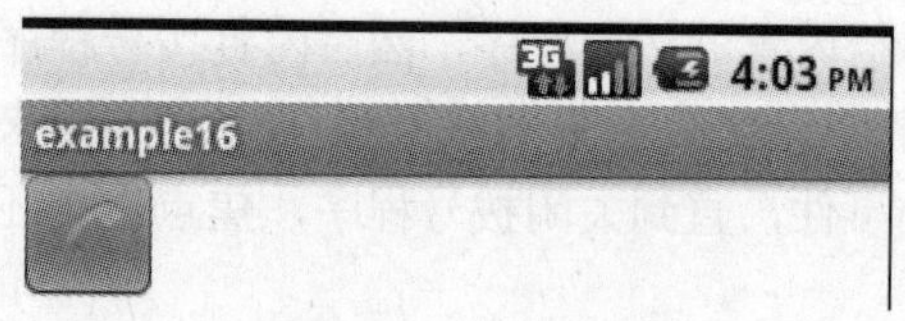

图 5-38　初始效果

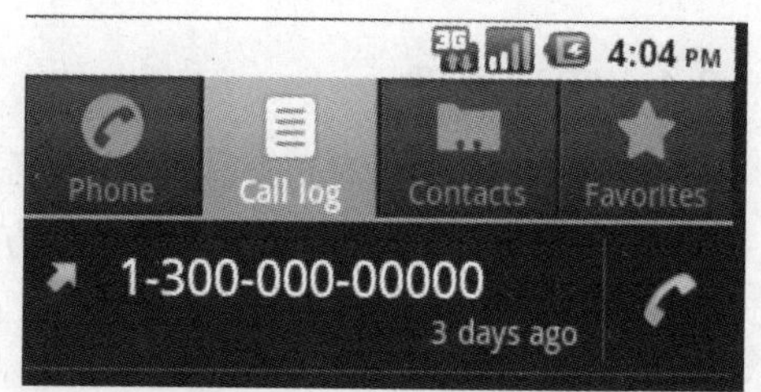

图 5-39　内置的拨号界面

5.17　查看正在运行的程序

电脑中的任务管理器大家都不会陌生，其中的进程管理器能够显示当前正在运行的程序。在本节的内容中，将通过一个具体实例的实现，介绍在查看 Android 中正在运行程序的基本流程。本实例的源代码保存在“光盘:\daima\5\example17”，下面开始讲解本实例的具体实现流程。

5.17.1　实现原理

在本实例中，首先插入了一个按钮，当单击按钮后会显示当前正在运行的程序。当前运行程序是通过 ActivityManager.getRunningTasks( )方法获取的，然后通过 ListView 将获取的信息显示出来。

当单击按钮后，如果在 ListView 的工作已经结束或被操作系统回收，则是不会更新运行列表的。另外，如果不具有访问其他运行程序的权限，也不会显示在 ListView 列表。

注意：为了保证 Android 程序的高效运行，建议限制获取运行程序的数量，在本实例中设置了最多获取 30 个进程。

5.17.2　具体实现

本实例的主程序文件是 example17.java，下面开始讲解其具体实现流程。

1）设置类成员最多能够获取 30 个 Task，具体代码如下。

```
public class example17 extends Activity
{
  private Button mButton01;
  private ListView mListView01;
  private ArrayAdapter<String> aryAdapter1;
  private ArrayList<String> arylistTask;

  /* 类成员设置最多几笔的 Task 数量 */
  private int intGetTastCounter=30;
```

2）设置类成员 ActivityManager 的对象，然后定义当单击按钮后取得正在后台运行的工作程序。具体代码如下。

```
/* 类成员 ActivityManager 对象 */
private ActivityManager mActivityManager;

/** Called when the activity is first created. */
@Override
public void onCreate(Bundle savedInstanceState)
{
  super.onCreate(savedInstanceState);
  setContentView(R.layout.main);

  mButton01 = (Button)findViewById(R.id.myButton1);
  mListView01 = (ListView)findViewById(R.id.myListView1);

  /* 单击按钮取得正在后台运行的工作程序 */
  mButton01.setOnClickListener(new Button.OnClickListener()
  {
    @Override
    public void onClick(View v)
    {
      // TODO Auto-generated method stub
      try
      {
        /* ActivityManager 对象向系统取得 ACTIVITY_SERVICE */
        mActivityManager = (ActivityManager)
        example17.this.getSystemService(ACTIVITY_SERVICE);

        arylistTask = new ArrayList<String>();

        /* 以 getRunningTasks 方法取回正在运行中的程序 TaskInfo */
        List<ActivityManager.RunningTaskInfo> mRunningTasks =
        mActivityManager.getRunningTasks(intGetTastCounter);

        int i = 1;
        /* 以循环及 baseActivity 方式取得工作名称与 ID */
        for (ActivityManager.RunningTaskInfo amTask : mRunningTasks)
        {
          /* baseActivity.getClassName 取出运行工作名称 */
          arylistTask.add("" + (i++) + ": "+
          amTask.baseActivity.getClassName()+
          "(ID=" + amTask.id +")");
        }
        aryAdapter1 = new ArrayAdapter<String>
        (example17.this, R.layout.simple_list_item_1, arylistTask);

        if(aryAdapter1.getCount()==0)
        {
```

```
                /* 当没有任何运行的工作，则提示信息 */
                mMakeTextToast
                (
                  getResources().getText
                  (R.string.str_err_no_running_task).toString(),
                  true
                );
              }
              else
              {
                /* 发现后台运行的工作程序，以 ListView Widget 条列呈现 */
                mListView01.setAdapter(aryAdapter1);
              }
            }
            catch(SecurityException e)
            {
              /* 当无 GET_TASKS 权限时(SecurityException 异常)提示信息 */
              mMakeTextToast
              (
                getResources().getText
                (R.string.str_err_permission).toString(),
                true
              );
            }
          }
        });
```

3）设置用户在运行工作选择时的事件处理，具体代码如下。

```
mListView01.setOnItemSelectedListener
(new ListView.OnItemSelectedListener()
{
  @Override
  public void onItemSelected
  (AdapterView<?> parent, View v, int id, long arg3)
  {
    /* 由于将运行工作以数组存放，所以使用 id 取出数组元素名称 */
    mMakeTextToast(arylistTask.get(id).toString(),false);
  }

  @Override
  public void onNothingSelected(AdapterView<?> arg0)
  {

  }
});
```

4）设置当用户在运行工作单击时的事件处理，具体代码如下。

```
/* 当用户在运行工作上单击时的事件处理 */
mListView01.setOnItemClickListener
(new ListView.OnItemClickListener()
{
  @Override
  public void onItemClick
  (AdapterView<?> parent, View v, int id,   long arg3)
  {
    /* 由于将运行工作以数组存放，故以 id 取出数组元素名称 */
    mMakeTextToast(arylistTask.get(id).toString(), false);
  }
});
}
```

5）定义 mMakeTextToast(String str, boolean isLong)方法，用于实现提醒。具体代码如下。

```
public void mMakeTextToast(String str, boolean isLong)
{
  if(isLong==true)
  {
    Toast.makeText(example17.this, str, Toast.LENGTH_LONG).show();
  }
  else
  {
    Toast.makeText(example17.this, str, Toast.LENGTH_SHORT).show();
  }
}
}
```

最后还需要在文件 AndroidManifest.xml 中设置权限，具体代码如下。

```
<uses-permission android:name="android.permission.GET_TASKS"></uses-permission>
```

执行后会显示对应的按钮，如图 5-40 所示。单击“正在获取运行的程序”按钮后，会显示当前正在运行的程序，如图 5-41 所示。

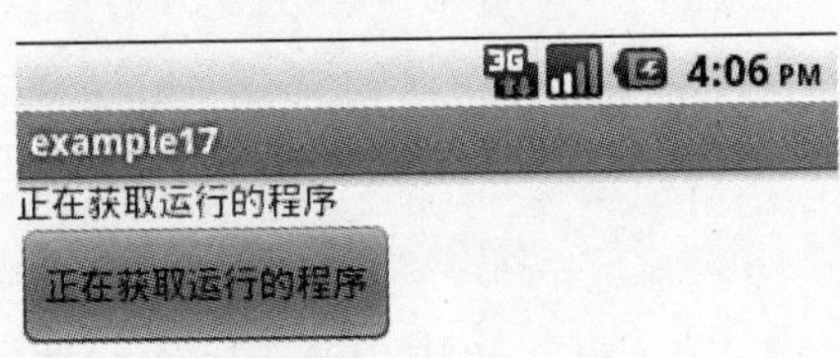

图 5-40　初始效果

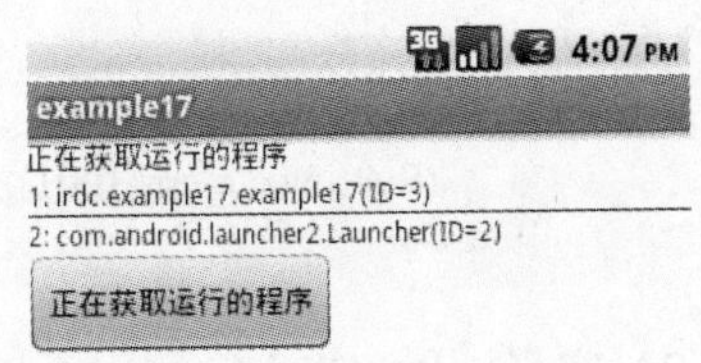

图 5-41　当前运行程序

5.18 更改屏幕方向

屏幕旋转对广大读者来说应该不会陌生，很多智能手机都能根据用户拿手机的方式，动态的横向或纵向显示屏幕中的内容。在本节的内容中，将通过一个具体实例的实现，介绍实现更改手机屏幕方向的基本流程。本实例的源代码保存在“光盘:\daima\5\example18”，下面开始讲解本实例的具体实现流程。

5.18.1 实现原理

在本实例中，首先插入了一个按钮，当单击按钮后会先判断当前屏幕的方向，即如果是横向显示则改为纵向显示，如果是纵向显示则改为横向显示。

在具体实现上，如果要改变屏幕方向，则必须要覆盖 setRequestedOrientation()方法。如果要获取当前的屏幕方向，则需要访问 getRequestedOrientation()方法。

5.18.2 具体实现

本实例的主程序文件是 example18.java，下面开始讲解其具体实现流程。

1）在 onCreate 方法中判断 getRequestedOrientation()是否为-1，如果是-1 则表示在 Activity 属性中没有设置 Android:screenOrientation（固定屏幕显示模式）的值，即表示即使单击了按钮，也无法判断出屏幕的方向，不会实现屏幕方向更改。具体代码如下。

```
public class example18 extends Activity
{
  private TextView mTextView01;
  private Button mButton01;

  /** Called when the activity is first created. */
  @Override
  public void onCreate(Bundle savedInstanceState)
  {
    super.onCreate(savedInstanceState);
    setContentView(R.layout.main);

    mButton01 = (Button)findViewById(R.id.myButton1);
    mTextView01 = (TextView)findViewById(R.id.myTextView1);

    if(getRequestedOrientation()==-1)
    {
      mTextView01.setText(getResources().getText
      (R.string.str_err_1001));
    }
```

2）定义 setOnClickListener(new Button.OnClickListener()方法，当用户单击按钮后开始旋转屏幕画面。具体代码如下。

```
mButton01.setOnClickListener(new Button.OnClickListener()
{
  @Override
  public void onClick(View arg0)
  {
    /* 方法一：重写 getRequestedOrientation( )方法 */

    /* 若无法取得 screenOrientation 属性 */
    if(getRequestedOrientation()==-1)
    {
      /* 提示无法进行画面旋转功能，因无法判别 Orientation */
      mTextView01.setText(getResources().getText
      (R.string.str_err_1001));
    }
    else
    {
      if(getRequestedOrientation()==
          ActivityInfo.SCREEN_ORIENTATION_LANDSCAPE)
      {
        /* 若当下为横排，则更改为竖排呈现 */
        setRequestedOrientation
        (ActivityInfo.SCREEN_ORIENTATION_PORTRAIT);
      }
      else if(getRequestedOrientation()==
                ActivityInfo.SCREEN_ORIENTATION_PORTRAIT)
      {
        /* 若当下为竖排，则更改为横排呈现 */
        setRequestedOrientation
        (ActivityInfo.SCREEN_ORIENTATION_LANDSCAPE);
      }
    }
  }
});
}
```

3）定义 setRequestedOrientation(int requestedOrientation)方法，判断当前要更改的屏幕方向，实现旋转。具体代码如下。

```
@Override
public void setRequestedOrientation(int requestedOrientation)
{

  /* 判断要更改的方向，以 Toast 提示 */
  switch(requestedOrientation)
  {
    /* 更改为 LANDSCAPE（横向显示） */
```

```
        case (ActivityInfo.SCREEN_ORIENTATION_LANDSCAPE):
          mMakeTextToast
          (
            getResources().getText(R.string.str_msg1).toString(),
            false
          );
          break;
        /* 更改为 PORTRAIT（直立显示） */
        case (ActivityInfo.SCREEN_ORIENTATION_PORTRAIT):
          mMakeTextToast
          (
            getResources().getText(R.string.str_msg2).toString(),
            false
          );
          break;
      }
      super.setRequestedOrientation(requestedOrientation);
    }
```

4）定义 getRequestedOrientation(int requestedOrientation)方法，获取当前屏幕的方向。具体代码如下。

```
    @Override
    public int getRequestedOrientation()
    {

      /* 此重写 getRequestedOrientation()方法，可取得当下屏幕的方向 */
      return super.getRequestedOrientation();
    }

    public void mMakeTextToast(String str, boolean isLong)
    {
      if(isLong==true)
      {
        Toast.makeText(example18.this, str, Toast.LENGTH_LONG).show();
      }
      else
      {
        Toast.makeText(example18.this, str, Toast.LENGTH_SHORT).show();
      }
    }
  }
```

接下来需要在文件 AndroidManifest.xml 中设置 Activity 的 Android:screenOrientation 属性，具体代码如下。

```
      <activity
```

```
        android:name=".example18"
        android:label="@string/app_name"
        android:screenOrientation="portrait"
    >
```

执行后会显示对应的按钮，如图 5-42 所示。单击“旋转处理”按钮后会实现屏幕的旋转，如图 5-43 所示。

图 5-42　初始效果

图 5-43　实现旋转

5.19　获取网络和手机相关信息

在手机应用中，为了特殊需求而需要获取手机和网络的相关信息。在本节的内容中，将通过一个具体实例的实现，介绍获取网络和手机信息的基本流程。本实例的源代码保存在“光盘:\daima\5\example19”，下面开始讲解本实例的具体实现流程。

5.19.1　实现原理

在本实例中，将使用 TelephonyManager 类来获取网络信息。另外，还可以通过 Android API 提供的 Android.provider.Setting.System 类来获取蓝牙和网络的相关信息。将以 getSystem Service()方法来获取 TelephonyManage 对象，然后通过 TelephonyManage 的方法来获取和电信有关的网络信息。然后通过 Android.provider.Setting. System.getString()来获取手机的相关设置信息，并将获取的信息存入自定义的 MyAdapter 中，最后将 setListAdapter（设置数据源方法）内的信息显示在 ListView 中。

5.19.2　具体实现

本实例的主程序文件是 example19.java 和 MyAdapter.java，下面开始讲解其具体实现流程。

1. 文件 example19.java

下面先讲解文件 example19.java 的具体实现流程。

1）先引入相关类，然后载入 main.xml 布局。具体代码如下。

```
public class example19 extends ListActivity
{
  private TelephonyManager telMgr;
  private List<String> item=new ArrayList<String>();
  private List<String> value=new ArrayList<String>();

  @SuppressWarnings("static-access")
  @Override
  public void onCreate(Bundle savedInstanceState)
  {
    super.onCreate(savedInstanceState);
    /*  载入 main.xml Layout */
    setContentView(R.layout.main);

    telMgr = (TelephonyManager)getSystemService(TELEPHONY_SERVICE);
```

2）将取得的信息写入 List（列表）中，具体包含下面的信息。

- 取得手机电话号码。
- 取得电信网络国别。
- 取得电信公司代码。
- 取得电信公司名称。
- 取得通信类型。
- 取得网络类型。
- 取得漫游状态。
- 取得手机 IMEI。
- 取得手机 IMEI。
- 取得 IMEI SV。
- 取得 IMSI。
- 取得蓝牙状态。
- 取得 Wi-Fi 状态。
- 飞行模式是否打开。
- 取得数据漫游是否打开。

如果上述信息都无法获得，则输出“无法取得”的提示。具体代码如下。

```
/* 将取得的信息写入 List 中 */
item.add(getResources().getText(R.string.str_list0).toString());
if(telMgr.getLine1Number()!=null)
{
```

```
    value.add(telMgr.getLine1Number());
  }
  else
  {
    value.add("无法取得");
  }

  /* 取得电信网络国别 */
  item.add(getResources().getText(R.string.str_list1).toString());
  if(telMgr.getNetworkCountryIso().equals(""))
  {
    value.add("无法取得");
  }
  else
  {
    value.add(""+telMgr.getNetworkCountryIso());
  }

  /* 取得电信公司代码 */
  item.add(getResources().getText(R.string.str_list2).toString());
  if(telMgr.getNetworkOperator().equals(""))
  {
    value.add("无法取得");
  }
  else
  {
    value.add(telMgr.getNetworkOperator());
  }

  /* 取得电信公司名称 */
  item.add(getResources().getText(R.string.str_list3).toString());
  if(telMgr.getNetworkOperatorName().equals(""))
  {
    value.add("无法取得");
  }
  else
  {
    value.add(telMgr.getNetworkOperatorName());
  }

  /* 取得行动通信类型 */
  item.add(getResources().getText(R.string.str_list4).toString());
  if(telMgr.getPhoneType()==telMgr.PHONE_TYPE_GSM)
  {
    value.add("GSM");
  }
```

```
else
{
  value.add("未知");
}

/* 取得网络类型 */
item.add(getResources().getText(R.string.str_list5).toString());
if(telMgr.getNetworkType()==telMgr.NETWORK_TYPE_EDGE)
{
  value.add("EDGE");
}
else if(telMgr.getNetworkType()==telMgr.NETWORK_TYPE_GPRS)
{
  value.add("GPRS");
}
else if(telMgr.getNetworkType()==telMgr.NETWORK_TYPE_UMTS)
{
  value.add("UMTS");
}
else if(telMgr.getNetworkType()==4)
{
  value.add("HSDPA");
}
else
{
  value.add("未知");
}

/* 取得漫游状态 */
item.add(getResources().getText(R.string.str_list6).toString());
if(telMgr.isNetworkRoaming())
{
  value.add("漫游中");
}
else{
  value.add("无漫游");
}

/* 取得手机 IMEI */
item.add(getResources().getText(R.string.str_list7).toString());
value.add(telMgr.getDeviceId());

/* 取得 IMEI SV */
item.add(getResources().getText(R.string.str_list8).toString());
if(telMgr.getDeviceSoftwareVersion()!=null)
{
```

```
    value.add(telMgr.getDeviceSoftwareVersion());
  }
  else
  {
    value.add("无法取得");
  }

  /* 取得手机 IMSI */
  item.add(getResources().getText(R.string.str_list9).toString());
  if(telMgr.getSubscriberId()!=null)
  {
    value.add(telMgr.getSubscriberId());
  }
  else
  {
    value.add("无法取得");
  }

  /* 取得 ContentResolver */
  ContentResolver cv = example19.this.getContentResolver();
  String tmpS="";

  /* 取得蓝牙状态 */
  item.add(getResources().getText(R.string.str_list10)
            .toString());
  tmpS=android.provider.Settings.System.getString(cv,
       android.provider.Settings.System.BLUETOOTH_ON);
  if(tmpS.equals("1"))
  {
    value.add("已打开");
  }
  else{
    value.add("未打开");
  }

  /* 取得 WIFI 状态 */
  item.add(getResources().getText(R.string.str_list11)
            .toString());
  tmpS=android.provider.Settings.System.getString(cv,
       android.provider.Settings.System.WIFI_ON);
  if(tmpS.equals("1"))
  {
    value.add("已打开");
```

```
}
else{
    value.add("未打开");
}

/* 取得飞行模式是否打开 */
item.add(getResources().getText(R.string.str_list12)
            .toString());
tmpS=android.provider.Settings.System.getString(cv,
        android.provider.Settings.System.AIRPLANE_MODE_ON);
if(tmpS.equals("1"))
{
    value.add("打开中");
}
else{
    value.add("未打开");
}

/* 取得数据漫游是否打开 */
item.add(getResources().getText(R.string.str_list13)
            .toString());
tmpS=android.provider.Settings.System.getString(cv,
        android.provider.Settings.System.DATA_ROAMING);
if(tmpS.equals("1"))
{
    value.add("打开中");
}
else{
    value.add("未打开");
}
```

3）使用自定义的 MyAdapter 来将数据传入 ListActivity，具体代码如下。

```
        setListAdapter(new MyAdapter(this,item,value));
    }
}
```

2．文件 MyAdapter.java

下面先讲解文件 MyAdapter.java 的具体实现流程。

1）先加载相关类，然后自定义 Adapter，并继承于 android.widget.BaseAdapter。具体代码如下。

```
/* 自定义 Adapter，继承于 android.widget.BaseAdapter */
public class MyAdapter extends BaseAdapter
{
```

```
/* 变量声明 */
private LayoutInflater mInflater;
private List<String> items;
private List<String> values;
/* MyAdapter 的构造器，传入三个参数  */
public MyAdapter(Context context,List<String> item,
                    List<String> value)
{
  /* 参数初始化 */
  mInflater = LayoutInflater.from(context);
  items = item;
  values = value;
}
```

2）通过@Override 分别覆盖方法 getCount()、getItem(int position)、getItemId(int position)和 getView(int position,View convertView,ViewGroup par)。具体代码如下。

```
/* 因继承 BaseAdapter，需覆盖以下方法 */
@Override
public int getCount()
{
  return items.size();
}

@Override
public Object getItem(int position)
{
  return items.get(position);
}

@Override
public long getItemId(int position)
{
  return position;
}

@Override
public View getView(int position,View convertView,ViewGroup par)
{
  ViewHolder holder;

  if(convertView == null)
  {
    /* 使用自定义的 file_row 作为 Layout */
    convertView = mInflater.inflate(R.layout.row_layout,null);
    /* 初始化 holder 的 text 与 icon */
    holder = new ViewHolder();
```

```
        holder.text1=(TextView)convertView.findViewById(R.id.myText1);
        holder.text2=(TextView)convertView.findViewById(R.id.myText2);

        convertView.setTag(holder);
      }
      else
      {
        holder = (ViewHolder) convertView.getTag();
      }
      /* 设置要显示的信息 */
      holder.text1.setText(items.get(position).toString());
      holder.text2.setText(values.get(position).toString());

      return convertView;
    }

    /* class ViewHolder */
    private class ViewHolder
    {
      /* text1：信息名称
       * text2：信息内容 */
      TextView text1;
      TextView text2;
    }
  }
```

执行后会按照指定样式显示获取的网络信息和手机信息，如图 5-44 所示。

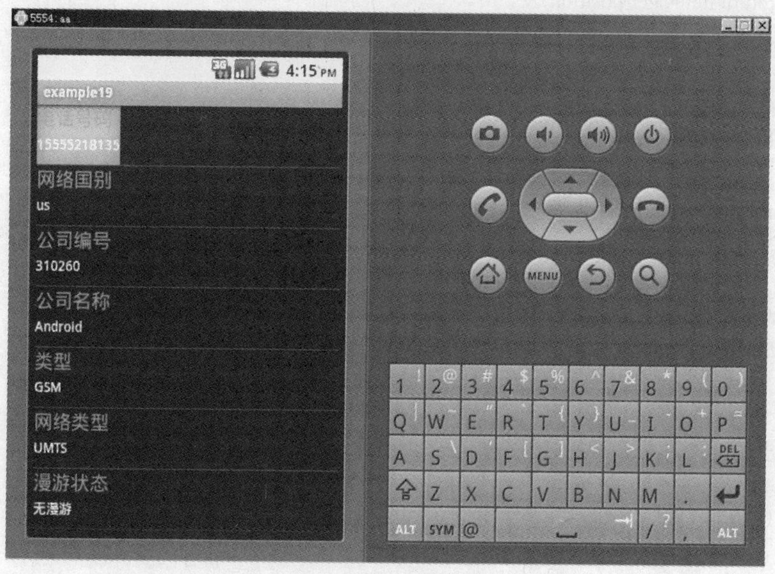

图 5-44　获取的信息

第6章　手机自动服务

在手机应用中，有很多的自动服务功能。例如，剩余电量提示、存储卡容量提示和黑名单自动屏蔽等。通过这些自动服务功能，很好的为用户提供了人性化的服务，整个操作过程更加方便。在本节的内容中，将通过几个典型实例的实现，来详细介绍这些自动手机服务的实现流程。

6.1　短信提醒

在当前手机应用中，如果收到了短信，系统将自动在手机上显示“有短信”的提示。在本节的内容中，将通过一个具体实例的实现，介绍短信提醒的基本流程。本实例的源代码保存在“光盘:\daima\6\example1”，下面开始讲解本实例的具体实现流程。

6.1.1　实现原理

在手机系统中，存在广播系统 BroadcastReceiver（广播接收者），它实时监听手机内的短信状况。当手机收到短信后，会通过 Notification 在状态栏中显示短信的摘要信息。

在本实例中，会将接收到的短信对象解析为可以识别发信人号码和短信正文的字符串。本实例的难点是如何向系统注册一个常驻的 BroadcastReceiver 对象，然后在后台中监听短信事件，最后将短信内容编译出来。

6.1.2　具体实现

本实例的主程序文件是 example1.java 和 example1_SMSreceiver.java，下面开始讲解其具体实现代码。

1．文件 example1.java

下面先讲解文件 example1.java，其功能是在文本框中显示“正在等待接收短信...”的提示。具体代码如下。

```
public class example1 extends Activity
{
  private TextView mTextView1;
  @Override
  public void onCreate(Bundle savedInstanceState)
  {
    super.onCreate(savedInstanceState);
    setContentView(R.layout.main);
    /*通过 findViewById 构造器创建 TextView 对象*/
```

```
        mTextView1 = (TextView) findViewById(R.id.myTextView1);
        mTextView1.setText("现在正在等待接收信息...");
    }
}
```

2. 文件 example1_SMSreceiver.java

接下来讲解文件 example1_SMSreceiver.java，下面开始讲解其具体实现流程。

1）先引用 BroadcastReceiver 类，然后引用 telephoney.gsm.SmsMessage 对象来收取短信，接着引用 Toast 类来告知用户收到短信。具体代码如下。

```
package irdc.example1;

/*必须引用 BroadcastReceiver 类*/
import android.content.BroadcastReceiver;
import android.content.Context;
import android.content.Intent;
import android.os.Bundle;
/*必须引用 telephoney.gsm.SmsMessage 来收取短信*/
import android.telephony.gsm.SmsMessage;
/*必须引用 Toast 类来告知用户收到短信*/
import android.widget.Toast;
```

2）自定义继承自 BroadcastReceiver 的类，用于监听系统服务广播的信息。具体实现流程如下。

第一步：声明静态字符串，并使用 android.provider.Telephony.SMS_RECEIVED 作为 Action 为短信的依据。

第二步：通过 if 语句判断传来的 Intent 是否为短信，如果是，则先建构一字符串集合变量 sb，然后接收由 Intent 传来的数据。

第三步：通过 if 语句判断 Intent 是有数据。

具体代码如下。

```
public class example1_SMSreceiver extends BroadcastReceiver
{
  /*声明静态字符串，并使用 android.provider.Telephony.SMS_RECEIVED
  作为 Action 为短信的依据*/
  private static final String mACTION =
  "android.provider.Telephony.SMS_RECEIVED";

  @Override
  public void onReceive(Context context, Intent intent)
  {
    /* 判断传来 Intent 是否为短信*/
    if (intent.getAction().equals(mACTION))
    {
      /*建构一字符串集合变量 sb*/
```

```
        StringBuilder sb = new StringBuilder();
        /*接收由 Intent 传来的数据*/
        Bundle bundle = intent.getExtras();
        /*判断 Intent 是有数据*/
        if (bundle != null)
        {
          /* pdus 为 android 内置短信参数 identifier
           * 通过 bundle.get("")返回一包含 pdus 的对象*/
          Object[] myOBJpdus = (Object[]) bundle.get("pdus");
          /*构建短信对象 array，并依据收到的对象长度来创建 array 的大小*/
          SmsMessage[] messages = new SmsMessage[myOBJpdus.length];
          for (int i = 0; i<myOBJpdus.length; i++)
          {
            messages[i] =
            SmsMessage.createFromPdu((byte[]) myOBJpdus[i]);
          }

          /* 将送来的短信合并，自定义信息于 StringBuilder 当中 */
          for (SmsMessage currentMessage : messages)
          {
            sb.append("正在接收到来自:\n");
            /* 来电者的电话号码 */
            sb.append(currentMessage.getDisplayOriginatingAddress());
            sb.append("\n------发来的短信------\n");
            /* 取得传来信息的 BODY */
            sb.append(currentMessage.getDisplayMessageBody());
          }
        }
```

3）以 Notification(Toase)显示短信信息，然后返回主 Activity 界面，并使其以一个全新的 task（进程）来运行。具体代码如下。

```
        Toast.makeText
        (
          context, sb.toString(), Toast.LENGTH_LONG
        ).show();

        /* 返回主 Activity */
        Intent i = new Intent(context, example1.class);
        /*使其以一个全新的进程来运行*/
        i.addFlags(Intent.FLAG_ACTIVITY_NEW_TASK);
        context.startActivity(i);
      }
    }
  }
```

接下来还需要在文件 AndroidManifest.xml 中向系统注册常驻的 receiver，并设置这个

receiver 的 intent-filter 名为“android.provider.Telephony.SMS_RECEIVED”。另外，还需要设置 permission.RECEIVE_SMS 权限。具体代码如下。

```
        <!-- 建立 receiver 来监听系统广播信息 -->
        <receiver android:name="example1_SMSreceiver">
        <!—设定要捕捉的讯息名称为 provider 中 Telephony.SMS_RECEIVED -->
    <intent-filter>
        <action
          android:name="android.provider.Telephony.SMS_RECEIVED" />
    </intent-filter>
        </receiver>
  </application>
  <uses-permission android:name="android.permission.RECEIVE_SMS"></uses-permission>
```

执行后的初始效果如图 6-1 所示。当接收到短信后会在屏幕中显示对应的提示信息，如图 6-2 所示。

图 6-1　初始效果

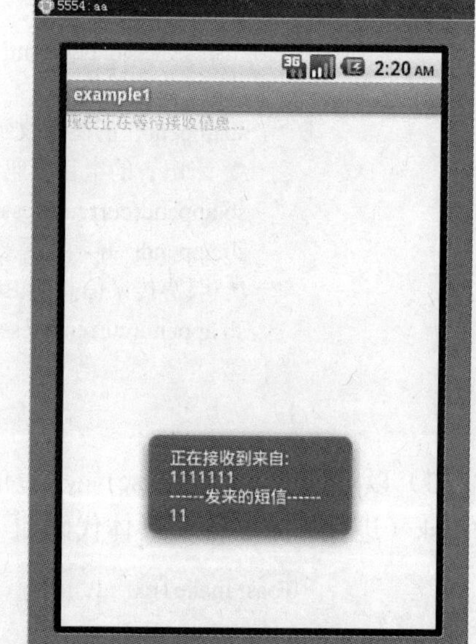

图 6-2　短信提示

同时在系统的短信栏目中会显示收到的短信，如图 6-3 所示。

注意：在具体测试时，需要同时运行两个模拟器，一个用于发送短信，另外一个接收。但是如果机器太慢，无法启动两个模拟器。也可以只启动一个模拟器。然后在 Eclipse 菜单中依次单击“windows”→“show view”→“other”→“Android”→“Emulator Control”，打开“Emulator Control”面板。在 Telephony Actions 分组框中，Voice 是呼叫，SMS 是发送短信。Incoming number 是模拟器的端口号，也可以使用这个功能给的模拟器拨打电话或发送短信。如图 6-4 所示。

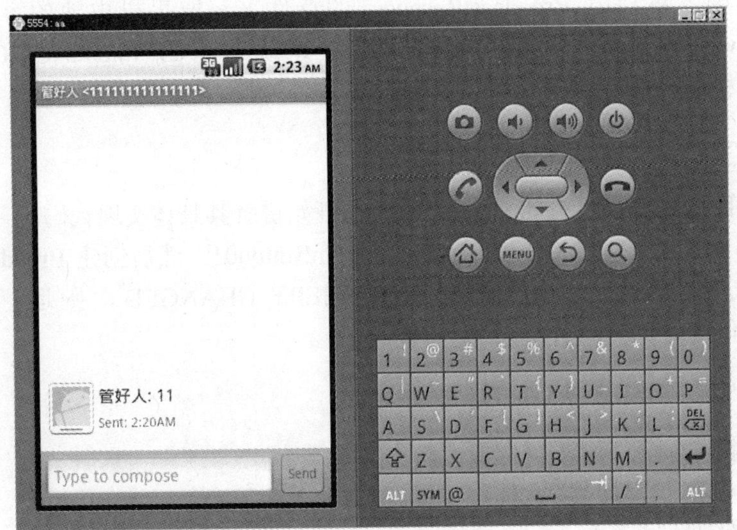

图 6-3　收到的短信

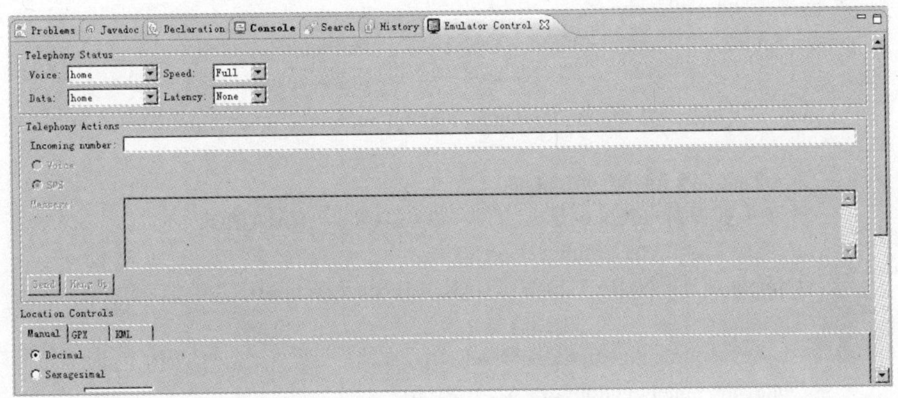

图 6-4　“Emulator Control”面板

6.2　电池容量提醒

在选购手机时，电池容量是一个十分重要的因素。在使用过程中，最害怕手机没电而影响业务。为此，显示电池容量的应用变得愈发重要了。在本节的内容中，将通过一个具体实例的实现，介绍实现查看手机电池容量的基本流程。本实例的源代码保存在“光盘:\daima\6\example2”，下面开始讲解本实例的具体实现流程。

6.2.1　实现原理

Android API 里面的 BroadcastReceiver 类和 Button 的 Listener 类似，当 Receiver 被注册后会在后台等待被其他程序调用。当指定要捕捉的 Action 发生时，Receiver 就会被调用，并运行 onReseiver 来实现内部的程序。

在本实例中，将利用 BroadcastReceiver 的特性来获取手机电池的容量。通过注册 BroadcastReceiver 时设置的 IntentFilter 来获取系统发出的 Intent.ACTION_BATTERY_CHANGED，然后获取电池的容量。

6.2.2 具体实现

本实例的主程序文件是 example2.java，下面开始讲解其具体实现代码。

1）分别声明 3 个变量 intLevel、intScale 和 mButton01，然后创建 BroadcastReceiver 对象，如果捕捉到的 Action 是 ACTION_BATTERY_CHANGED，则运行 onBatteryInfoReceiver()。具体代码如下。

```
public class example2 extends Activity
{
  /* 变量声明 */
  private int intLevel;
  private int intScale;
  private Button mButton01;

  /* 创建 BroadcastReceiver */
  private BroadcastReceiver mBatInfoReceiver=new BroadcastReceiver()
  {
    public void onReceive(Context context, Intent intent)
    {
      String action = intent.getAction();
      /* 如果捕捉到的 Action 是 ACTION_BATTERY_CHANGED,
       * 就运行 onBatteryInfoReceiver()方法 */
      if (Intent.ACTION_BATTERY_CHANGED.equals(action))
      {
        intLevel = intent.getIntExtra("level", 0);
        intScale = intent.getIntExtra("scale", 100);
        onBatteryInfoReceiver(intLevel,intScale);
      }
    }
  };
```

2）在 onCreate()中载入主布局文件 main.xml，然后初始化 Button 和设置单击后的动作，并注册一个系统 BroadcastReceiver，用于访问电池容量。具体代码如下。

```
/** Called when the activity is first created. */
@Override
public void onCreate(Bundle savedInstanceState)
{
  super.onCreate(savedInstanceState);
  /* 载入布局文件 main.xml */
  setContentView(R.layout.main);
```

```
        /* 初始化 Button，并设置单击后的动作 */
        mButton01 = (Button)findViewById(R.id.myButton1);
        mButton01.setOnClickListener(new Button.OnClickListener()
        {
          @Override
          public void onClick(View v)
          {
            /* 注册一个系统 BroadcastReceiver，作为访问电池容量之用 */
            registerReceiver
            (
              mBatInfoReceiver,
              new IntentFilter(Intent.ACTION_BATTERY_CHANGED)
            );
          }
        });
      }
```

3）定义方法 onBatteryInfoReceiver()，当捕捉到 ACTION_BATTERY_CHANGED 时要运行这个方法，首先创建一个背景模糊的 Window 窗口，且将对话窗口放在前景，然后将取得的电池计量显示于 Dialog 对话框中，最后设置返回主画面的按钮。具体代码如下。

```
        /* 捕捉到 ACTION_BATTERY_CHANGED 时要运行的方法 */
        public void onBatteryInfoReceiver(int intLevel, int intScale)
        {
          /* 创建跳出的对话窗口 */
          final Dialog d = new Dialog(example2.this);
          d.setTitle(R.string.str_dialog_title);
          d.setContentView(R.layout.mydialog);

          /* 创建一个背景模糊的 Window 窗口，且将对话窗口放在前景 */
          Window window = d.getWindow();
          window.setFlags
          (
            WindowManager.LayoutParams.FLAG_BLUR_BEHIND,
            WindowManager.LayoutParams.FLAG_BLUR_BEHIND
          );

          /* 将取得的电池容量显示于对话框中 */
          TextView mTextView02=(TextView)d.findViewById(R.id.myTextView2);
          mTextView02.setText
          (
            getResources().getText(R.string.str_dialog_body)+
            String.valueOf(intLevel * 100 / intScale) + "%"
          );

          /* 设置返回主画面的按钮 */
```

```
            Button mButton02 = (Button)d.findViewById(R.id.myButton2);
            mButton02.setOnClickListener(new Button.OnClickListener()
            {
              @Override
              public void onClick(View v)
              {
                /* 反注册 Receiver，并关闭对话窗口 */
                unregisterReceiver(mBatInfoReceiver);
                d.dismiss();
              }
            });
            d.show();
          }
        }
```

执行后的初始效果如图 6-5 所示。当单击“获取”按钮后会显示当前电池的容量，如图 6-6 所示。

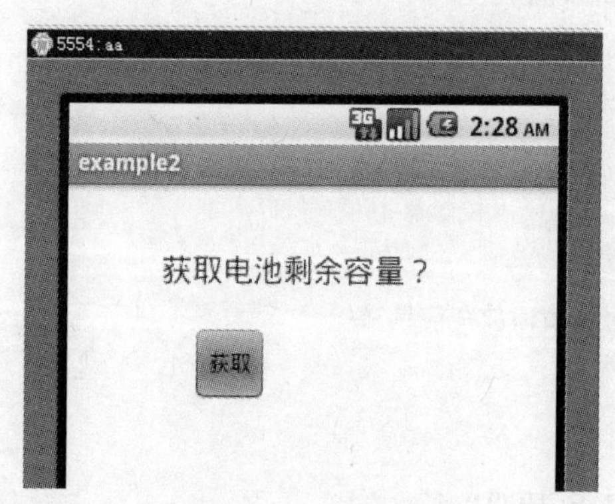

图 6-5　初始效果

图 6-6　显示容量

6.3　短信群发

在当前手机应用中，短信群发是一个十分重要的功能。在本节的内容中，将通过一个具体实例的实现，介绍短信群发功能的基本实现流程。本实例的源代码保存在“光盘:\daima\6\example3”，下面开始讲解本实例的具体实现流程。

6.3.1　实现原理

在本实例中，当单击“发送”按钮后，会首先获取手机通讯录的信息，让用户选择短信

接收者。选好后返回主程序，然后实现短信群发功能。

在使用本实例前，首先在通讯录添加一些联系人信息，这些联系人作为接收短信者。当用户选择接收者后，会将短信发送到目标者。

6.3.2　具体实现

具体流程如下。

1）先定义 5 个变量 mTextView01、mTextView3、mTextView5、mButton01 和 mButton 01，然后分别为上述变量赋值。具体代码如下。

```
public class example3 extends Activity
{
  private TextView mTextView01;
  private TextView mTextView3;
  private TextView mTextView5;
  private Button mButton01;
  private Button mButton02;
  /*先声明 strMessage 为 String*/
  String strMessage;

  private static final int PICK_CONTACT_SUBACTIVITY = 2;

  @Override
  public void onCreate(Bundle savedInstanceState)
  {
    super.onCreate(savedInstanceState);
    setContentView(R.layout.main);

    mTextView01 = (TextView)findViewById(R.id.myTextView1);
    mButton01 = (Button)findViewById(R.id.myButton1);
    mTextView3 = (TextView)findViewById(R.id.myTextView3);
    mButton02 = (Button)findViewById(R.id.myButton2);
    mTextView5= (TextView)findViewById(R.id.myTextView5);
```

2）分别设置 2 个 Button 的处理事件，其中 mButton01 用于获取 mTextView3 里的内容，mButton02 用于获取 mTextView5 里的内容。具体代码如下。

```
    /*设置第一个按钮的单击事件*/
    mButton01.setOnClickListener(new Button.OnClickListener()
    {
      @Override
      public void onClick(View v)
      {

        Uri uri = Uri.parse("content://contacts/people");
        Intent intent = new Intent(Intent.ACTION_PICK, uri);
        /*获取 mTextView3 里的内容*/
        strMessage = mTextView3.getText().toString();
```

```
            startActivityForResult(intent, PICK_CONTACT_SUBACTIVITY);
          }
        });

        /*设置第二个按钮的单击事件*/
        mButton02.setOnClickListener(new Button.OnClickListener()
        {
          @Override
          public void onClick(View v)
          {
            Uri uri = Uri.parse("content://contacts/people");
            Intent intent = new Intent(Intent.ACTION_PICK, uri);
            /*获取 mTextView5 里的内容*/
            strMessage = mTextView5.getText().toString();

            startActivityForResult(intent, PICK_CONTACT_SUBACTIVITY);
          }
        });
      }
```

3）在获取 android.permission.READ_CONTACTS 的权限下，通过 try 语句来获取通讯录中姓名和电话，然后设置要寄给通讯录中的哪一个号码，接着用 smsManager. SendText Message()方法发送短信，最后用 Toast 来显示正在传送的提示。具体代码如下。

```
@Override
protected void onActivityResult
(int requestCode, int resultCode, Intent data)
{
  switch (requestCode)
  {
    case PICK_CONTACT_SUBACTIVITY:
      final Uri uriRet = data.getData();
      if(uriRet != null)
      {
        try
        {
          /* 必须要有 android.permission.READ_CONTACTS 权限 */
          Cursor c = managedQuery(uriRet, null, null, null, null);
          c.moveToFirst();
          /*获取通讯录的姓名*/
          String strName =
          c.getString(c.getColumnIndexOrThrow(People.NAME));
          /*获取通讯录的电话*/
          String strPhone =
          c.getString(c.getColumnIndexOrThrow(People.NUMBER));
```

```
                    /*设置要寄给通讯录里的电话*/
                    String strDestAddress = strPhone;
                    System.out.println(strMessage);
                    SmsManager smsManager = SmsManager.getDefault();

                    PendingIntent mPI = PendingIntent.getBroadcast
                    (example3.this, 0, new Intent(), 0);
                    /*发出短信*/
                    smsManager.sendTextMessage
                    (
                      strDestAddress, null, strMessage, mPI, null
                    );
                    /*用 Toast 显示传送中*/
                    Toast.makeText
                    (
                      example3.this,
                      getString(R.string.str_msg)+strName,
                      Toast.LENGTH_SHORT
                    ).show();

                  mTextView01.setText(strName+":"+strPhone);
                }
                catch(Exception e)
                {
                  mTextView01.setText(e.toString());
                  e.printStackTrace();
                }
              }
              break;
          }
          super.onActivityResult(requestCode, resultCode, data);
        }
      }
```

接下来还需要在文件 AndroidManifest.xml 中设置允许 READ_CONTACTS 权限和 SEND_SMS 权限。具体代码如下。

```
<uses-permission android:name="android.permission.READ_CONTACTS"></uses-permission>
<uses-permission android:name="android.permission.SEND_SMS"></uses-permission>
```

执行后的初始效果如图 6-7 所示。当单击“发送”按钮后会显示联系人界面，如图 6-8 所示。

单击其中的一个联系人后，会将短信发送给他，并输出发送提示，如图 6-9 所示。

在上述实例中，虽然实现了短信发送功能，但是还不能算是群组发送。实际上 Android 系统允许用户创建若干个群组。可以把你的联系人们很轻松的放入各种不同的群组里面。Android 系统之所以提供了这样的一个特性，是为了让用户给整组联系人群发邮件或者短信。如图 6-10 所示。

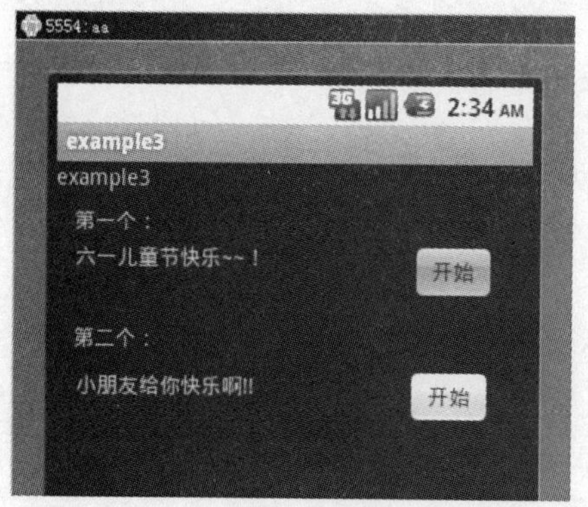

图 6-7　初始效果

图 6-8　联系人界面

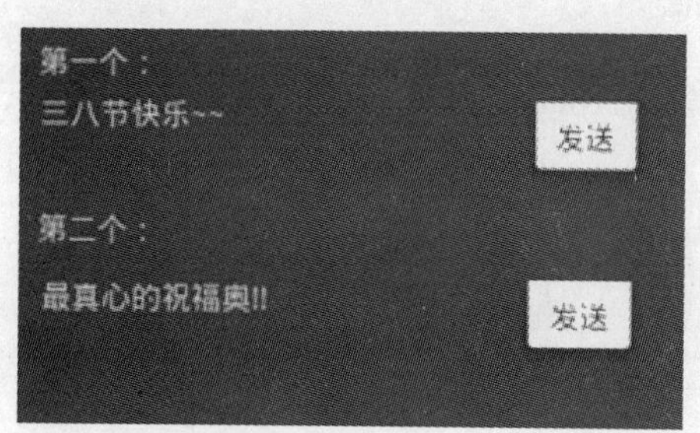

图 6-9　已发送

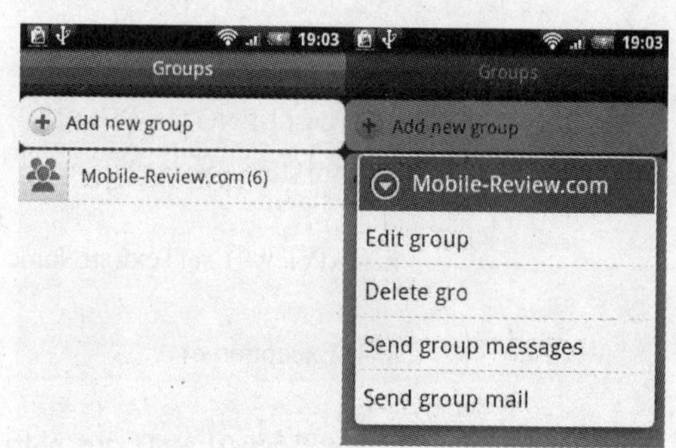

图 6-10　Android 的群组

当然也可以用编程的方式实现群发短信。我们可以用字符串数组的形式保存联系人数据，也可以通过 cursor 为对象，然后通过循环的方式，在取得了联系人数据的同时就传出指定的短信内容。

6.4　发送短信实现 Email 通知

短信和邮件是两个十分重要的应用。在本节的内容中，将通过一个具体实例的实现，介绍通过短信来发送 Email 通知的基本实现流程。本实例的源代码保存在“光盘:\daima\6\example4”，下面开始讲解本实例的具体实现流程。

6.4.1　实现原理

在本实例中，当收到一条短信后，先用 Toast 来提示获取了短信，然后再通过 Email 发送到用户的邮箱中，这样就可以将重要的短信在邮箱中保存，而不用担心手机容量的问题了。

在具体实现上，先在后台设计一个 BroadcastReceiver 类，用于等待接收短信，当收到短信后，以 Bundle（捆绑）的方式来封装短信的内容，然后通过 Intent 方式返回给主程序

Activity。因为 Receiver 无法直接发送 Email，所以需要将控制权给主程序，通过主程序来运行发送 Email。当主程序收到 Bundle 后，会以 Bundle.getString()的方法来取得返回短信的内容，然后以 Intent.setType("plain/text")来设置要打开的 Intent 类型，并以关键程序 Intent.putExtra(android.content.Intent.EXTRA_EMAIL,strEmailReciver)来指定要打开的是 Email 所需要的 Extra 参数，当 Android 系统收到这些参数后，就会打开内置的 Email 发送程序。读者在此需要注意的是，在模拟器运行后会显示"No application can perform this action"的提示，而在真实机器上不会出现此问题。

6.4.2 具体实现

本实例的主程序文件是 example4.java 和 example4SMSreceiver.java，下面开始讲解其具体实现代码。

1. 文件 example1.java

下面先讲解文件 example1.java，其具体实现流程如下。

1）声明一个 TextView，String 数组与两个文本字符串变量，具体代码如下。

```
public class example4 extends Activity
{
  /*声明一个 TextView，String 数组与两个文本字符串变量*/
  private TextView mTextView1;
  public String[] strEmailReciver;
  public String strEmailSubject;
  public String strEmailBody;
```

2）在 onCreate()方法中通过 findViewById 构造器来创建 TextView 对象，并通过 TextView 来显示"等待接收短信..."的提示。

```
/** Called when the activity is first created. */
@Override
public void onCreate(Bundle savedInstanceState)
{
  super.onCreate(savedInstanceState);
  setContentView(R.layout.main);

  /*通过 findViewById 构造器创建 TextView 对象*/
  mTextView1 = (TextView) findViewById(R.id.myTextView1);
  mTextView1.setText("等待接收短信...");
```

3）通过 try 语句取得短信传来的 bundle，并取出 bundle 内的字符串，然后自定义 Intent 来运行发送 E-mail 的工作，同时设置邮件格式为"plain/text"，并分别取得 EditText01,02,03,04 的值作为收件人地址、附件、主题和正文，最后将取得的字符串放入 mEmailIntent 中。具体代码如下。

```
    try
    {
```

```
          /*取得短信传来的 bundle*/
          Bundle bundle = this.getIntent().getExtras();
          if (bundle!= null)
          {
            /*将 bundle 内的字符串取出*/
            String sb = bundle.getString("STR_INPUT");
            /*自定义一 Intent 来运行寄送 E-mail 的工作*/
            Intent mEmailIntent =
            new Intent(android.content.Intent.ACTION_SEND);
            /*设置邮件格式为"plain/text"*/
            mEmailIntent.setType("plain/text");

            /*
            * 取得 EditText01,02,03,04 的值作为
            * 收件人地址、附件、主题、正文
            */
            strEmailReciver =new String[]{"jay.mingchieh@gmail.com"};
            strEmailSubject = "你有一封短信!!";
            strEmailBody = sb.toString();

            /*将取得的字符串放入 mEmailIntent 中*/
            mEmailIntent.putExtra(android.content.Intent.EXTRA_EMAIL,
            strEmailReciver);
            mEmailIntent.putExtra(android.content.Intent.EXTRA_SUBJECT,
            strEmailSubject);
            mEmailIntent.putExtra(android.content.Intent.EXTRA_TEXT,
            strEmailBody);
            startActivity(Intent.createChooser(mEmailIntent,
            getResources().getString(R.string.str_message)));
          }
          else
          {
            finish();
          }
        }
        catch(Exception e)
        {
          e.printStackTrace();
        }
      }
    }
```

2．文件 example4SMSreceiver.java

接下来讲解文件 example4SMSreceiver.java，其具体实现流程如下。

1）先引用 BroadcastReceiver 类，然后引用 telephoney.gsm.SmsMessage 来收取短信，引用 Toast 类来告知用户收到短信，具体代码如下。

```
package irdc.example4;

/*引用 BroadcastReceiver 类*/
import android.content.BroadcastReceiver;
import android.content.Context;
import android.content.Intent;
import android.os.Bundle;
/*引用 telephoney.gsm.SmsMessage 来收取短信*/
import android.telephony.gsm.SmsMessage;
/*引用 Toast 类来告知用户收到短信*/
import android.widget.Toast;
```

2）自定义继承自 BroadcastReceiver 类，用于监听系统服务广播的信息，然后声明静态字符串并作为 Action 的依据，具体代码如下。

```
public class example4SMSreceiver extends BroadcastReceiver
{
  * android.provider.Telephony.SMS_RECEIVED
  private static final String mACTION =
  "android.provider.Telephony.SMS_RECEIVED";

  private String str_receive="收到短信!";
```

3）定义 onReceive(Context context, Intent intent)用于获取短信，先通过 if 语句判断传来 Intent 是否为短信。如果是，则建构一字符串集合变量 sb 并接收由 Intent 传来的数据。具体代码如下。

```
@Override
public void onReceive(Context context, Intent intent)
{
  // TODO Auto-generated method stub
  Toast.makeText(context, str_receive.toString(),
  Toast.LENGTH_LONG).show();

  /*判断传来 Intent 是否为短信*/
  if (intent.getAction().equals(mACTION))
  {
    /*建构一字符串集合变量 sb*/
    StringBuilder sb = new StringBuilder();
    /*接收由 Intent 传来的数据*/
    Bundle bundle = intent.getExtras();
```

4）用 if 语句判断 Intent 是否有数据，用 pdus 作为 android 内置短信参数 identifier（能够识别身份），并通过 bundle.get("")返回一包含 pdus 的对象。具体代码如下。

```
/*判断 Intent 是有数据*/
if (bundle != null)
```

```
{
   Object[] myOBJpdus = (Object[]) bundle.get("pdus");
```

5）构造短信对象 array，并依据收到的对象长度来创建 array 的大小。具体代码如下。

```
SmsMessage[] messages = new SmsMessage[myOBJpdus.length];

for (int i = 0; i<myOBJpdus.length; i++)
{
   messages[i] =
   SmsMessage.createFromPdu((byte[]) myOBJpdus[i]);
}
```

6）将传递来的短信合并，然后自定义信息于 StringBuilder 当中。即分别获取收信人的电话号码，传来信息的 BODY。具体代码如下。

```
for (SmsMessage currentMessage : messages)
{
   sb.append("接收到来自:\n");
   /* 收信人的电话号码 */
   sb.append(currentMessage.getDisplayOriginatingAddress());
   sb.append("\n------传来的短信------\n");
   /* 取得传来信息的内容 */
   sb.append(currentMessage.getDisplayMessageBody());
   Toast.makeText
   (
      context, sb.toString(), Toast.LENGTH_LONG
   ).show();
 }
}
```

7）用 Notification(Toase)显示来讯信息，具体代码如下。

```
Toast.makeText
(
   context, sb.toString(), Toast.LENGTH_LONG
).show();
```

8）返回主 Activity，然后自定义一 Bundle，用 putString()方法将短信存入自定义的 bundle 内，最后设置 Intent 的 Flag 以一个全新的 task（进程）来运行。具体代码如下。

```
Intent i = new Intent(context, example4.class);
/*自定义一 Bundle*/
Bundle mbundle = new Bundle();
/*将短信信息以 putString()方法存入自定义的 bundle 内*/
mbundle.putString("STR_INPUT",    sb.toString());
/*将自定义 bundle 写入 Intent 中*/
i.putExtras(mbundle);
/*设置 Intent 的 Flag 以一个全新的进程来运行*/
```

```
            i.addFlags(Intent.FLAG_ACTIVITY_NEW_TASK);
            context.startActivity(i);
        }
      }
    }
```

最后，还需要在文件 AndroidManifest.xml 中向系统注册一个常驻的 BroadcastReceiver，并设置这个 Receiver 的 intent-filter，让其 SMSreceiver 针对收到短信事件做出反应，并添加 android.permission.RECEIVE_SMS 权限。具体代码如下。

```
        <receiver android:name="example4SMSreceiver">
        <!-- 设定要捕捉的讯息名称为 provider 中 Telephony.SMS_RECEIVED -->
      <intent-filter>
        <action
          android:name="android.provider.Telephony.SMS_RECEIVED" />
      </intent-filter>
      </receiver>
        </application>
    <uses-permission android:name="android.permission.RECEIVE_SMS"></uses-permission>
```

执行后可以向其发送一条短信，收到短信后会显示提示信息，并生成邮件提示。如图 6-11 所示。

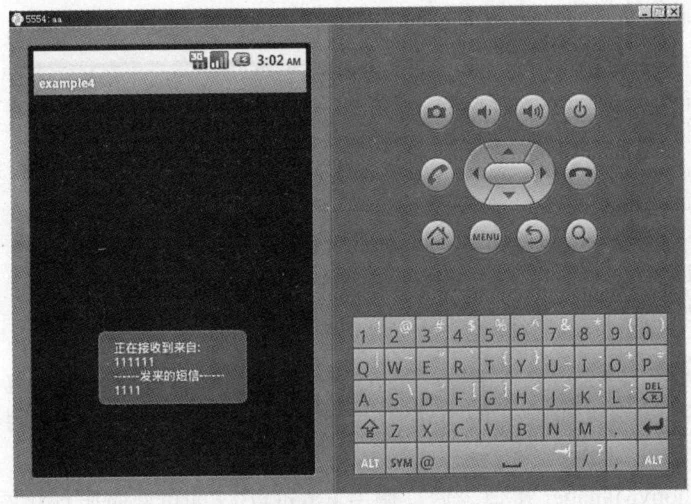

图 6-11　运行效果

6.5　来电的信息提醒

在当前手机应用中，如果有来电，则会在屏幕中显示拨打用户的姓名等基本信息。在本节的内容中，将通过一个具体实例的实现，介绍实现来电屏幕提醒的基本实现流程。本实例的源代码保存在“光盘:\daima\6\example5”，下面开始讲解本实例的具体实现流程。

6.5.1 实现原理

在 Android 中，可以通过 PhoneStateListener（监听电话状态的方法）提供的方法来监听来电状态。在具体实施时，需要创建 PhoneStateListener 对象，并重写其中的 onCallStateChanged()方法，并通过传入的“state”来判断来电状态。

要获取来电状态，需要用户具有读取电话状态的权限，否则不能成功获取状态。在具体实施上，需要在模拟器中先添加一个联系人记录，并为其起名。这样当电话进来后，会在屏幕中显示他的名字。如果是非通讯录的来电，则在屏幕中显示 Unknown Caller。

6.5.2 TelephonyManager 和 PhoneStateListener

开发应用程序的时候，希望能够监听电话的呼入，以便执行暂停音乐播放器等操作，当电话结束之后，再次恢复播放。Android 平台可以通过 TelephonyManager 和 PhoneStateListener 来完成此任务。

TelephonyManager 作为一个 Service 接口提供给用户查询电话相关的内容，例如 IMEI、LineNumber1 等。通过下面的代码即可获得 TelephonyManager 的实例。

```
TelephonyManager mTelephonyMgr = (TelephonyManager) this
.getSystemService(Context.TELEPHONY_SERVICE);
```

在 Android 平台中，PhoneStateListener 是个很有用的监听器，用来监听电话的状态，如呼叫状态和连接服务等。其方法如下。

```
public void onCallForwardingIndicatorChanged(boolean cfi)
public void onCallStateChanged(int state, String incomingNumber)
public void onCellLocationChanged(CellLocation location)
public void onDataActivity(int direction)
public void onDataConnectionStateChanged(int state)
public void onMessageWaitingIndicatorChanged(boolean mwi)
public void onServiceStateChanged(ServiceState serviceState)
public void onSignalStrengthChanged(int asu)
```

这里只需要覆盖 onCallStateChanged()方法即可监听呼叫状态。在 TelephonyManager 中定义了三种状态，分别是振铃（RINGING）、摘机（OFFHOOK）和空闲（IDLE），通过“state”的值就知道现在的电话状态了。

获得了 TelephonyManager 接口之后，调用 listen()方法即可监听电话状态。

```
mTelephonyMgr.listen(new TeleListener(),
PhoneStateListener.LISTEN_CALL_STATE);
```

6.5.3 具体实现

本实例的主程序文件是 example5.java，下面开始讲解其具体实现流程。

1）通过方法 setContentView 来引用主布局文件 main.xml，然后通过 myTextView1 显示提示。具体代码如下。

```
public class example5 extends Activity
{
  private TextView myTextView1;

  /** Called when the activity is first created. */
  @Override
  public void onCreate(Bundle savedInstanceState)
  {
    super.onCreate(savedInstanceState);
    setContentView(R.layout.main);

    myTextView1 = (TextView) findViewById(R.id.myTextView1);
```

2）定义 TelephonyManager 对象 tm，用于获取电话服务，然后通过 tm.listen 来注册电话通信 Listener。具体代码如下。

```
    /* 添加自己实现的 PhoneStateListener */
    exPhoneCallListener myPhoneCallListener =
    new exPhoneCallListener();

    /* 取得电话服务 */
    TelephonyManager tm =
    (TelephonyManager) this.getSystemService
    (Context.TELEPHONY_SERVICE);

    /* 注册电话通信 Listener */
    tm.listen
    (
      myPhoneCallListener,
      PhoneStateListener.LISTEN_CALL_STATE
    );

  }
```

3）使用内部类来继承 PhoneStateListener，重写 onCallStateChanged 方法，这样当状态改变时改变 myTextView1 的文字及颜色。然后分别设置无任何状态、接起电话和电话呼入的显示。具体代码如下。

```
  public class exPhoneCallListener extends PhoneStateListener
  {
    /* 重写 onCallStateChanged
    当状态改变时改变 myTextView1 的文字及颜色 */
    public void onCallStateChanged(int state, String incomingNumber)
    {
      switch (state)
      {
```

```
        /* 无任何状态时 */
        case TelephonyManager.CALL_STATE_IDLE:
          myTextView1.setTextColor
          (
            getResources().getColor(R.drawable.red)
          );
          myTextView1.setText("CALL_STATE_IDLE");
          break;
        /* 接起电话时 */
        case TelephonyManager.CALL_STATE_OFFHOOK:
          myTextView1.setTextColor
          (
            getResources().getColor(R.drawable.green)
          );
          myTextView1.setText("CALL_STATE_OFFHOOK");
          break;
        /* 电话呼入时 */
        case TelephonyManager.CALL_STATE_RINGING:
          getContactPeople(incomingNumber);
          break;
        default:
          break;
      }
      super.onCallStateChanged(state, incomingNumber);
    }
  }
```

4）通过方法 getContactPeople(String incomingNumber)来获取机器内的联系人信息，然后在光标里存放字段名称。具体代码如下。

```
private void getContactPeople(String incomingNumber)
{
  myTextView1.setTextColor(Color.BLUE);
  ContentResolver contentResolver = getContentResolver();
  Cursor cursor = null;

  /* cursor 里要放的字段名称 */
  String[] projection = new String[]
  {
    Contacts.People._ID,
    Contacts.People.NAME,
    Contacts.People.NUMBER
  };
```

5）通过来电号码查找对应的联系人，查找到则显示姓名，没有查找到则只显示号码。具体代码如下。

```
/* 用来电电话号码去找该联系人 */
cursor = contentResolver.query
(
  Contacts.People.CONTENT_URI, projection,
  Contacts.People.NUMBER + "=?",
  new String[]
  {
    incomingNumber
  },
  Contacts.People.DEFAULT_SORT_ORDER
);

/* 找不到联系人 */
if (cursor.getCount() == 0)
{
  myTextView1.setText("unknown Number:" + incomingNumber);
}
else if (cursor.getCount() > 0)
{
  cursor.moveToFirst();
  /* 在 projection 这个数组里名字是放在第 1 个位置 */
  String name = cursor.getString(1);
  myTextView1.setText(name + ":" + incomingNumber);
}
}
}
```

接下来还需要在文件 AndroidManifest.xml 中获取如下两个权限。

- 读取通讯录权限：android.permission.READ_CONTACTS。
- 获取电话状态权限：android.permission.READ_PHONE_STATE。

具体代码如下。

```
<uses-permission android:name="android.permission.READ_CONTACTS"></uses-permission>
<uses-permission android:name="android.permission.READ_PHONE_STATE"></uses-permission>
```

执行后的效果如图 6-12 所示，当打来电话后会显示来电的基本信息，如图 6-13 所示。

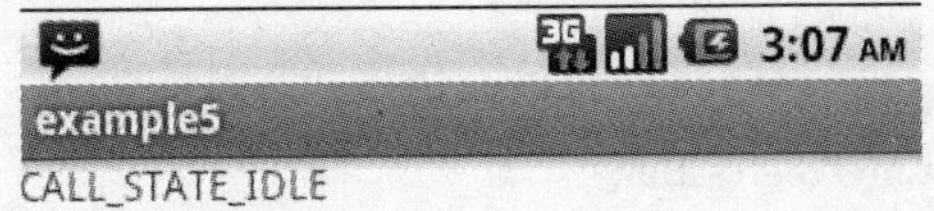

图 6-12　初始效果

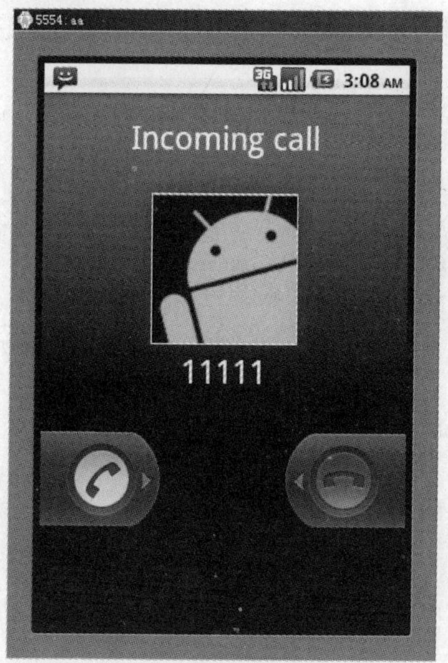

图 6-13　来电后界面

6.6　获取存储卡容量

在手机应用中，需要及时了解存储卡的容量信息。在本节的内容中，将通过一个具体实例的实现，介绍编程实现获取存储卡容量的基本流程。本实例的源代码保存在“光盘:\daima\6\example6”，下面开始讲解本实例的具体实现流程。

6.6.1　实现原理

存储卡是可以随时插拔的，每次插拔时会对操作系统进行 ACTION broadcast。在本实例中，将通过 StatFs 文件系统的方法来取得 MicroSD 存储卡的剩余容量。并且在具体实施时，需要首先判断是否安装存储卡，如果不存在则直接不计算。为了更好的显示容量，在布局中插入了一个 ProgressBar 进度条组件，这样将一目了然。

6.6.2　具体实现

本实例的主程序文件是 example6.java，下面开始讲解其具体实现流程。

1）先分别定义 3 个参数 myButton、myProgressBar 和 myTextView，然后通过 findViewById 构造 3 个对象 myButton、myProgressBar 和 myTextView。具体代码如下。

```
public class example6 extends Activity
{
  private Button myButton;
  private ProgressBar myProgressBar;
  private TextView myTextView;
```

```
@Override
public void onCreate(Bundle savedInstanceState)
{
  super.onCreate(savedInstanceState);
  setContentView(R.layout.main);

  myButton = (Button) findViewById(R.id.myButton);
  myProgressBar = (ProgressBar) findViewById(R.id.myProgressBar);
  myTextView = (TextView) findViewById(R.id.myTextView);
```

2）定义 setOnClickListener 事件，用于触发按钮单击事件处理程序。具体代码如下。

```
myButton.setOnClickListener(new Button.OnClickListener()
{

  @Override
  public void onClick(View arg0)
  {
    showSize();
  }
});

}
```

3）定义方法 showSize()，用于显示存储卡的容量大小。流程如下。

- 第一步：将 TextView 及 ProgressBar 设置为空值及 0。
- 第二步：获取 SD CARD 文件路径。
- 第三步：通过 StatFs 方法来查看文件系统空间使用状况。
- 第四步：分别获取全部的 Block 数量和已使用的 Block 数量。
- 第五步：通过 getMax()方法获取在 main.xml 里进度条设置的最大值。
- 第六步：显示出容量信息。
- 第七部：如果没有 SD 卡则输出“SD CARD 已删除”的提示。

具体代码如下所示：

```
private void showSize()
{
  /*  将 TextView 及 ProgressBar 设置为空值及 0 */
  myTextView.setText("");
  myProgressBar.setProgress(0);
  /* 判断存储卡是否插入 */
  if (Environment.getExternalStorageState().equals(Environment.MEDIA_MOUNTED))
  {
    /*    取得 SD 卡文件路径，一般是/sdcard */
    File path = Environment.getExternalStorageDirectory();

    /* StatFs 查看文件系统空间使用状况 */
```

```
        StatFs statFs = new StatFs(path.getPath());
        /* Block 的 size*/
        long blockSize = statFs.getBlockSize();
        /* 总的 Block 数量 */
        long totalBlocks = statFs.getBlockCount();
        /*    已使用的 Block 数量 */
        long availableBlocks = statFs.getAvailableBlocks();

        String[] total = fileSize(totalBlocks * blockSize);
        String[] available = fileSize(availableBlocks * blockSize);

        /* getMax 取得在 main.xml 里 ProgressBar 设置的最大值 */
        int ss = Integer.parseInt(available[0]) * myProgressBar.getMax()
            / Integer.parseInt(total[0]);

        myProgressBar.setProgress(ss);
        String text = "总共" + total[0] + total[1] + "\n";
        text += "可用" + available[0] + available[1];
        myTextView.setText(text);

    } else if (Environment.getExternalStorageState().equals(
            Environment.MEDIA_REMOVED))
    {
        String text = "SD CARD 已删除";
        myTextView.setText(text);
    }
}

/*返回为字符串数组[0]为大小[1]为单位 KB 或 MB*/
private String[] fileSize(long size)
{
    String str = "";
    if (size >= 1024)
    {
        str = "KB";
        size /= 1024;
        if (size >= 1024)
        {
            str = "MB";
            size /= 1024;
        }
    }
    DecimalFormat formatter = new DecimalFormat();
    /* 每 3 个数字用,分隔如：1,000 */
    formatter.setGroupingSize(3);
    String result[] = new String[2];
```

```
            result[0] = formatter.format(size);
            result[1] = str;
            return result;
        }
    }
```

执行后的效果如图 6-14 所示。

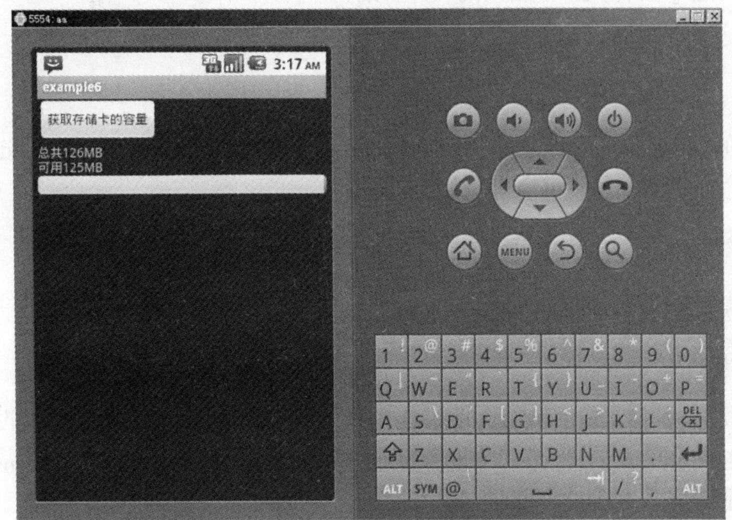

图 6-14　执行效果

Android 模拟器能够让我们使用 FAT32 格式的磁盘镜像作为 SD 卡的模拟，具体过程如下。

1）进入 Android SDK 目录下的 tools 子目录，运行：

```
mksdcard -l sdcard 512M /your_path_for_img/sdcard.img
```

这样就创建了一个 512M 的 SD 卡镜像文件。

在使用 mksdcard 命令时要注意如下 6 点。

- mksdcard 命令可以使用三种尺寸：字节、千字节和兆字节。如果只使用数字，表示字节。后面还可以跟 K，如 262144K，也表示 256M。
- mksdcard 建立的虚拟文件最小为 8M，也就是说，模拟器只支持大于 8M 的虚拟文件。
- –l 命令行参数表示虚拟磁盘的卷标，可以没有该参数。
- 虚拟文件的扩展名可以是任意的，如 mycard.abc。
- mksdcard 命令不会自动建立不存在的目录，因此，在执行上面命令之前，要先在当前目录中建立一个 card 目录。
- mksdcard 命令是按实际大小生成的 sdcard 虚拟文件。也就是说，生成 256M 的虚拟文件的尺寸就是 256M，如果生成较大的虚拟文件，要看看自己的硬盘空间够不够。

2）运行模拟器的时候指定路径（注意需要完整路径）。

```
emulator -sdcard /your_path_for_img/sdcard.img
```

这样模拟器中就可以使用“/sdcard”这个路径来指向模拟的 SD 卡了。

那么如何复制本机文件到 SD 卡，或者管理 SD 卡上的内容呢？有如下 2 种方案。

第一种：mount 是 Linux 下的一个命令，它可以将 Windows 分区作为 Linux 的一个“文件”挂接到 Linux 的一个空文件夹下，从而将 Windows 的分区和/mnt 这个目录联系起来，因此我们只要访问这个文件夹，就相当于访问该分区了。

```
mount -o loop sdcard.img android_sdcard
```

这样管理这个目录就是管理 sdcard 内容了。

第二种：在 Windows 可是环境下可以用 mtools 来做管理，也可以用 android SDK 自带的命令（这个命令在 linux 下面也可以用）。

```
adb push local_file sdcard/remote_file
```

在执行完上面的命令后，执行下面的命令启动 Android 模拟器。

```
emulator -avd avd1 -sdcard card/mycard.img
```

如果在开发环境（Eclipse）中，可以在“Run Configuration”对话框中设置启动参数，当然，也可以在 Preferences 对话框中设置默认启动参数。这样在新建立的 Android 工程中就自动加入了装载 sdcard 虚拟文件的命令行参数。

如果读者使用 OPhone 虚拟机，设置的方法完全一样的。在虚拟机中的 Setting 里看看 sdcard，是否找到。那么如何查看 sdcard 虚拟设备中的内容呢？方法很多，最简单的就是使用 Android Eclipse 插件带的 DDMS 透视图。

6.7 来电邮件通知你

在前面的实例中，介绍了短信、邮件通知的实现过程。同理，也可以编写一个程序，当来电时用邮件实现通知。在本节的内容中，将通过一个具体实例的实现，介绍来电邮件通知的基本流程。本实例的源代码保存在“光盘:\daima\6\example7”，下面开始讲解本实例的具体实现流程。

6.7.1 实现原理

在本实例中，通过 TelephoneManage 来判断来电状态，并实现来电通知。程序通过 Email 来通知来电记录，本实例继承了前面的实例，并再次对 PhoneCallListener 类实现电话事件判断，并根据来电状态发送 Email。和前面的实例一样，在模拟器运行后会显示“No application can perform this action”的提示，而在真实机器上不会出现此问题。

6.7.2 具体实现

本实例的主程序文件是 example7.java 和 example1_SMSreceiver.java，下面开始讲解其具体实现代码。

1）先定义三个参数 mTextView1、mEditText01 和 strEmailSubject，并分别为后两个赋值。具体代码如下。

```
public class example7 extends Activity
{
  private TextView mTextView1;
  private String mEditText01 ="IRDC@gmail.com";
  private String strEmailSubject = "You have phone!!";
```

2）定义 TelephonyManager 类的对象 telMgr，用于获取 TELEPHONY_SERVICE 系统信息。具体代码如下。

```
  /** Called when the activity is first created. */
 @Override
 public void onCreate(Bundle savedInstanceState)
 {
  super.onCreate(savedInstanceState);
  setContentView(R.layout.main);

  mPhoneCallListener phoneListener=new mPhoneCallListener();
  /*对象 telMgr，用于获取 TELEPHONY_SERVICE 系统*/
  TelephonyManager telMgr = (TelephonyManager)getSystemService
             (TELEPHONY_SERVICE);
  telMgr.listen(phoneListener, mPhoneCallListener.
         LISTEN_CALL_STATE);

  mTextView1 = (TextView)findViewById(R.id.myTextView1);
 }
```

3）使用 PhoneCallListener 来监听电话状态更改事件，onCallStateChanged()方法的具体实现流程如下。

第一步：获取电话待机状态。

第二步：获取电话通话状态。

第三步：获取电话来电状态。

第四步：显示号码。

第五步：有电话时发送邮件。

第六步：设置收信人邮箱地址。

第七步：设置邮件标题。

第八步：设置邮件内容

第九步：实现发信处理。

上述功能的具体代码如下。

```
public class mPhoneCallListener extends PhoneStateListener
{
```

```
        @Override
        public void onCallStateChanged(int state, String incomingNumber)
        {
          switch(state)
          {
            /*  获取电话待机状态*/
            case TelephonyManager.CALL_STATE_IDLE:
              mTextView1.setText(R.string.str_CALL_STATE_IDLE);
              break;
            /*  获取电话通话状态*/
            case TelephonyManager.CALL_STATE_OFFHOOK:
              mTextView1.setText(R.string.str_CALL_STATE_OFFHOOK);
              break;
            /*  获取电话来电状态*/
            case TelephonyManager.CALL_STATE_RINGING:
              mTextView1.setText
              (
                /*显示号码*/
                getResources().getText(R.string.str_CALL_STATE_RINGING)+
                incomingNumber
              );

              /*有电话时发送邮件*/
              Intent mEmailIntent = new Intent(android.content.Intent
                          .ACTION_SEND);
              mEmailIntent.setType("plain/text");
              /*设置收信人邮箱地址*/
              mEmailIntent.putExtra(android.content.Intent.EXTRA_EMAIL,
                       new String[]{mEditText01.toString()});
              /*设置邮件标题*/
              mEmailIntent.putExtra(android.content.Intent.EXTRA_SUBJECT,
                       strEmailSubject);
              /*设置邮件内容*/
              mEmailIntent.putExtra(android.content.Intent.EXTRA_TEXT,
                       R.string.str_EmailBody+incomingNumber);
              /*实现发信处理*/
              startActivity(Intent.createChooser(mEmailIntent,
                     getResources().getString(R.string.str_message)));

              break;
            default:
              break;
          }
          super.onCallStateChanged(state, incomingNumber);
        }
      }
    }
```

执行后的初始效果如图 6-15 所示，当拨打电话后会显示来电信息，如图 6-16 所示。

图 6-15　初始效果

图 6-16　来电后界面

当挂机后会发送邮件，当然因为是模拟器，所以显示“No application can perform this action”的提示。如图 6-17 所示。

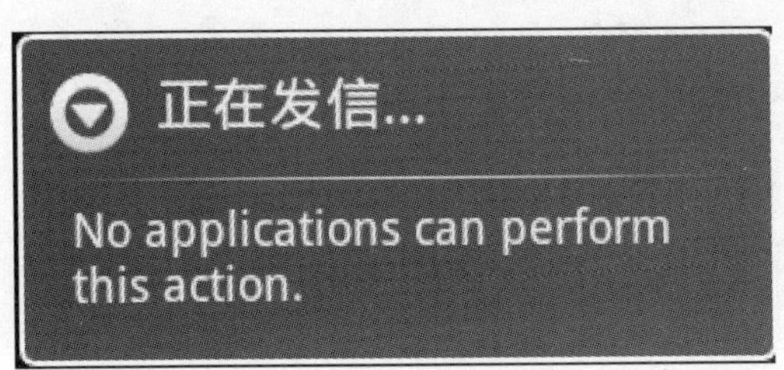

图 6-17　发邮件

6.8　内存和存储卡控制

在前面的实例中，介绍了和存储卡有关的操作。在本节的内容中，将进一步讲解对 SD 存储卡的操作方法，通过一个具体实例的实现，介绍对内存和内存卡中文件进行操作的基本流程。本实例的源代码保存在“光盘:\daima\6\example8”，下面开始讲解本实例的具体实现流程。

6.8.1　实现原理

移动手机的存储控件分为内存控件和存储卡控件，在本实例中将添加两个按钮，分别用于添加和删除内存或存储卡内的文件。并且在实例中使用了 3 个 Activity，主程序的是 Entry Activity，另外 2 个分别用于处理内存卡和存储卡。

当用户选择内存或存储卡后，将以列表形式显示里面的所有的目录和文件名，并在

Android 主菜单中显示“添加”或“删除”按钮。单击“添加”按钮后会显示一个添加菜单，实现添加文件功能。当单击“删除”按钮后，可以删除指定的文件。

6.8.2 具体实现

本实例的主程序文件是 example8.java、example8_1.java 和 example8_2.java，下面开始讲解其具体实现流程。

1．文件 example8.java

下面先看文件 example8.java 的具体实现流程。

1）先定义 4 个参数对象 myButton1、myButton2、fileDir 和 sdcardDir，然后通过 findViewById 构造两个对象 myButton1 和 myButton2。具体代码如下。

```
public class example8 extends Activity
{
  private Button myButton1;
  private Button myButton2;
  private File fileDir;
  private File sdcardDir;

  /** Called when the activity is first created. */
  @Override
  public void onCreate(Bundle savedInstanceState)
  {
    super.onCreate(savedInstanceState);
    setContentView(R.layout.main);

    myButton1 = (Button) findViewById(R.id.myButton1);
    myButton2 = (Button) findViewById(R.id.myButton2);
```

2）通过 getFilesDir()方法取得 SD 卡的目录，设置当 SD Card 无插入时，myButton2 处于不可用状态。具体代码如下。

```
/* 取得目前 File 目录 */
fileDir = this.getFilesDir();

/* 取得 SD 卡的目录 */
sdcardDir = Environment.getExternalStorageDirectory();

/* 当 SD Card 无插入时，将 myButton2 设成不可用 */
if (Environment.getExternalStorageState().equals(Environment.MEDIA_REMOVED))
{
  myButton2.setEnabled(false);
}
```

3）分别定义按钮单击处理事件 setOnClickListener 和 setOnClickListener，具体代码如下。

```
myButton1.setOnClickListener(new Button.OnClickListener()
{

   @Override
   public void onClick(View arg0)
   {
      String path = fileDir.getParent() + java.io.File.separator
            + fileDir.getName();
      showListActivity(path);
   }
});
myButton2.setOnClickListener(new Button.OnClickListener()
{

   @Override
   public void onClick(View arg0)
   {
      String path = sdcardDir.getParent() + sdcardDir.getName();
      showListActivity(path);
   }
});
}
```

4）定义方法 showListActivity(String path)，定义一个 Intent 类对象 intent，然后将路径传到 example_1。具体代码如下。

```
private void showListActivity(String path)
{
   Intent intent = new Intent();
   intent.setClass(example8.this, example8_1.class);
   Bundle bundle = new Bundle();
   /* 将路径传到 example_1 */
   bundle.putString("path", path);
   intent.putExtras(bundle);
   startActivity(intent);
}
}
```

2．文件 example8_1.java

接下来讲解文件 example8_1.java，其具体实现流程如下。

1）主 Activity 传来的 path 字符串作为传入路径，如果路径不存在，则使用 java.io.File 来创建。具体代码如下。

```
public class example8_1 extends ListActivity
{
   private List<String> items = null;
   private String path;
```

```
protected final static int MENU_NEW = Menu.FIRST;
protected final static int MENU_DELETE = Menu.FIRST + 1;

@Override
public void onCreate(Bundle savedInstanceState)
{
  super.onCreate(savedInstanceState);
  setContentView(R.layout.ex06_09_1);

  Bundle bunde = this.getIntent().getExtras();
  path = bunde.getString("path");

  java.io.File file = new java.io.File(path);
  /* 当该目录不存在时将目录创建 */
  if (!file.exists())
  {
    file.mkdir();
  }
  fill(file.listFiles());
}
```

2）定义 onOptionsItemSelected 方法，根据单击的菜单实现添加或删除。具体代码如下。

```
@Override
public boolean onOptionsItemSelected(MenuItem item)
{
  super.onOptionsItemSelected(item);
  switch (item.getItemId())
  {
    case MENU_NEW:
      /* 单击添加菜单 */
      showListActivity(path, "", "");
      break;
    case MENU_DELETE:
      /* 单击删除菜单 */
      deleteFile();
      break;
  }
  return true;
}
```

3）定义 onCreateOptionsMenu(Menu menu)方法，用于添加需要的菜单项。具体代码如下。

```
@Override
public boolean onCreateOptionsMenu(Menu menu)
{
```

```
    super.onCreateOptionsMenu(menu);
    /* 添加菜单 */
    menu.add(Menu.NONE, MENU_NEW, 0, R.string.strNewMenu);
    menu.add(Menu.NONE, MENU_DELETE, 0, R.string.strDeleteMenu);
    return true;
}
```

4）当单击文件名后取得文件内容，具体代码如下。

```
@Override
protected void onListItemClick
(ListView l, View v, int position, long id)
{
  File file = new File
  (path + java.io.File.separator + items.get(position));

  /* 单击文件取得文件内容 */
  if (file.isFile())
  {
    String data = "";
    try
    {
      FileInputStream stream = new FileInputStream(file);
      StringBuffer sb = new StringBuffer();
      int c;
      while ((c = stream.read()) != -1)
      {
        sb.append((char) c);
      }
      stream.close();
      data = sb.toString();
    }
    catch (Exception e)
    {
      e.printStackTrace();
    }
    showListActivity(path, file.getName(), data);
  }
}
```

5）定义 fill(File[] files)方法，用于填充内容到文件。具体代码如下。

```
private void fill(File[] files)
{
  items = new ArrayList<String>();
  if (files == null)
  {
```

```
        return;
    }
    for (File file : files)
    {
        items.add(file.getName());
    }
    ArrayAdapter<String> fileList = new ArrayAdapter<String>
    (this,android.R.layout.simple_list_item_1, items);
    setListAdapter(fileList);
}
```

6）定义 showListActivity()，用于显示已经存在的文件列表。具体代码如下。

```
private void showListActivity
(String path, String   ilename, String data)
{
    Intent intent = new Intent();
    intent.setClass(example8_1.this, example8_2.class);

    Bundle bundle = new Bundle();
    /* 文件路径 */
    bundle.putString("path", path);
    /* 文件名 */
    bundle.putString("   ilename",   ilename);
    /* 文件内容 */
    bundle.putString("data", data);
    intent.putExtras(bundle);

    startActivity(intent);
}
```

7）定义 deleteFile()方法，用于删除选定的文件。具体代码如下所示：

```
private void deleteFile()
{
    int position = this.getSelectedItemPosition();
    if (position >= 0)
    {
        File file = new File(path + java.io.File.separator +
        items.get(position));

        /* 删除文件 */
        file.delete();
        items.remove(position);
        getListView().invalidateViews();
    }
}
}
```

3．文件 example8_2.java

当单击“添加”按钮后会转到 example8_2.java，接下来讲解文件 example8_2.java，其具体实现流程如下。

1）先定义三个变量：myDialogEditText、MENU_SAVE 和 fileName。具体代码如下。

```
public class example8_2 extends Activity
{
  private String path;
  private String data;
  private EditText myEditText1;

  private EditText myDialogEditText;
  protected final static int MENU_SAVE = Menu.FIRST;
  private String fileName;
```

2）设置 myEditText1 用于放置文件内容，然后定义 Bundle 对象 bunde，用于获取路径和数据。具体代码如下。

```
@Override
public void onCreate(Bundle savedInstanceState)
{
  super.onCreate(savedInstanceState);
  setContentView(R.layout.ex06_09_2);

  /* 放置文件内容的 EditText */
  myEditText1 = (EditText) findViewById(R.id.myEditText1);

  Bundle bunde = this.getIntent().getExtras();
  path = bunde.getString("path");
  data = bunde.getString("data");
  fileName = bunde.getString("fileName");
  myEditText1.setText(data);
}
```

3）使用 onOptionsItemSelected 方法根据用户选择而进行操作。当选择 MENU_SAVE 时，保存这个文件。具体代码如下。

```
@Override
public boolean onOptionsItemSelected(MenuItem item)
{
  super.onOptionsItemSelected(item);
  switch (item.getItemId())
  {
    case MENU_SAVE:
      saveFile();
      break;
```

```
    }
    return true;
}
```

4）定义 onCreateOptionsMenu(Menu menu)方法，用于添加一个菜单。具体代码如下。

```
@Override
public boolean onCreateOptionsMenu(Menu menu)
{
  super.onCreateOptionsMenu(menu);
  /* 添加菜单 */
  menu.add(Menu.NONE, MENU_SAVE, 0, R.string.strSaveMenu);
  return true;
}
```

5）定义 saveFile()方法，用于保存文件。先定义 LayoutInflater 对象 factory，用于跳出存档。然后通过 myDialogEditText 取得 Dialog 里的 EditText，最后通过实现存档处理。具体代码如下。

```
private void saveFile()
{
  /* 跳出存档的对话框 */
  LayoutInflater factory = LayoutInflater.from(this);

  final View textEntryView = factory.inflate
  (R.layout.save_dialog, null);

  Builder mBuilder1 = new AlertDialog.Builder(example8_2.this);

  mBuilder1.setView(textEntryView);
  /* 取得对话框里的 EditText */
  myDialogEditText = (EditText) textEntryView.findViewById
                                          (R.id.myDialogEditText);

  myDialogEditText.setText(fileName);

  mBuilder1.setPositiveButton
  (
    R.string.str_alert_ok,new DialogInterface.OnClickListener()
    {
      public void onClick(DialogInterface dialoginterface, int i)
      {
        /* 存档 */
        String Filename = path + java.io.File.separator
```

```
            + myDialogEditText.getText().toString();
        java.io.BufferedWriter bw;
        try
        {
          bw = new java.io.BufferedWriter(new java.io.FileWriter(
            new java.io.File(Filename)));
          String str = myEditText1.getText().toString();
          bw.write(str, 0, str.length());
          bw.newLine();
          bw.close();
        }
        catch (IOException e)
        {
          e.printStackTrace();
        }
        /* 回到 example8_1 */
        Intent intent = new Intent();
        intent.setClass(example8_2.this, example8_1.class);
        Bundle bundle = new Bundle();
        /* 将路径传到 example8_1 */
        bundle.putString("path", path);
        intent.putExtras(bundle);
        startActivity(intent);

        finish();
      }
    });
  mBuilder1.setNegativeButton(R.string.str_alert_cancel, null);
  mBuilder1.show();
 }
}
```

执行后的初始效果如图 6-18 所示，当单击一个按钮后会显示对应的存储信息，如图 6-19 所示。当单击图 6-19 中的“menu”后，会弹出两个 menu 选项，如图 6-20 所示。此时，可以通过这两个选项分别对存储卡中数据实现管理。

图 6-18　初始效果

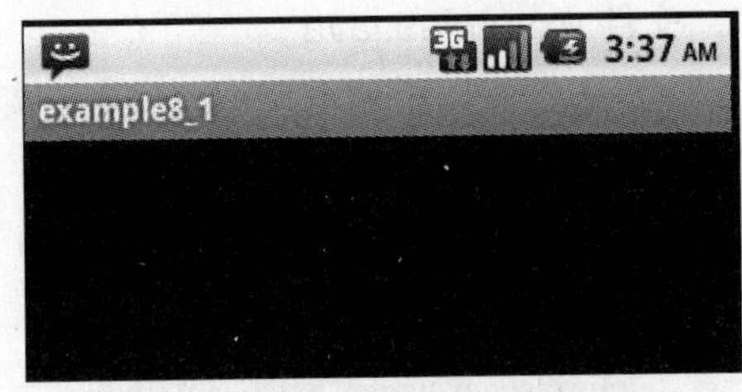

图 6-19　SD 卡的文件信息

图 6-20　管理 menu

6.9　实现定时闹钟

手机闹钟，大家都不陌生。在本节的内容中，将通过一个具体实例的实现，介绍在 Android 中实现定时闹钟的具体过程。本实例的源代码保存在“光盘:\daima\6\example9”，下面开始讲解本实例的具体实现流程。

6.9.1　实现原理

在 Android 中提供了 AlarmManager（闹钟组件）类，通过此类可以设置在指定时间运行某些动作。并且，在 Android 中内置了 Alarm Clock（闹钟），通过 AlarmManager 可以实现闹钟功能。

6.9.2　具体实现

本实例的主程序文件是 example9.java、example9_1.java 和 example9_2.java，下面开始讲解其具体实现流程。

1．文件 example9.java

下面先看文件 example9.java，其具体实现流程如下。

1）先分别定义 6 个变量 setTime1、setTime2、mButton1、mButton2、mButton3 及 mButton4，然后通过 findViewById 构造 3 个对象 myButton、myProgressBar 和 myTextView。具体代码如下。

```
public class example9 extends Activity
{
```

```
/* 声明变量 */
TextView setTime1;
TextView setTime2;
Button mButton1;
Button mButton2;
Button mButton3;
Button mButton4;
Calendar c=Calendar.getInstance();
```

2）载入主布局文件 main.xml，通过 setTime1()方法设置闹钟只响一次，然后设置只响一次闹钟的 Button1。具体代码如下。

```
@Override
public void onCreate(Bundle savedInstanceState)
{
  super.onCreate(savedInstanceState);
  /* 载入 main.xml Layout */
  setContentView(R.layout.main);

  /* 以下为只响一次的闹钟的设置 */
  setTime1=(TextView) findViewById(R.id.setTime1);
  /* 只响一次的闹钟的设置 Button */
  mButton1=(Button)findViewById(R.id.mButton1);
  mButton1.setOnClickListener(new View.OnClickListener()
  {
    public void onClick(View v)
    {
      /* 取得单击按钮时的时间作为 TimePickerDialog 的默认值 */
      c.setTimeInMillis(System.currentTimeMillis());
      int mHour=c.get(Calendar.HOUR_OF_DAY);
      int mMinute=c.get(Calendar.MINUTE);
```

3）通过 TimePickerDialog（时间对话框）来弹出一个对话框，供用户来设置时间。

- 第一步：获取设置后的时间。
- 第二步：指定闹钟设置时间到时要运行 CallAlarm.class。
- 第三步：创建 PendingIntent。
- 第四步：通过 AlarmManager.RTC_WAKEUP 设置服务在系统休眠时同样会运行。
- 第五步：定义 tmpS 对象，用于更新显示的闹钟时间。
- 第六步：以 Toast 提示设置已完成。

具体代码如下。

```
/* 跳出 TimePickerDialog 来设置时间 */
new TimePickerDialog(example9.this,
  new TimePickerDialog.OnTimeSetListener()
  {
```

```
                public void onTimeSet(TimePicker view,int hourOfDay,
                                      int minute)
                {
                  /* 取得设置后的时间，秒和毫秒设为 0 */
                  c.setTimeInMillis(System.currentTimeMillis());
                  c.set(Calendar.HOUR_OF_DAY,hourOfDay);
                  c.set(Calendar.MINUTE,minute);
                  c.set(Calendar.SECOND,0);
                  c.set(Calendar.MILLISECOND,0);

                  /* 指定闹钟设置时间到时要运行 CallAlarm.class */
                  Intent intent = new Intent(example9.this, example9_2.class);
                  /* 创建 PendingIntent */
                  PendingIntent sender=PendingIntent.getBroadcast(
                                        example9.this,0, intent, 0);
                  /* AlarmManager.RTC_WAKEUP 设置服务在系统休眠时同样会运行
                   * 以 set()设置的 PendingIntent 只会运行一次
                   * */
                  AlarmManager am;
                  am = (AlarmManager)getSystemService(ALARM_SERVICE);
                  am.set(AlarmManager.RTC_WAKEUP,
                      c.getTimeInMillis(),
                          sender
                          );
                  /* 更新显示的闹钟时间 */
                  String tmpS=format(hourOfDay)+"："+format(minute);
                  setTime1.setText(tmpS);
                  /* 以 Toast 提示设置已完成 */
                  Toast.makeText(example9.this,"设置闹钟时间为"+tmpS,
                    Toast.LENGTH_SHORT)
                    .show();
                }
              },mHour,mMinute,true).show();
        }
    });
```

4）设置按钮 mButton2，用于实现只响一次的闹钟的删除。

- 第一步：在 AlarmManager 中实现删除。
- 第二步：通过 Toast 提示已删除设置，并更新显示的闹钟时间。

具体代码如下。

```
    /* 只响一次的闹钟的删除 Button */
    mButton2=(Button) findViewById(R.id.mButton2);
    mButton2.setOnClickListener(new View.OnClickListener()
    {
      public void onClick(View v)
```

```
{
    Intent intent = new Intent(example9.this, example9_2.class);
    PendingIntent sender=PendingIntent.getBroadcast(
                              example9.this,0, intent, 0);
    /* 由 AlarmManager 中删除 */
    AlarmManager am;
    am =(AlarmManager)getSystemService(ALARM_SERVICE);
    am.cancel(sender);
    /* 以 Toast 提示已删除设置，并更新显示的闹钟时间 */
    Toast.makeText(example9.this,"闹钟时间解除",
                    Toast.LENGTH_SHORT).show();
    setTime1.setText("目前无设置");
  }
});
```

5）开始设置重复响起的闹钟。

- 第一步：建立重复响起的闹钟的设置画面，并引用 timeset.xml 为布局文件。
- 第二步：建立重复响起闹钟的设置 Dialog 对话框。
- 第三步：获取设置的间隔秒数。
- 第四步：获取设置的开始时间，秒及毫秒都设为 0。
- 第五步：指定闹钟设置时间到时要运行 CallAlarm.class。
- 第六步：通过 setRepeating()方法让闹钟重复运行。
- 第七步：通过 dmpS 更新显示的设置闹钟时间。
- 第八步：通过以 Toast 提示设置已完成。

具体代码如下。

```
/* 以下为重复响起的闹钟的设置 */
setTime2=(TextView) findViewById(R.id.setTime2);
/* 建立重复响起的闹钟的设置画面 */
/* 引用 timeset.xml 为 Layout */
LayoutInflater factory = LayoutInflater.from(this);
final View setView = factory.inflate(R.layout.timeset,null);
final TimePicker tPicker=(TimePicker)setView
                          .findViewById(R.id.tPicker);
tPicker.setIs24HourView(true);

/* 建立重复响起闹钟的设置 Dialog */
final AlertDialog di=new AlertDialog.Builder(example9.this)
        .setIcon(R.drawable.clock)
        .setTitle("设置")
        .setView(setView)
        .setPositiveButton("确定",
          new DialogInterface.OnClickListener()
          {
            public void onClick(DialogInterface dialog, int which)
```

```
        {
          /* 取得设置的间隔秒数 */
          EditText ed=(EditText)setView.findViewById(R.id.mEdit);
          int times=Integer.parseInt(ed.getText().toString())
                  *1000;
          /* 取得设置的开始时间，秒及毫秒设为 0 */
          c.setTimeInMillis(System.currentTimeMillis());
          c.set(Calendar.HOUR_OF_DAY,tPicker.getCurrentHour());
          c.set(Calendar.MINUTE,tPicker.getCurrentMinute());
          c.set(Calendar.SECOND,0);
          c.set(Calendar.MILLISECOND,0);

          /* 指定闹钟设置时间到时要运行 CallAlarm.class */
          Intent intent = new Intent(example9.this,
                                  example9_2.class);
          PendingIntent sender = PendingIntent.getBroadcast(
                                  example9.this,1, intent, 0);
          /* setRepeating()方法可让闹钟重复运行 */
          AlarmManager am;
          am = (AlarmManager)getSystemService(ALARM_SERVICE);
          am.setRepeating(AlarmManager.RTC_WAKEUP,
                  c.getTimeInMillis(),times,sender);
          /* 更新显示的设置闹钟时间 */
          String tmpS=format(tPicker.getCurrentHour())+"："+
                  format(tPicker.getCurrentMinute());
          setTime2.setText("设置闹钟时间为"+tmpS+
                      "开始，重复间隔为"+times/1000+"秒");
          /* 以 Toast 提示设置已完成 */
          Toast.makeText(example9.this,"设置闹钟时间为"+tmpS+
                    "开始，重复间隔为"+times/1000+"秒",
                    Toast.LENGTH_SHORT).show();
        }
      })
    .setNegativeButton("取消",
      new DialogInterface.OnClickListener()
      {
        public void onClick(DialogInterface dialog, int which)
        {
        }
      }).create();
```

6）实现重复响起的闹钟的设置 Button，具体代码如下。

```
/* 重复响起的闹钟的设置 Button */
mButton3=(Button) findViewById(R.id.mButton3);
mButton3.setOnClickListener(new View.OnClickListener()
{
```

```
        public void onClick(View v)
        {
          /* 取得单击按钮时的时间作为 tPicker 的默认值 */
          c.setTimeInMillis(System.currentTimeMillis());
          tPicker.setCurrentHour(c.get(Calendar.HOUR_OF_DAY));
          tPicker.setCurrentMinute(c.get(Calendar.MINUTE));
          /* 跳出设置画面 di */
          di.show();
        }
      });
```

7）开始设置重复响起的闹钟的删除 Button。

❑ 第一步：在 AlarmManager 中删除。

❑ 第二步：以 Toast 提示已删除设置，并更新显示的闹钟时间。

具体代码如下。

```
      /* 重复响起的闹钟的删除 Button */
      mButton4=(Button) findViewById(R.id.mButton4);
      mButton4.setOnClickListener(new View.OnClickListener()
      {
        public void onClick(View v)
        {
          Intent intent = new Intent(example9.this, example9_2.class);
          PendingIntent sender = PendingIntent.getBroadcast(
                                   example9.this,1, intent, 0);
          /* 由 AlarmManager 中删除 */
          AlarmManager am;
          am = (AlarmManager)getSystemService(ALARM_SERVICE);
          am.cancel(sender);
          /* 以 Toast 提示已删除设置，并更新显示的闹钟时间 */
          Toast.makeText(example9.this,"闹钟时间解除",
                      Toast.LENGTH_SHORT).show();
          setTime2.setText("目前无设置");
        }
      });
    }
```

8）格式设置方法 format(int x)，用于设置日期时间示两位数的显示格式。具体代码如下。

```
    /* 日期时间显示两位数的方法 */
    private String format(int x)
    {
      String s=""+x;
      if(s.length()==1) s="0"+s;
      return s;
```

```
    }
  }
```

2．文件 example9_1.java

下面先看文件 example9_1.java，其具体实现流程如下。

1）加载相关类。

2）实现实际跳出闹铃 Dialog 的 Activity。

3）通过 AlertDialog.Builder(example9_1.this)实现弹出闹钟警示框。

4）通过 onClick(DialogInterface dialog, int whichButton) 关闭 Activity。

具体代码如下。

```
/* 加载相关类 */
import irdc.example9.R;
import android.app.Activity;
import android.app.AlertDialog;
import android.content.DialogInterface;
import android.os.Bundle;

/* 实际跳出闹铃 Dialog 的 Activity */
public class example9_1 extends Activity
{
  @Override
  protected void onCreate(Bundle savedInstanceState)
  {
    super.onCreate(savedInstanceState);
    /* 跳出的闹铃警示   */
    new AlertDialog.Builder(example9_1.this)
        .setIcon(R.drawable.clock)
        .setTitle("闹钟响了!!")
        .setMessage("赶快起床吧!!!")
        .setPositiveButton("关掉他",
         new DialogInterface.OnClickListener()
        {
          public void onClick(DialogInterface dialog, int whichButton)
          {
            /* 关闭 Activity */
            example9_1.this.finish();
          }
        })
        .show();
  }
}
```

3．文件 example9_2.java

下面先看文件 example9_2.java，其具体实现流程如下。

1）加载相关类。

2）调用闹钟 Alert 对象的 Receiver。

3）创建 Intent，用于调用 AlarmAlert.class。

具体代码如下。

```
package irdc.example9;

/* 加载相关类 */
import android.content.Context;
import android.content.Intent;
import android.content.BroadcastReceiver;
import android.os.Bundle;

/* 调用闹钟 Alert 的 Receiver */
public class example9_2 extends BroadcastReceiver
{
  @Override
  public void onReceive(Context context, Intent intent)
  {
    /* create Intent，调用 AlarmAlert.class */
    Intent i = new Intent(context, example9_1.class);

    Bundle bundleRet = new Bundle();
    bundleRet.putString("STR_CALLER", "");
    i.putExtras(bundleRet);
    i.addFlags(Intent.FLAG_ACTIVITY_NEW_TASK);
    context.startActivity(i);
  }
}
```

最后，还需要在文件 AndroidManifest.xml 中添加对 CallAlarm（呼叫闹钟）的 receiver 设置。具体代码如下。

```
<!—注册 receiver CallAlarm -->
<receiver android:name=".example9_2" android:process=":remote" />
<activity android:name=".example9_1" ndroid:label="@string/app_name">
</activity>
```

执行后的初始效果如图 6-21 所示，单击第一个“设置”按钮后在弹出的界面中可以设置闹钟时间，如图 6-22 所示。单击第二个按钮可以设置重复响起的时间，如图 6-23 所示。闹钟响起后的界面如图 6-24 所示。

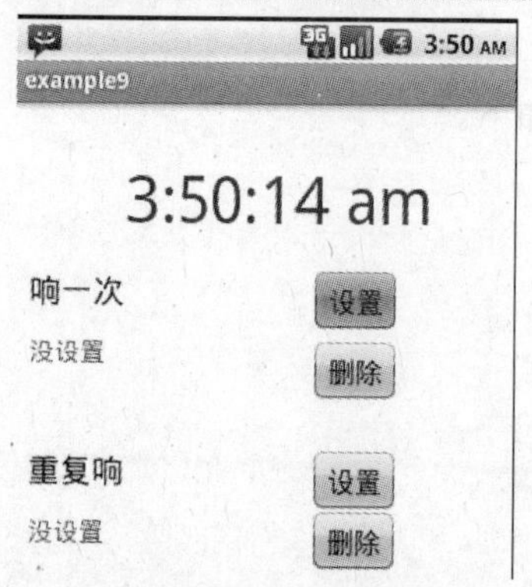

图 6-21　初始效果

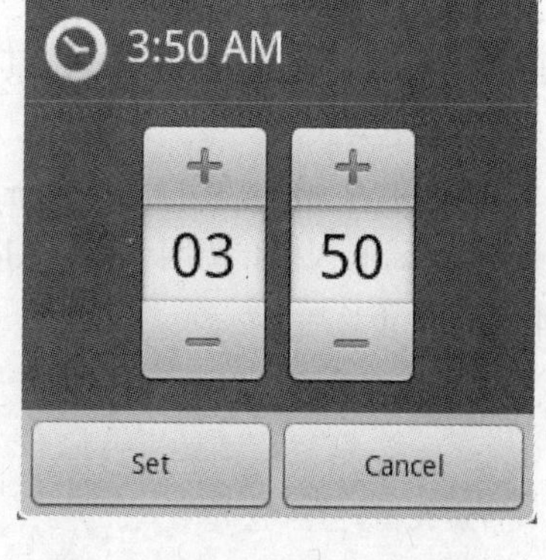

图 6-22　响一次的设置界面

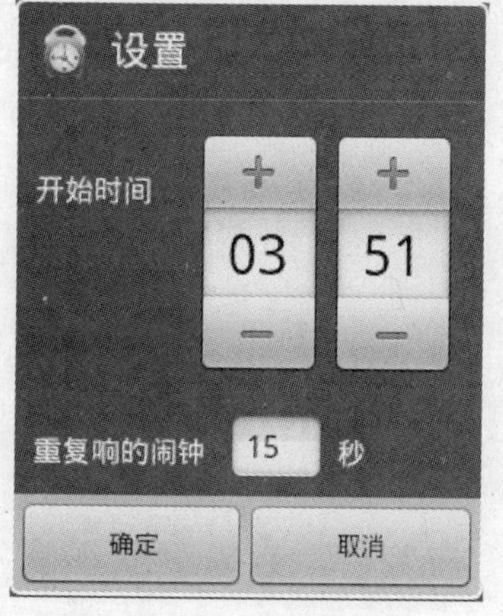

图 6-23　重复响的设置界面

图 6-24　闹钟响起后界面

6.10　黑名单来电自动静音

几乎每个手机中都有黑名单功能，被列入黑名单的用户不能打进电话和发送短信。在本节的内容中，将通过一个具体实例的实现，介绍在 Android 中实现黑名单用户来电静音的具体过程。本实例的源代码保存在“光盘:\daima\6\example10”，下面开始讲解本实例的具体实现流程。

6.10.1　实现原理

在本实例中，首先添加一个 EditText，用户在里面可以输入黑名单用户的号码。当此号码来电时，系统会自动切换为静音模式。当对方挂机后，再自动转换为正常模式，并使用 Toast 实现提示。

在具体实现上，转换铃声模式功能是通过 setRingerMode 参数实现的，正常模式是 RINGER_MODE_NORMAL，静音模式是 RINGER_MODE_SILENT，震动模式是 RINGER_MODE_VIBRATE。

6.10.2　具体实现

本实例的主程序文件是 example10.java，下面开始讲解其具体实现流程。

1）先分别定义 3 个变量 mTextView01、mEditText1 和 mEditText03。具体代码如下。

```
public class example10 extends Activity
{
  private TextView mTextView01;
  private TextView mTextView03;
  private EditText mEditText1;
```

2）设置 PhoneCallListener 对象 phoneListener，用 TelephonyManager 抓取 Telephony Severice，然后设置 Listen Call()方法，并查找 TextView 和 EditText 的数据。具体代码如下。

```
/** Called when the activity is first created. */
@Override
public void onCreate(Bundle savedInstanceState)
{
  super.onCreate(savedInstanceState);
  setContentView(R.layout.main);

  /*设置 PhoneCallListener*/
  mPhoneCallListener phoneListener=new mPhoneCallListener();
  /*用 TelephonyManager 抓取 Telephony Severice*/
  TelephonyManager telMgr = (TelephonyManager)getSystemService(
      TELEPHONY_SERVICE);
  /*设置 Listen Call*/
  telMgr.listen(phoneListener, mPhoneCallListener.
      LISTEN_CALL_STATE);

  /*查找 TextView 和 EditText 中的数据*/
  mTextView01 = (TextView)findViewById(R.id.myTextView1);
  mTextView03 = (TextView)findViewById(R.id.myTextView3);
  mEditText1 = (EditText)findViewById(R.id.myEditText1);

}
```

3）判断 PhoneStateListener 当前状态。

- 第一步：获取手机待机状态。
- 第二步：设置手机为待机时响铃正常。
- 第三步：如果获取手机状态为通话中则显示对应信息。
- 第四步：如果获取手机状态为来电则显示来电信息。
- 第五步：判断输入电话是否一致，如果一致时用静音。

具体代码如下。

```
/*  判断 PhoneStateListener 当前状态*/
public class mPhoneCallListener extends PhoneStateListener
{
  @Override
  public void onCallStateChanged(int state, String incomingNumber)
  {
    switch(state)
    {
      /*  获取手机待机状态*/
      case TelephonyManager.CALL_STATE_IDLE:
        mTextView01.setText(R.string.str_CALL_STATE_IDLE);

        try
        {
          AudioManager audioManager = (AudioManager)
                      getSystemService(Context.AUDIO_SERVICE);
          if (audioManager != null)
          {
            /*设置手机为待机时响铃正常*/
            audioManager.setRingerMode(AudioManager.
                RINGER_MODE_NORMAL);
            audioManager.getStreamVolume(
                AudioManager.STREAM_RING);
          }
        }
        catch(Exception e)
        {
         mTextView01.setText(e.toString());
         e.printStackTrace();
        }
        break;

      /*  获取手机状态为通话中 */
      case TelephonyManager.CALL_STATE_OFFHOOK:
        mTextView01.setText(R.string.str_CALL_STATE_OFFHOOK);
        break;

      /*  获取手机状态为来电 */
      case TelephonyManager.CALL_STATE_RINGING:
        /*显示来电信息  */
        mTextView01.setText(
            getResources().getText(R.string.str_CALL_STATE_RINGING)+
          incomingNumber);

        /*  判断输入电话是否一致，一致时用静音  */
        if(incomingNumber.equals(mTextView03.getText().toString()))
```

```
        {
        try
        {
          AudioManager audioManager = (AudioManager)
                          getSystemService(Context.AUDIO_SERVICE);
        if (audioManager != null)
          {
          /*设置响铃为静音  */
          audioManager.setRingerMode(AudioManager.
               RINGER_MODE_SILENT);
          audioManager.getStreamVolume(
               AudioManager.STREAM_RING);
          Toast.makeText(example10.this, getString(R.string.str_msg)
               ,Toast.LENGTH_SHORT).show();
          }
        }
        catch(Exception e)
        {
         mTextView01.setText(e.toString());
         e.printStackTrace();
        break;
        }
      }
    }

    super.onCallStateChanged(state, incomingNumber);

    mEditText1.setOnKeyListener(new EditText.OnKeyListener()
    {
```

4）定义 onKey(View v, int keyCode, KeyEvent event)方法，用于将 EditText 的输入数据显示在 TextView 对象中。具体代码如下。

```
        @Override
        public boolean onKey(View v, int keyCode, KeyEvent event)
        {
          /*设置 EditText 的输入数据显示在 TextView*/
          mTextView03.setText(mEditText1.getText());
          return false;
        }
      });
    }
  }
}
```

执行后的初始效果如图 6-25 所示，在输入框中可以输入黑名单号码，如图 6-26 所示。当此号码来电时会是静音模式，如图 6-27 所示。

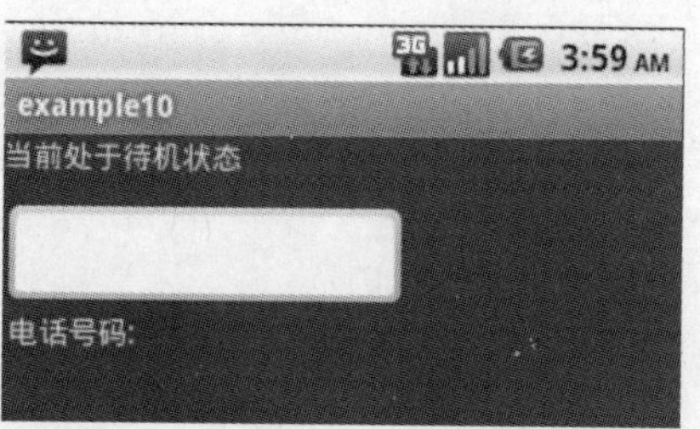

图 6-25　初始效果

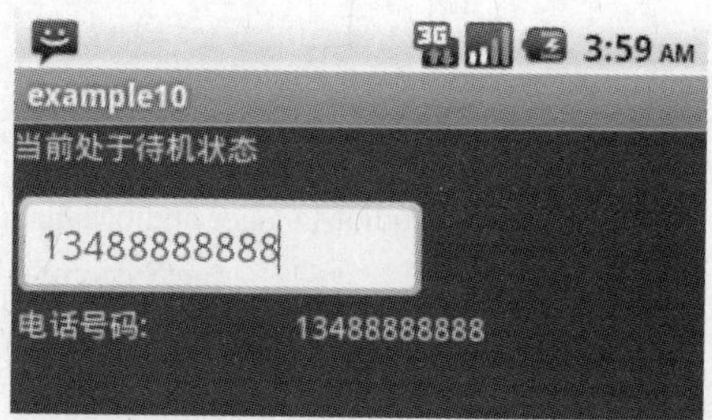

图 6-26　黑名单号码

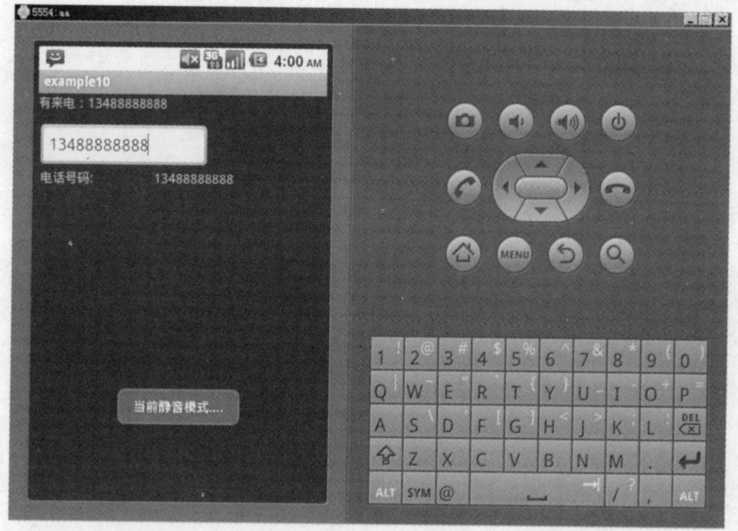

图 6-27　来电静音

注意：上述实例在模拟器中不会显示静音模式图片，但是在真实机器上会显示。

6.11　指定时间置换桌面背景

在移动手机设备中，通常为用户提供了在不同时间显示不同屏幕背景照片的功能。在本节的内容中，将通过一个具体实例的实现，介绍在指定时间置换桌面背景的具体过程。本实例的源代码保存在“光盘:\daima\6\example11”，下面开始讲解本实例的具体实现流程。

6.11.1　实现原理

本实例结合了前面的闹钟实例，实际上 AlarmManage 类并不是只能用作闹钟，它还可以自行设置在什么时间运行什么样的动作。

在本实例中，预先准备了 7 张素材图片供用户选择，放在了“res\drawable”目录下。

6.11.2　具体实现

本实例的主程序文件是 example11.java、ChangeBgImage.java、DailyBgDB.java 和 MyReceiver.java，下面开始讲解其具体实现流程。

1．文件 example11.java

下面先看文件 example11.java，其具体实现流程如下。

1）先加载相关类，然后声明设置图片的 7 个 Button 按钮变量及启动与终止的两个 Button 变量，并声明显示 7 个图文件名称的 TextView 变量。具体代码如下。

```
public class example11 extends Activity
{
  /* 声明设置图片的七个 Button 及启动与终止的两个 Button */
  private Button mButton1;
  private Button mButton2;
  private Button setButton1;
  private Button setButton2;
  private Button setButton3;
  private Button setButton4;
  private Button setButton5;
  private Button setButton6;
  private Button setButton7;
  /* 声明显示图文件名称的七个 TextView */
  private TextView mySet1;
  private TextView mySet2;
  private TextView mySet3;
  private TextView mySet4;
  private TextView mySet5;
  private TextView mySet6;
  private TextView mySet7;
```

2）分别声明自定义的数据库变量 DailyBgDB，存放设置值的 Map，存放图文件 id 的数组 bg 与存放图文件名称的数组 bgName，具体代码如下。

```
/* 声明自定义的数据库变量 DailyBgDB */
private DailyBgDB db;
/* 声明存放设置值的 Map */
private Map<Integer,Integer> map;
private LayoutInflater inflater;
private int tmpWhich=0;
/* 声明存放图文件 id 的数组 bg 与存放图文件名称的数组 bgName */
private final int[] bg =
{R.drawable.b01,R.drawable.b02,R.drawable.b03,R.drawable.b04,
R.drawable.b05,R.drawable.b06,R.drawable.b07};
private final String[] bgName =
{"b01.png","b02.png","b03.png","b04.png","b05.png","b06.png",
"b07.png"};
```

3）载入主布局文件 main.xml，并将数据库存放的设置值放入 map 中，然后初始化各个 TextView 对象，具体代码如下。

```
    @Override
    public void onCreate(Bundle savedInstanceState) {
      super.onCreate(savedInstanceState);
      /* 载入 main.xml Layout */
      setContentView(R.layout.main);
      inflater=(LayoutInflater)getSystemService(
        Context.LAYOUT_INFLATER_SERVICE);

      /* 将数据库存放的设置值放入 map 中 */
      initSettingData();
      /* 初始化 TextView 对象 */
      mySet1=(TextView) findViewById(R.id.mySet1);
      mySet2=(TextView) findViewById(R.id.mySet2);
      mySet3=(TextView) findViewById(R.id.mySet3);
      mySet4=(TextView) findViewById(R.id.mySet4);
      mySet5=(TextView) findViewById(R.id.mySet5);
      mySet6=(TextView) findViewById(R.id.mySet6);
      mySet7=(TextView) findViewById(R.id.mySet7);
```

4）根据获取的图像设置显示的图文件名称，具体代码如下。

```
    /* 设置显示的图文件名称 */
    if(!map.get(0).equals(99))
    {
      mySet1.setText(bgName[map.get(0)]);
    }
    if(!map.get(1).equals(99))
    {
      mySet2.setText(bgName[map.get(1)]);
    }
    if(!map.get(2).equals(99))
    {
      mySet3.setText(bgName[map.get(2)]);
    }
    if(!map.get(3).equals(99))
    {
      mySet4.setText(bgName[map.get(3)]);
    }
    if(!map.get(4).equals(99))
    {
      mySet5.setText(bgName[map.get(4)]);
    }
    if(!map.get(5).equals(99))
    {
      mySet6.setText(bgName[map.get(5)]);
    }
    if(!map.get(6).equals(99))
```

```
{
  mySet7.setText(bgName[map.get(6)]);
}
```

5）初始化各个 Button 对象，然后以 initButton()方法来设置 OnClickListener，具体代码如下。

```
/* 初始化 Button 对象 */
setButton1=(Button) findViewById(R.id.setButton1);
setButton2=(Button) findViewById(R.id.setButton2);
setButton3=(Button) findViewById(R.id.setButton3);
setButton4=(Button) findViewById(R.id.setButton4);
setButton5=(Button) findViewById(R.id.setButton5);
setButton6=(Button) findViewById(R.id.setButton6);
setButton7=(Button) findViewById(R.id.setButton7);
/* 以 initButton()来设置 OnClickListener */
setButton1=initButton(setButton1,mySet1,0);
setButton2=initButton(setButton2,mySet2,1);
setButton3=initButton(setButton3,mySet3,2);
setButton4=initButton(setButton4,mySet4,3);
setButton5=initButton(setButton5,mySet5,4);
setButton6=initButton(setButton6,mySet6,5);
setButton7=initButton(setButton7,mySet7,6);
```

6）设置启动服务的 Button，通过 setOnClickListener 实现单击事件的监听。

- 第一步：取得服务启动后一天的 0 点 0 分 0 秒的 millsTime。
- 第二步：重复运行的间隔时间。
- 第三步：将更换背景的进程添加 AlarmManager 中。
- 第四步：通过 setRepeating()方法让进程处理重复运行。
- 第五步：使用 Toast 提示已启动。
- 第六步：启动后马上先运行一次换背景的程序以更换今天的背景。

具体代码如下。

```
/* 设置启动服务的 Button */
mButton1=(Button)findViewById(R.id.myButton1);
mButton1.setOnClickListener(new View.OnClickListener()
{
  public void onClick(View v)
  {
    /* 取得服务启动后一天的 0 点 0 分 0 秒的 millsTime */
    Calendar calendar=Calendar.getInstance();
    calendar.add(Calendar.DATE,1);
    calendar.set(Calendar.HOUR_OF_DAY,0);
    calendar.set(Calendar.MINUTE,0);
    calendar.set(Calendar.SECOND,0);
    calendar.set(Calendar.MILLISECOND,0);
```

```
        long startTime=calendar.getTimeInMillis();
        /* 重复运行的间隔时间 */
        long repeatTime=24*60*60*1000;
        /* 将更换桌布的进程添加 AlarmManager 中 */
        Intent intent = new Intent(example11.this,MyReceiver.class);
        PendingIntent sender = PendingIntent.getBroadcast(
                        example11.this, 0, intent, 0);
        AlarmManager am = (AlarmManager)getSystemService(
          ALARM_SERVICE);
        /* setRepeating()可让进程重复运行
           startTime 为开始运行时间
           repeatTime 为重复运行间隔
           AlarmManager.RTC 可使服务休眠时仍然会运行 */
        am.setRepeating(AlarmManager.RTC,startTime,repeatTime,
                sender);
        /* 以 Toast 提示已启动 */
        Toast.makeText(example11.this,"服务已启动",Toast.LENGTH_SHORT)
          .show();
        /* 启动后马上先运行一次换背景的程序以更换今天的背景 */
        Intent i = new Intent(example11.this,ChangeBgImage.class);
        startActivity(i);
      }
    });
```

7）设置终止服务的 Button，定义 onClick(View v)方法实现对单击按钮事件的监听。

- 定义 Intent 对象 intent。
- 在 AlarmManager 中删除调度。
- 通过 Toast 提示已终止。

具体代码如下。

```
    /* 设置终止服务的 Button */
    mButton2=(Button) findViewById(R.id.myButton2);
    mButton2.setOnClickListener(new View.OnClickListener()
    {
      public void onClick(View v)
      {
        Intent intent = new Intent(example11.this,MyReceiver.class);
        PendingIntent sender = PendingIntent.getBroadcast(
                        example11.this, 0, intent, 0);
        /* 由 AlarmManager 中删除调度 */
        AlarmManager am = (AlarmManager)getSystemService(
          ALARM_SERVICE);
        am.cancel(sender);
        /* 通过 Toast 提示已终止 */
        Toast.makeText(example11.this,"服务已终止",Toast.LENGTH_SHORT)
          .show();
```

```
        }
    });
}
```

8）定义 initSettingData()，用于从数据库中取得设置值的方法。具体代码如下。

```
/* 由数据库中取得设置值的方法 */
private void initSettingData()
{
    map=new LinkedHashMap<Integer,Integer>();
    db=new DailyBgDB(example11.this);
    Cursor cur=db.select();
    while(cur.moveToNext()){
        map.put(cur.getInt(0),cur.getInt(1));
    }
    cur.close();
    db.close();
}
```

9）设置 Button 的 OnClickListener 的方法，首先设置单击 Button 后跳出的选择图片的 Dialog，然后设置预览画面的文件名与 ImageView（图片组件），最后改变画面显示的图文件文件名，并将更改的信息存入数据库。具体代码如下。

```
/* 设置 Button 的 OnClickListener 的方法 */
private Button initButton(Button b,final TextView t,final int id)
{
    b.setOnClickListener(new View.OnClickListener()
    {
        public void onClick(View v)
        {
            /* 设置单击 Button 后跳出的选择图片的 Dialog */
            new AlertDialog.Builder(example11.this)
                .setIcon(R.drawable.pic_icon)
                .setTitle("请选择图片！")
                .setSingleChoiceItems(bgName,map.get(id),
                 new DialogInterface.OnClickListener()
                 {
                    public void onClick(DialogInterface dialog,int which)
                    {
                        tmpWhich=which;
                        /* 选择图片后跳出预览图文件的窗口 */
                        View view=inflater.inflate(R.layout.preview, null);
                        TextView message=(TextView) view.findViewById(
                            R.id.bgName);
                        /* 设置预览画面的文件名与 ImageView */
                        message.setText(bgName[which]);
                        ImageView mView01 = (ImageView)view.findViewById(
```

```
                    R.id.bgImage);
                mView01.setImageResource(bg[which]);

                Toast toast=Toast.makeText(example11.this,"",
                    Toast.LENGTH_SHORT);
                toast.setView(view);
                toast.show();
            }
        })
        .setPositiveButton("确定",
         new DialogInterface.OnClickListener()
        {
          public void onClick(DialogInterface dialog,int which1)
          {
            /* 改变画面显示的设置图文件文件名 */
            t.setText(bgName[tmpWhich]);
            /* 改变 map 对象里的值 */
            map.put(id,tmpWhich);
            /* 将更改的设置存入数据库 */
            saveData(id,tmpWhich);
          }
        })
        .setNegativeButton("取消",
         new DialogInterface.OnClickListener()
        {
          public void onClick(DialogInterface dialog,int which)
          {
          }
        }).show();
    }
  });
  return b;
}
```

10）定义 saveData(int id,int value)方法，用于将设置值存储到数据库。具体代码如下。

```
    /* 存储设置值至 DB 的方法 */
    private void saveData(int id,int value)
    {
      db=new DailyBgDB(example11.this);
      db.update(id,value);
      db.close();
    }
}
```

2. 文件 MyReceiver.java

接下来看文件 MyReceiver.java，其功能是引入相关类，然后运行更换桌面背景的

Receiver。具体代码如下。

```
package irdc.example11;

/*  加载相关类  */
import android.content.Context;
import android.content.Intent;
import android.content.BroadcastReceiver;
import android.os.Bundle;

/*  运行更换桌面背景的 Receiver */
public class MyReceiver extends BroadcastReceiver
{
  @Override
  public void onReceive(Context context, Intent intent)
  {
    /* create Intent，调用 ChangeBgImage.class */
    Intent i = new Intent(context, ChangeBgImage.class);

    Bundle bundleRet = new Bundle();
    bundleRet.putString("STR_CALLER", "");
    i.putExtras(bundleRet);
    i.addFlags(Intent.FLAG_ACTIVITY_NEW_TASK);
    context.startActivity(i);
  }
}
```

3. 文件 ChangeBgImage.java

接下来看文件 ChangeBgImage.java，其具体实现流程如下。

1）先加载相关类，具体代码如下。

```
package irdc.example11;

/* import 相关 class */

import irdc.example11.R;

import java.io.IOException;
import java.util.Calendar;
import android.app.Activity;
import android.database.Cursor;
import android.graphics.Bitmap;
import android.graphics.BitmapFactory;
import android.os.Bundle;
```

2）实际运行更换桌面背景的 Activity，并声明存放图文件 id 的数组 bg，具体代码如下。

```
/* 实际运行更换桌面背景的 Activity */
public class ChangeBgImage extends Activity
{
  /* 声明存放图文件 id 的数组 bg */
  private static final int[] bg =
    {R.drawable.b01,R.drawable.b02,R.drawable.b03,R.drawable.b04,
    R.drawable.b05,R.drawable.b06,R.drawable.b07};
```

3）载入布局文件 progress.xml，并获取今天是星期几，然后从数据库中取得今天应该换哪一张背景，如果“DailyBg==99”代表没设置，则不运行。具体代码如下。

```
@Override
protected void onCreate(Bundle savedInstanceState)
{
  super.onCreate(savedInstanceState);
  /* 载入 progress.xml Layout */
  setContentView(R.layout.progress);
  /* 取得今天是星期几 */
  Calendar ca=Calendar.getInstance();
  int dayOfWeek=ca.get(Calendar.DAY_OF_WEEK)-1;

  /* 从数据库中取得今天应该换哪一张背景 */
  int DailyBg=0;
  String selection = "DailyId=?";
  String[] selectionArgs = new String[]{""+dayOfWeek};
  DailyBgDB db=new DailyBgDB(ChangeBgImage.this);
  Cursor cur=db.select(selection,selectionArgs);
  while(cur.moveToNext())
  {
    DailyBg=cur.getInt(0);
  }
  cur.close();
  db.close();

  /* 如果 DailyBg==99 代表没设置，所以不运行 */
  if(DailyBg!=99)
  {
    Bitmap bmp=BitmapFactory.decodeResource
    (getResources(), bg[DailyBg]);
    try
    {
      super.setWallpaper(bmp);
    }
    catch (IOException e)
    {
      e.printStackTrace();
    }
```

```
        }
        finish();
    }
}
```

4．文件 DailyBgDB.java

接下来看文件 DailyBgDB.java，其具体实现流程如下。

1）先进行变量声明，然后定义构造器 DailyBgDB(Context context)。具体代码如下。

```
package irdc.example11;

/* 加载相关类 */
import android.content.ContentValues;
import android.content.Context;
import android.database.Cursor;
import android.database.sqlite.SQLiteDatabase;
import android.database.sqlite.SQLiteOpenHelper;

public class DailyBgDB extends SQLiteOpenHelper
{
  /* 变量声明 */
  private final static String DATABASE_NAME = "dailyBG_db";
  private final static int DATABASE_VERSION = 1;
  private final static String TABLE_NAME = "dailySetting_table";
  public final static String FIELD1 = "DailyId";
  public final static String FIELD2 = "DailyBg";
  public SQLiteDatabase sdb;

  /* 构造器 */
  public DailyBgDB(Context context)
  {
    super(context, DATABASE_NAME, null, DATABASE_VERSION);
    sdb= this.getWritableDatabase();
  }
```

2）开始数据库处理，如果表格不存在就创建表对象，并存入初始的数据到数据库。具体代码如下。

```
  @Override
  public void onCreate(SQLiteDatabase db)
  {
    /* 表不存在就创建表 */
    String sql = "CREATE TABLE IF NOT EXISTS "+TABLE_NAME+"("+FIELD1
        +" INTEGER primary key, "+FIELD2+" INTEGER)";
    db.execSQL(sql);

    /* 存入初始的数据到数据库 */
```

```
        sdb=db;
        insert(0,99);
        insert(1,99);
        insert(2,99);
        insert(3,99);
        insert(4,99);
        insert(5,99);
        insert(6,99);
    }

    @Override
    public void onUpgrade(SQLiteDatabase db,int oldVersion,
                    int newVersion)
    {
    }

    public Cursor select()
    {
      Cursor cursor=sdb.query(TABLE_NAME,null,null,
                        null,null,null,null);
      return cursor;
    }
```

3）定义方法 select(String selection,String[] selectionArgs)，当 select 时，有 where 条件时要用此方法，用于检索数据。具体代码如下。

```
/* select 时有 where 条件要用此方法 */
public Cursor select(String selection,String[] selectionArgs)
{
  String[] columns = new String[] { FIELD2 };
  Cursor cursor=sdb.query(TABLE_NAME,columns,selection,selectionArgs,null,null,null);
  return cursor;
}
```

4）定义方法 insert(int value1,int value2)，用于将添加的值放入 ContentValues。具体代码如下。

```
public long insert(int value1,int value2)
{
  /* 将添加的值放入 ContentValues */
  ContentValues cv = new ContentValues();
  cv.put(FIELD1, value1);
  cv.put(FIELD2, value2);
  long row = sdb.insert(TABLE_NAME, null, cv);
  return row;
}
```

5）定义方法 delete(int id)，用于删除设置。具体代码如下。

```
public void delete(int id)
{
  String where = FIELD1 + " = ?";
  String[] whereValue ={ Integer.toString(id) };
  sdb.delete(TABLE_NAME, where, whereValue);
}
```

6）定义 update(int id, int value)方法，用于修改设置。具体代码如下。

```
public void update(int id, int value)
{
  String where = FIELD1 + " = ?";
  String[] whereValue ={ Integer.toString(id) };
  /* 将修改的值放入 ContentValues */
  ContentValues cv = new ContentValues();
  cv.put(FIELD2, value);
  sdb.update(TABLE_NAME, cv, where, whereValue);
 }
}
```

最后还需要在文件 AndroidManifest.xml 中添加 MyReceiver 的 receiver 设置，并添加背景图像权限 android.permission.SET_WALLPAPER。具体代码如下。

```
<receiver android:name=".MyReceiver" android:process=":remote"/>
<!-- 设定 SET_WALLPAPER 权限 -->
<uses-permission android:name="android.permission.SET_WALLPAPER" />
```

执行后的初始效果如图 6-28 所示，选择星期数，单击后面的“设置”按钮后，可以在弹出界面中设置这天的背景图片，如图 6-29 所示。

图 6-28 初始效果

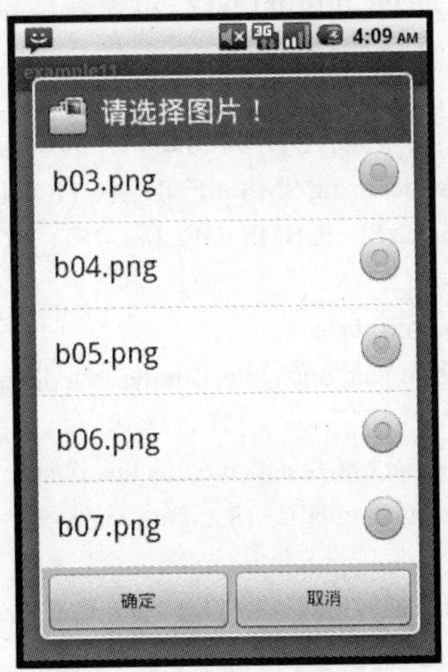

图 6-29 设置背景图片

6.12 监听短信状态

当发送短信后，用户往往对是否发送成功比较关心。在本节的内容中，将通过一个具体实例的实现，介绍监听短信发送的具体过程。本实例的源代码保存在“光盘:\daima\6\example12”，下面开始讲解本实例的具体实现流程。

6.12.1 实现原理

本实例中，当发送一条短信后，会及时提供一条信息，说明短信是发送成功还是失败。手机的默认程序可以捕捉到发送状态，这是因为经过系统广播的信息，程序可以捕捉到发送结果。本实例的学习重点是如何衍生广播类 mServiceReceiver，并在这个 Receiver 对象中判断短信的发送结果。

6.12.2 具体实现

本实例的主程序文件是 example12.java，下面开始讲解其具体实现流程。

1）先创建两个 mServiceReceiver 对象，作为类成员变量。然后分别创建 5 个变量：mButton1、mButton2、mTextView01、mEditText1 和 mEditText2。具体代码如下。

```
public class example12 extends Activity
{
  /* 创建两个 mServiceReceiver 对象，作为类成员变量 */
  private mServiceReceiver mReceiver01, mReceiver02;
  private Button mButton1;
  private TextView mTextView01;
  private EditText mEditText1, mEditText2;
```

2）自定义 ACTION 常数，作为广播的 Intent Filter 识别常数。然后通过 mEditText1 获取电话号码，通过 mEditText2 对象获取信息内容，然后设置默认值为 5556，表示第二个模拟器的 Port。具体代码如下。

```
/* 自定义 ACTION 常数，作为广播的 Intent Filter 识别常数 */
private String SMS_SEND_ACTIOIN = "SMS_SEND_ACTIOIN";
private String SMS_DELIVERED_ACTION = "SMS_DELIVERED_ACTION";

@Override
public void onCreate(Bundle savedInstanceState)
{
  super.onCreate(savedInstanceState);
  setContentView(R.layout.main);

  mTextView01 = (TextView)findViewById(R.id.myTextView1);

  /* 电话号码 */
  mEditText1 = (EditText) findViewById(R.id.myEditText1);
```

```
/* 短信内容 */
mEditText2 = (EditText) findViewById(R.id.myEditText2);
mButton1 = (Button) findViewById(R.id.myButton1);

//mEditText1.setText("+12345678");
/* 设置默认值为 5556 表示第二个模拟器的 Port */
mEditText1.setText("5556");
mEditText2.setText("Hello DavidLanz!");
```

3）定义 setOnClickListener 作为发送 SMS 短信按钮事件处理程序，strDestAddress 对象是欲发送的电话号码，strMessage 对象是要发送的内容。具体代码如下。

```
/* 发送 SMS 短信按钮事件处理 */
mButton1.setOnClickListener(new Button.OnClickListener()
{
  @Override
  public void onClick(View arg0)
  {
    // TODO Auto-generated method stub

    /* 欲发送的电话号码 */
    String strDestAddress = mEditText1.getText().toString();

    /* 欲发送的短信内容 */
    String strMessage = mEditText2.getText().toString();
```

4）创建 SmsManager 对象 smsManager，用于发送短信。具体流程如下。

- 第一步：创建自定义 Action 常数的 Intent。
- 第二步：用 sentIntent 参数为传送后接受的广播信息 PendingIntent。
- 第三步：用 deliveryIntent 参数为送达后接受的广播信息 PendingIntent。
- 第四步：发送 SMS 短信。
- 第五步：有异常则用 mTextView01.setText(e.toString())输出异常。

具体代码如下。

```
/* 创建 SmsManager 对象 */
SmsManager smsManager = SmsManager.getDefault();

try
{
  /* 创建自定义 Action 常数的 Intent（给 PendingIntent 参数之用）*/
  Intent itSend = new Intent(SMS_SEND_ACTIOIN);
  Intent itDeliver = new Intent(SMS_DELIVERED_ACTION);

  /* sentIntent 参数为传送后接受的广播信息 PendingIntent */
  PendingIntent mSendPI = PendingIntent.getBroadcast
```

```
                (getApplicationContext(), 0, itSend, 0);

                /* deliveryIntent 参数为送达后接受的广播信息 PendingIntent */
                PendingIntent mDeliverPI = PendingIntent.getBroadcast
                (getApplicationContext(), 0, itDeliver, 0);

                /* 发送 SMS 短信，注意倒数的两个 PendingIntent 参数 */
                smsManager.sendTextMessage
                (strDestAddress, null, strMessage, mSendPI, mDeliverPI);

                mTextView01.setText(R.string.str_sms_sending);
              }
              catch(Exception e)
              {
                mTextView01.setText(e.toString());
                e.printStackTrace();
              }
            }
          });
        }
```

5）自定义 mServiceReceiver 对象，用于覆盖 BroadcastReceiver 监听短信状态信息。如果发送短信成功则输出“成功发送”提示，如果发送短信失败则输出“发送失败”提示。具体代码如下。

```
/* 自定义 mServiceReceiver 覆盖 BroadcastReceiver 聆听短信状态信息 */
public class mServiceReceiver extends BroadcastReceiver
{
  @Override
  public void onReceive(Context context, Intent intent)
  {

    try
    {
      /* android.content.BroadcastReceiver.getResultCode()方法 */
      switch(getResultCode())
      {
        case Activity.RESULT_OK:
          /* 发送短信成功 */
          mTextView01.setText(R.string.str_sms_sent_success);
          mMakeTextToast
          (
            getResources().getText
            (R.string.str_sms_sent_success).toString(),
            true
          );
```

```
            break;
        case SmsManager.RESULT_ERROR_GENERIC_FAILURE:
            /* 发送短信失败 */
            mTextView01.setText(R.string.str_sms_sent_failed);
            mMakeTextToast
            (
                getResources().getText

                (R.string.str_sms_sent_failed).toString(),
                true
            );
            break;
        case SmsManager.RESULT_ERROR_RADIO_OFF:
            break;
        case SmsManager.RESULT_ERROR_NULL_PDU:
            break;
        }
    }
    catch(Exception e)
    {
        mTextView01.setText(e.toString());
        e.getStackTrace();
    }
  }
}
```

6）定义方法 mMakeTextToast(String str, boolean isLong)，用于输出发送成功还是失败的提示。具体代码如下。

```
public void mMakeTextToast(String str, boolean isLong)
{
  if(isLong==true)
  {
    Toast.makeText(example12.this, str, Toast.LENGTH_LONG).show();
  }
  else
  {
    Toast.makeText(example12.this, str, Toast.LENGTH_SHORT).show();
  }
}
```

7）定义 onResume()方法，此时 Activity 会被重建。具体代码如下。

```
@Override
protected void onResume()
{
  /* 自定义 IntentFilter 为 SENT_SMS_ACTIOIN Receiver */
```

```
    IntentFilter mFilter01;
    mFilter01 = new IntentFilter(SMS_SEND_ACTIOIN);
    mReceiver01 = new mServiceReceiver();
    registerReceiver(mReceiver01, mFilter01);

    /*  自定义 IntentFilter 为 DELIVERED_SMS_ACTION Receiver */
    mFilter01 = new IntentFilter(SMS_DELIVERED_ACTION);
    mReceiver02 = new mServiceReceiver();
    registerReceiver(mReceiver02, mFilter01);

    super.onResume();
  }
```

8）定义 onPause()方法，此时 Activity 会被暂停。具体代码如下。

```
    @Override
    protected void onPause()
    {

      /*  取消注册自定义 Receiver */
      unregisterReceiver(mReceiver01);
      unregisterReceiver(mReceiver02);

      super.onPause();
    }
  }
```

发送短信后能够显示短信是否成功的提示，如图 6-30 所示。

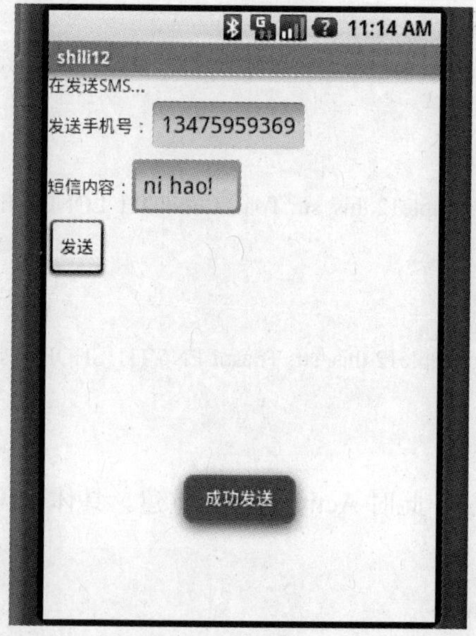

图 6-30　短信提示

6.13　设计开机显示程序

在移动手机设备中，开机后一般都会显示一个特定的开机界面。在本节的内容中，将通过一个具体实例的实现，介绍设计开机显示程序的具体过程。本实例的源代码保存在“光盘:\daima\6\example13”，下面开始讲解本实例的具体实现流程。

6.13.1　实现原理

前面讲解的 Activity、Service 和 Broadcast Receiver，都是在开机之后运行的。实际上此时开机事件会发送出一个 Android.intent.action.BOOT_COMPLETED 广播信息。只要接收到此 Action，就能在 Receiver 中打开自己的程序。

在本实例中，包含了一个主程序 Activity 和一个 Broadcast Receiver 类，只要此程序运行一次，以后只要一开机就会运行这个程序。

6.13.2　具体实现

本实例的主程序文件是 example13.java 和 StartupIntentReceiver.java，下面开始讲解其具体实现流程。

1．文件 example13.java

文件 example13.java 的功能是定义主程序 Activity，本程序只要运行一次，在以后开机时，就会自动运行，并以欢迎的 TextView 文字作为提示文本。具体代码如下。

```
package irdc.example13;

import irdc.example13.R;
import android.app.Activity;
import android.os.Bundle;
import android.widget.TextView;

public class example13 extends Activity
{
  /* 本程序只要运行一次，在以后开机时，就会自动运行 */
  private TextView mTextView01;

  /** Called when the activity is first created. */
  @Override
  public void onCreate(Bundle savedInstanceState)
  {
    super.onCreate(savedInstanceState);
    setContentView(R.layout.main);

    /* 为了快速提示，程序仅以欢迎的 TextView 文字作为展示 */
    mTextView01 = (TextView)findViewById(R.id.myTextView1);
    mTextView01.setText(R.string.str_welcome);
```

```
    }
}
```

2．文件 StartupIntentReceiver.java

在文件 StartupIntentReceiver.java 中，添加了一个继承 BroadcastReceiver 类的 HippoStartup-IntentReceiver 类，在其内部覆盖了 onReceive()方法，此方法会接收来自系统的广播。此 onReceive()方法的唯一任务是把自己唤醒，所以在传入 Intent 的参数中的第二个参数是指定 Activity 的类，最后以 start-Activity()方法打开并运行程序。具体代码如下。

```
package irdc.example13;

import android.content.BroadcastReceiver;
import android.content.Context;
import android.content.Intent;

/* 捕捉 android.intent.action.BOOT_COMPLETED 的 Receiver 类 */
public class StartupIntentReceiver extends BroadcastReceiver
{
  @Override
  public void onReceive(Context context, Intent intent)
  {

    /* 当收到 Receiver 时，指定打开此程序（example13.class） */
    Intent mBootIntent = new Intent(context, example13.class);

    /* 设置 Intent 打开为 FLAG_ACTIVITY_NEW_TASK */
    mBootIntent.setFlags(Intent.FLAG_ACTIVITY_NEW_TASK);

    /* 将 Intent 以 startActivity 传送给操作系统 */
    context.startActivity(mBootIntent);
  }
}
```

执行后将会显示预先设置的开机欢迎语，如图 6-31 所示。

图 6-31　执行效果

第 7 章　娱乐和多媒体编程

在移动手机应用中，娱乐和多媒体是一个重要的构成模块。主要包含了屏保、图片、MP3 播放、影片播放和拍照等应用。在本节的内容中，将通过几个典型实例的实现，来详细介绍 Android 中娱乐和多媒体编程的基本知识。

7.1　获取图片的宽高

在当前手机应用中，通常需要获取一副图片的宽和高。在本节的内容中，将通过一个具体实例的实现，介绍获取指定图片宽和高的基本流程。本实例的源代码保存在“光盘:\daima\7\example1”，下面开始讲解本实例的具体实现流程。

7.1.1　实现原理

在本实例中，通过一个 ListView 对象并设置了 setOnCreateContextMenuListener 方法，当用户单击一个选项后，能够弹出多个 ContextView 菜单，并以 onContextItemSelected 来判断用户是单击了哪个选项，最后分别获取图片的宽和高。

在具体实现上，通过 Bitmap 对象的 BitmapFactory.decodeResource()方法来获取预先设定的图片“123.png”，然后再通过 Bitmap 对象的 getHeight()方法和 getWidth()方法来获取图片的宽和高。

7.1.2　具体实现

本实例的主程序文件是 example1.java，下面开始讲解其具体实现代码。

1）先声明一个 TextView 变量 mTextView01 和一个 ImageView 变量 mImageView01，然后分别声明 Context 菜单的选项常数 CONTEXT_ITEM1、CONTEXT_ITEM2 和 CONTEXT_ITEM3。具体代码如下。

```
package irdc.example1;


import irdc.example1.R;
import android.app.Activity;
import android.graphics.Bitmap;
import android.graphics.BitmapFactory;
import android.os.Bundle;
import android.view.ContextMenu;
import android.view.Menu;
import android.view.MenuItem;
```

```
import android.view.View;
import android.view.ContextMenu.ContextMenuInfo;
import android.widget.ImageView;
import android.widget.ListView;
import android.widget.TextView;
public class example1 extends Activity
{
  /*声明一个 TextView 变量与一个 ImageView 变量*/
  private TextView mTextView01;
  private ImageView mImageView01;
  /*声明 Context 菜单的选项常数*/
  protected static final int CONTEXT_ITEM1 = Menu.FIRST;
  protected static final int CONTEXT_ITEM2 = Menu.FIRST+1;
  protected static final int CONTEXT_ITEM3 = Menu.FIRST+2;
```

2）先通过 findViewById 构造器来创建 TextView 和 ImageView 对象，然后将 Drawable 中的图片“baby.png”放入自定义的 ImageView 中。具体代码如下。

```
@Override
public void onCreate(Bundle savedInstanceState)
{
  super.onCreate(savedInstanceState);
  setContentView(R.layout.main);

  /*通过 findViewById 构造器创建 TextView 与 ImageView 对象*/
  mTextView01 = (TextView)findViewById(R.id.myTextView1);
  mImageView01= (ImageView)findViewById(R.id.myImageView1);
  /*将 Drawable 中的图片 baby.png 放入自定义的 ImageView 中*/
  mImageView01.setImageDrawable(getResources().
              getDrawable(R.drawable.baby));
```

3）先设置 OnCreateContextMenuListener 给 TextView，这样在图片上可以使用 ContextMenu，然后覆盖 OnCreateContextMenu 方法来创建 ContextMenu 类的选项。具体代码如下。

```
/*设置 OnCreateContextMenuListener 给 TextView
 * 在图片上可以使用 ContextMenu*/
mImageView01.setOnCreateContextMenuListener
(new ListView.OnCreateContextMenuListener()
{
  @Override
  /*覆盖 OnCreateContextMenu 方法来创建 ContextMenu 类的选项*/
  public void onCreateContextMenu
  (ContextMenu menu, View v, ContextMenuInfo menuInfo)
  {
    menu.add(Menu.NONE, CONTEXT_ITEM1, 0, R.string.str_context1);
    menu.add(Menu.NONE, CONTEXT_ITEM2, 0, R.string.str_context2);
```

```
            menu.add(Menu.NONE, CONTEXT_ITEM3, 0, R.string.str_context3);
        }
    });
}
```

4）覆盖 OnContextItemSelected 方法来定义用户单击菜单后的动作，然后通过自定义 Bitmap 对象 BitmapFactory.decodeResource 来取得预设的图片资源。具体代码如下。

```
@Override
/*覆盖 OnContextItemSelected 方法来定义用户单击菜单后的动作*/
public boolean onContextItemSelected(MenuItem item)
{
  /*自定义 Bitmap 对象并通过 BitmapFactory.decodeResource 取得
   *预先 Import 至 Drawable 的 baby.png 图像*/
  Bitmap myBmp = BitmapFactory.decodeResource
    (getResources(), R.drawable.baby);
  /*通过 Bitmap 对象的 getHight 与 getWidth 来取得图片宽高*/
  int intHeight = myBmp.getHeight();
  int intWidth = myBmp.getWidth();
```

5）根据用户选择的选项，分别通过 getHeight()方法和 getWidth()方法获取对应图片宽度和高度。具体代码如下。

```
try
{
  /*菜单选项与动作*/
  switch(item.getItemId())
  {
    /*将图片宽度显示在 TextView 中*/
    case CONTEXT_ITEM1:
      String strOpt =
      getResources().getString(R.string.str_width)
      +"="+Integer.toString(intWidth);
      mTextView01.setText(strOpt);
      break;
    /*将图片高度显示在 TextView 中*/
    case CONTEXT_ITEM2:
      String strOpt2 =
      getResources().getString(R.string.str_height)
      +"="+Integer.toString(intHeight);
      mTextView01.setText(strOpt2);
      break;
    /*将图片宽高显示在 TextView 中*/
    case CONTEXT_ITEM3:
      String strOpt3 =
      getResources().getString(R.string.str_width)
```

```
                +"="+Integer.toString(intWidth)+"\n"
                +getResources().getString(R.string.str_height)
                +"="+Integer.toString(intHeight);
                mTextView01.setText(strOpt3);
                break;
        }
    }
    catch(Exception e)
    {
        e.printStackTrace();
    }
    return super.onContextItemSelected(item);
  }
}
```

执行后的初始效果如图 7-1 所示，当长时间选中图片后会弹出用户选项，如图 7-2 所示。当选择一个选项后，会弹出对应的获取数值。

图 7-1　初始效果

获取宽度

获取高度

获取宽和高

图 7-2　弹出选项

7.2　几何图形绘制

当前手机应用中，很多游戏程序中经常需要绘制几何图形。在本节的内容中，将通过一个具体实例的实现，介绍在 Android 中绘制几何图形的实现流程。本实例的源代码保存在“光盘:\daima\7\example2”，下面开始讲解本实例的具体实现流程。

7.2.1　实现原理

在本实例中，通过 Android.graphics 的类来绘制 2D 向量图。在其包中提供了很多在手机

上绘制图形的类和方法。例如，Canvas 类相当于一个图纸，所有的图形都在它上面绘制并显示出来。而 Paint 类则像铅笔，可以设置为不同的颜色，从而绘制不同颜色的图形。

7.2.2　具体实现

本实例的主程序文件是 example2.java，下面开始讲解其具体实现代码。

1）先加载相关类，然后设置 ContentView 为自定义的 MyView。具体代码如下。

```
public class example2 extends Activity
{
  @Override
  public void onCreate(Bundle savedInstanceState)
  {
    super.onCreate(savedInstanceState);
    /* 设置 ContentView 为自定义的 MyView */
    MyView myView = new MyView(this);
    setContentView(myView);
  }
```

2）自定义继承 View 类的 MyView 对象，具体代码如下。

```
/* 自定义继承 View 的 MyView */
private class MyView extends View
{
  public MyView(Context context)
  {
    super(context);
  }
```

3）分别设置背景和消除锯齿，然后分别绘制空心圆形、空心正方形、空心长方形、空心椭圆形、空心三角形和空心梯形。具体代码如下。

```
/* 覆盖 onDraw()方法 */
@Override
protected void onDraw(Canvas canvas)
{
  super.onDraw(canvas);
  /* 设置背景为白色 */
  canvas.drawColor(Color.WHITE);

  Paint paint = new Paint();
  /* 去锯齿 */
  paint.setAntiAlias(true);
  /* 设置 paint 对象的颜色 */
  paint.setColor(Color.RED);
  /* 设置 paint 对象的 style 为 STROKE：空心的 */
  paint.setStyle(Paint.Style.STROKE);
```

```
/* 设置 paint 对象的外框宽度 */
paint.setStrokeWidth(3);

/* 画一个空心圆形 */
canvas.drawCircle(40,40,30, paint);
/* 画一个空心正方形 */
canvas.drawRect(10,90,70,150,paint);
/* 画一个空心长方形 */
canvas.drawRect(10,170,70,200,paint);
/* 画一个空心椭圆形 */
RectF re=new RectF(10,220,70,250);
canvas.drawOval(re, paint);
/* 画一个空心三角形 */
Path path = new Path();
path.moveTo(10,330);
path.lineTo(70,330);
path.lineTo(40,270);
path.close();
canvas.drawPath(path, paint);
/* 画一个空心梯形 */
Path path1 = new Path();
path1.moveTo(10,410);
path1.lineTo(70,410);
path1.lineTo(55,350);
path1.lineTo(25,350);
path1.close();
canvas.drawPath(path1, paint);
```

4）设置实心样式和颜色，然后分别绘制实心圆形、实心正方形、实心长方形、实心椭圆形、实心三角形和实心梯形。具体代码如下。

```
/* 设置 paint 对象的 style 为 FILL：实心 */
paint.setStyle(Paint.Style.FILL);
/* 设置 paint 的颜色 */
paint.setColor(Color.BLUE);

/* 画一个实心圆 */
canvas.drawCircle(120, 40, 30, paint);
/* 画一个实心正方形 */
canvas.drawRect(90,90,150,150,paint);
/* 画一个实心长方形 */
canvas.drawRect(90,170,150,200,paint);
/* 画一个实心椭圆形 */
RectF re2=new RectF(90,220,150,250);
canvas.drawOval(re2, paint);
/* 画一个实心三角形 */
Path path2 = new Path();
```

```
path2.moveTo(90,330);
path2.lineTo(150,330);
path2.lineTo(120,270);
path2.close();
canvas.drawPath(path2, paint);
/* 画一个实心梯形 */
Path path3 = new Path();
path3.moveTo(90,410);
path3.lineTo(150,410);
path3.lineTo(135,350);
path3.lineTo(105,350);
path3.close();
canvas.drawPath(path3, paint);
```

5）设置渐变样式和颜色，然后分别绘制渐变圆形、渐变正方形、渐变长方形、渐变椭圆形、渐变三角形和渐变梯形。具体代码如下。

```
/* 设置渐变色 */
Shader mShader=new LinearGradient(0, 0,100,100,
    new int[]{Color.RED, Color.GREEN,Color.BLUE,Color.YELLOW},
    null, Shader.TileMode.REPEAT);
paint.setShader(mShader);

/* 画一个渐变色的圆形 */
canvas.drawCircle(200,40, 30, paint);
/* 画一个渐变色的正方形 */
canvas.drawRect(170,90,230,150,paint);
/* 画一个渐变色的长方形 */
canvas.drawRect(170,170,230,200,paint);
/* 画一个渐变色的椭圆形 */
RectF re3=new RectF(170,220,230,250);
canvas.drawOval(re3, paint);
/* 画一个渐变色的三角形 */
Path path4 = new Path();
path4.moveTo(170,330);
path4.lineTo(230,330);
path4.lineTo(200,270);
path4.close();
canvas.drawPath(path4, paint);
/* 画一个渐变色的梯形 */
Path path5 = new Path();
path5.moveTo(170,410);
path5.lineTo(230,410);
path5.lineTo(215,350);
path5.lineTo(185,350);
path5.close();
canvas.drawPath(path5, paint);
```

6）通过 canvas.drawText()方法实现写字功能，具体代码如下。

```
        /* 写字 */
        paint.setTextSize(24);
        canvas.drawText(getResources().getString(R.string.str_text1),
                        240,50,paint);
        canvas.drawText(getResources().getString(R.string.str_text2),
                        240,120,paint);
        canvas.drawText(getResources().getString(R.string.str_text3),
                        240,190,paint);
        canvas.drawText(getResources().getString(R.string.str_text4),
                        240,250,paint);
        canvas.drawText(getResources().getString(R.string.str_text5),
                        240,320,paint);
        canvas.drawText(getResources().getString(R.string.str_text6),
                        240,390,paint);

      }
    }
}
```

执行后将会在屏幕内显示不同的图形，并在后面显示对应的文字描述。效果如图 7-3 所示。

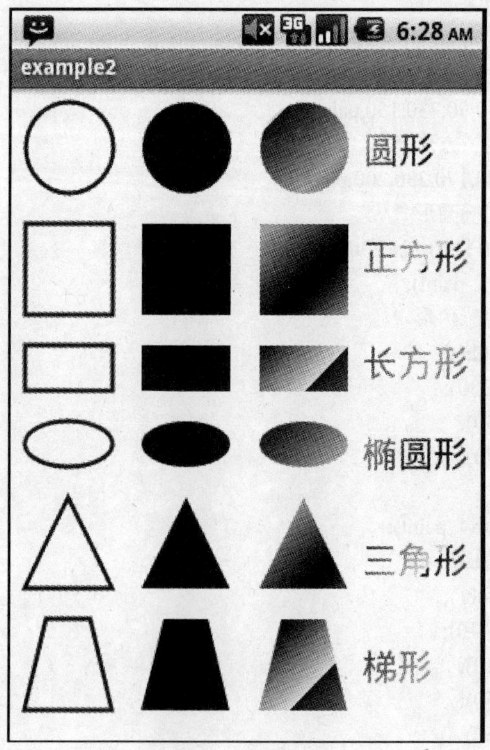

图 7-3　执行效果

7.3　手机屏幕保护程序

在手机应用中，屏幕保护程序是一个十分重要的功能。在本节的内容中，将通过一个具体实例的实现，介绍在 Android 中实现屏保功能的基本流程。本实例的源代码保存在“光盘:\daima\7\example3”，下面开始讲解本实例的具体实现流程。

7.3.1　实现原理

本实例的主要难点如下。

- 控制和判断用户静止未触动手机键盘或屏幕的时间及其事件。
- 动态全屏幕淡入、淡出和图片交换效果。

上述难点都是通过线程实现的，它以时间戳记录的方式，判断距离上一次单击键盘或屏幕的时间，并计算两次的时间间隔。当超过了设置了时间后，会进行屏保，本实例设置的时间间隔为 5 秒。

7.3.2　具体实现

本实例的主程序文件是 example3.java，下面开始讲解其具体实现代码。

1）先加载相关类，然后设置 LayoutInflater 对象供新建的 AlertDialog 使用。具体代码如下。

```
public class example3 extends Activity
{
  private TextView mTextView01;
  private ImageView mImageView01;

  /* LayoutInflater 对象作为新建 AlertDialog 之用 */
  private LayoutInflater mInflater01;
```

2）定义 mView01 对象，用于输入解锁的 View。通过 Menu 选项 identifier，用以识别对应的事件。具体代码如下。

```
/* 输入解锁的 View */
private View mView01;
private EditText mEditText01,mEditText02;

/* Menu 选项 identifier，用以识别事件 */
static final private int MENU_ABOUT = Menu.FIRST;
static final private int MENU_EXIT = Menu.FIRST+1;
private Handler mHandler01 = new Handler();
private Handler mHandler02 = new Handler();
private Handler mHandler03 = new Handler();
private Handler mHandler04 = new Handler();
```

3）分别定义控制 User 静止与否的 Counter，控制 Fade In 与 Fade Out 的 Counter，控制循序替换背景图 ID 的 Counter。具体代码如下。

```
/* 控制 User 静止与否的 Counter */
private int intCounter1, intCounter2;
/* 控制 Fade#In 与 Fade Out 的 Counter */
private int intCounter3, intCounter4;
/* 控制循序替换背景图 ID 的 Counter    */
private int intDrawable=0;
```

4）设置 timePeriod。设置静止超过 ns 将自动进入屏幕保护。具体代码如下。

```
/* 上一次用户有动作的时间戳 */
private Date lastUpdateTime;
/* 计算用户多少秒没有动作 */
private long timePeriod;
/* 静止超过 ns 将自动进入屏幕保护 */
private float fHoldStillSecond = (float) 5;
private boolean bIfRunScreenSaver;
private boolean bFadeFlagOut, bFadeFlagIn = false;
private long intervalScreenSaver = 1000;
private long intervalKeypadeSaver = 1000;
private long intervalFade = 100;
private int screenWidth, screenHeight;
```

5）设置每隔 5s 置换图片，用 Screen 保存需要用到的背景图。具体代码如下。

```
/* 每 n 秒置换图片 */
private int intSecondsToChange = 5;

/* 设置 Screen 保存需要用到的背景图 */
private static int[] screenDrawable = new int[]
{
  R.drawable.screen1,
  R.drawable.screen2,
  R.drawable.screen3,
  R.drawable.screen4,
  R.drawable.screen5
};
```

6）设置在 setContentView 之前调用全屏幕显示，通过 lastUpdateTime 方法取得用户触碰手机的时间，并用 recoverOriginalLayout()方法来初始化 Layout 上的 Widget 的可见性。具体代码如下。

```
@Override
public void onCreate(Bundle savedInstanceState)
{
  super.onCreate(savedInstanceState);

  /* 必须在 setContentView 之前调用全屏幕显示 */
```

```
    requestWindowFeature(Window.FEATURE_NO_TITLE);
    getWindow().setFlags
    (
      WindowManager.LayoutParams.FLAG_FULLSCREEN,
      WindowManager.LayoutParams.FLAG_FULLSCREEN
    );
    setContentView(R.layout.main);

    mTextView01 = (TextView)findViewById(R.id.myTextView1);
    mImageView01 = (ImageView)findViewById(R.id.myImageView1);
    mEditText01 = (EditText)findViewById(R.id.myEditText1);

    /* 取得用户触碰手机的时间 */
    lastUpdateTime = new Date(System.currentTimeMillis());

    /* 初始化 Layout 上的 Widget 可见性 */
    recoverOriginalLayout();
  }
```

7）设置菜单群组 ID，然后通过 menu.add 创建具有子项的菜单，最后创建退出菜单。具体代码如下。

```
  @Override
  public boolean onCreateOptionsMenu(Menu menu)
  {

    /* menu 群组 ID */
    int idGroup1 = 0;

    /* The order position of the item */
    int orderMenuItem1 = Menu.NONE;
    int orderMenuItem2 = Menu.NONE+1;

    /* 创建具有子项的菜单 */
    menu.add
    (
      idGroup1, MENU_ABOUT, orderMenuItem1, R.string.app_about
    );
    /* 创建退出菜单 */

    menu.add(idGroup1, MENU_EXIT, orderMenuItem2, R.string.str_exit);
    menu.setGroupCheckable(idGroup1, true, true);

    return super.onCreateOptionsMenu(menu);
  }
```

8）根据用户选择的菜单项，显示对应的 AlertDialog 提示框。具体代码如下。

```
@Override
public boolean onOptionsItemSelected(MenuItem item)
{
  switch(item.getItemId())
  {
    case (MENU_ABOUT):
      new AlertDialog.Builder
      (
        example3.this
      ).setTitle(R.string.app_about).setIcon
      (
        R.drawable.hippo
      ).setMessage
      (
        R.string.app_about_msg
      ).setPositiveButton(R.string.str_ok,
      new DialogInterface.OnClickListener()
      {
        public void onClick
        (DialogInterface dialoginterface, int i)
        {
        }
      }).show();
      break;
    case (MENU_EXIT):
      /* 离开程序 */
      finish();
      break;
  }
  return super.onOptionsItemSelected(item);
}
```

9）用 mTasks01 监控用户是否有动作的运行线程，通过 timePeriod 计算用户静止不动作的时间间距，如果静止不动，超过设置的 5s，则运行对应的线程。具体代码如下。

```
/* 监控 User 没有动作的运行线程 */
private Runnable mTasks01 = new Runnable()
{
  public void run()
  {
    intCounter1++;
    Date timeNow = new Date(System.currentTimeMillis());

    /* 计算用户静止不动作的时间间距 */
    timePeriod =
    (long)timeNow.getTime() - (long)lastUpdateTime.getTime();
```

```
float timePeriodSecond = ((float)timePeriod/1000);

/* 如果超过时间静止不动 */
if(timePeriodSecond>fHoldStillSecond)
{
  /* 静止超过时间第一次的标记 */
  if(bIfRunScreenSaver==false)
  {
    /* 启动运行线程 2 */
    mHandler02.postDelayed(mTasks02, intervalScreenSaver);

    /* Fade Out*/
    if(intCounter1%(intSecondsToChange)==0)
    {
      bFadeFlagOut=true;
      mHandler03.postDelayed(mTasks03, intervalFade);
    }
    else
    {
      /* 在 Fade Out 后立即 Fade In */
      if(bFadeFlagOut==true)
      {
        bFadeFlagIn=true;
        mHandler04.postDelayed(mTasks04, intervalFade);
      }
      else
      {
        bFadeFlagIn=false;
        intCounter4 = 0;
        mHandler04.removeCallbacks(mTasks04);
      }
      intCounter3 = 0;
      bFadeFlagOut = false;
    }
    bIfRunScreenSaver = true;
  }
  else
  {

    if(intCounter1%(intSecondsToChange)==0)
    {
      bFadeFlagOut=true;
      mHandler03.postDelayed(mTasks03, intervalFade);
    }
    else
    {
```

```
                /* 在 Fade Out（淡出）后立即 Fade In（淡入）*/
                if(bFadeFlagOut==true)
                {
                  bFadeFlagIn=true;
                  mHandler04.postDelayed(mTasks04, intervalFade);
                }
                else
                {
                  bFadeFlagIn=false;
                  intCounter4 = 0;
                  mHandler04.removeCallbacks(mTasks04);
                }
                intCounter3 = 0;
                bFadeFlagOut=false;
              }
            }
          }
          else
          {
            /* 当用户没有动作的间距未超过时间 */
            bIfRunScreenSaver = false;
            /* 恢复原来的 Layout Visible*/
            recoverOriginalLayout();
          }

          /* 以 LogCat 监看 User 静止不动的时间间距 */
          Log.i
          (
            "HIPPO",
            "Counter1:"+Integer.toString(intCounter1)+
            "/"+
            Float.toString(timePeriodSecond));

          /* 反复运行运行线程 1 */
          mHandler01.postDelayed(mTasks01, intervalKeypadeSaver);
        }
      };
```

10）定义 mTasks02，设置每 1 秒运行一次屏保程序，并隐藏原有 Layout 上面的 Widget，调用 ScreenSaver()方法加载图片，即轮换显示预设的 5 幅图片。具体代码如下。

```
      /* Screen Saver Runnable */
      private Runnable mTasks02 = new Runnable()
      {
        public void run()
        {
          if(bIfRunScreenSaver==true)
```

```
        {
          intCounter2++;

          hideOriginalLayout();
          showScreenSaver();

          Log.i("HIPPO", "Counter2:"+Integer.toString(intCounter2));
          mHandler02.postDelayed(mTasks02, intervalScreenSaver);
        }
        else
        {
          mHandler02.removeCallbacks(mTasks02);
        }
      }
    };
```

11）定义 mTasks03，通过 setAlpha 方法设置 ImageView（图像组件）的透明度渐暗下去。具体代码如下。

```
/* Fade Out 特效 Runnable */
private Runnable mTasks03 = new Runnable()
{
  public void run()
  {
    if(bIfRunScreenSaver==true && bFadeFlagOut==true)
    {
      intCounter3++;

      /* 设置 ImageView 的透明度渐暗下去 */
      mImageView01.setAlpha(255-intCounter3*28);

      Log.i("HIPPO", "Fade out:"+Integer.toString(intCounter3));
      mHandler03.postDelayed(mTasks03, intervalFade);
    }
    else
    {
      mHandler03.removeCallbacks(mTasks03);
    }
  }
};
```

12）定义 mTasks03，通过 setAlpha 方法设置 ImageView 的透明度渐亮起来。具体代码如下。

```
/* Fade In 特效 Runnable */
private Runnable mTasks04 = new Runnable()
{
  public void run()
```

```
    {
      if(bIfRunScreenSaver==true && bFadeFlagIn==true)
      {
        intCounter4++;

        /* 设置 ImageView 的透明度渐亮起来 */
        mImageView01.setAlpha(intCounter4*28);

        mHandler04.postDelayed(mTasks04, intervalFade);
        Log.i("HIPPO", "Fade In:"+Integer.toString(intCounter4));
      }
      else
      {
        mHandler04.removeCallbacks(mTasks04);
      }
    }
  };
```

13）先定义 recoverOriginalLayout()方法，用于恢复原有的 Layout 可视性；然后定义 hideOriginalLayout()方法，用于隐藏原有应用程序里的布局配置组件。具体代码如下。

```
/* 恢复原有的 Layout 可视性 */
private void recoverOriginalLayout()
{
  mTextView01.setVisibility(View.VISIBLE);
  mEditText01.setVisibility(View.VISIBLE);
  mImageView01.setVisibility(View.GONE);
}

/* 隐藏原有应用程序里的布局配置组件 */
private void hideOriginalLayout()
{
  /* 欲隐藏的 Widget */
  mTextView01.setVisibility(View.INVISIBLE);
  mEditText01.setVisibility(View.INVISIBLE);
}

/* 开始 Screen 保存 */
private void showScreenSaver()
{

  if(intDrawable>4)
  {
    intDrawable = 0;
  }

  DisplayMetrics dm=new DisplayMetrics();
```

```
    getWindowManager().getDefaultDisplay().getMetrics(dm);
    screenWidth = dm.widthPixels;
    screenHeight = dm.heightPixels;
    Bitmap                    bmp=BitmapFactory.decodeResource(getResources(),
screenDrawable[intDrawable]);
```

14）通过 Matrix 设置比例，使用 Matrix.postScale()方法设置新维度 ReSize，通过 resizedBitmap 对象设置图文件至屏幕分辨率，新建 Drawable 对象 myNewBitmapDrawable 用于放大图文件至全屏幕，通过 setVisibility(View.VISIBLE)使 ImageView 可见。具体代码如下。

```
    /* Matrix 比例 */
    float scaleWidth = ((float) screenWidth) / bmp.getWidth();
    float scaleHeight = ((float) screenHeight) / bmp.getHeight() ;

    Matrix matrix = new Matrix();
    /* 使用 Matrix.postScale 设置维度 ReSize */
    matrix.postScale(scaleWidth, scaleHeight);

    /* ReSize 图文件至屏幕分辨率 */
    Bitmap resizedBitmap = Bitmap.createBitmap
    (
      bmp,0,0,bmp.getWidth(),bmp.getHeight(),matrix,true
    );

    /* 新建 Drawable 放大图文件至全屏幕 */
    BitmapDrawable myNewBitmapDrawable =
        new BitmapDrawable(resizedBitmap);
    mImageView01.setImageDrawable(myNewBitmapDrawable);

    /* 使 ImageView 可见 */
    mImageView01.setVisibility(View.VISIBLE);

    /* 每间隔设置秒数置换图片 ID，于下一个 runnable2 才会生效 */
    if(intCounter2%intSecondsToChange==0)
    {
      intDrawable++;
    }
  }
```

15）定义方法 onUserWakeUpEvent()，实现解锁和加密处理。具体代码如下。

```
    public void onUserWakeUpEvent()
    {
      if(bIfRunScreenSaver==true)
      {
        try
        {
```

```
            /* LayoutInflater.from 取得此 Activity 的 context */
            mInflater01 = LayoutInflater.from(example3.this);

            /* 创建解锁密码使用 View 的 Layout */
            mView01 = mInflater01.inflate(R.layout.securescreen, null);

            /* 对话框中唯一的 EditText 等待输入解锁密码 */
            mEditText02 =
            (EditText) mView01.findViewById(R.id.myEditText2);

            /* 创建 AlertDialog */
            new AlertDialog.Builder(this)
            .setView(mView01)
            .setPositiveButton("OK",
            new DialogInterface.OnClickListener()
            {
              public void onClick(DialogInterface dialog, int whichButton)
              {
                /* 比较输入的密码与原 Activity 里的设置是否相符 */
                if(mEditText01.getText().toString().equals
                  (mEditText02.getText().toString()))
                {
                  /* 当密码正确才真的解锁屏幕保护装置 */
                  resetScreenSaverListener();
                }
              }
            }).show();
        }
        catch(Exception e)
        {
          e.printStackTrace();
        }
      }
    }
```

16）定义 updateUserActionTime()，用于统计用户单击键盘或屏幕的时间间隔。具体流程如下。

- 第一步：取得单击按键事件时的系统 Time Millis。
- 第二步：重新计算单击按键距离上一次静止的时间间距。

具体代码如下。

```
public void updateUserActionTime()
{
  /* 取得单击按键事件时的系统 Time Millis（时间波动） */
  Date timeNow = new Date(System.currentTimeMillis());
```

```
    /* 重新计算单击按键距离上一次静止的时间间距 */
    timePeriod =
    (long)timeNow.getTime() - (long)lastUpdateTime.getTime();
    lastUpdateTime.setTime(timeNow.getTime());
  }
```

17）定义方法 resetScreenSaverListener()，用于重新设置屏幕。流程如下。

- 第一步：删除现有的 Runnable。
- 第二步：取得单击按键事件时的系统 Time Millis。
- 第三步：重新计算单击按键距离上一次静止的时间间距。
- 第四步：通过 bIfRunScreenSaver 取消屏保。
- 第五步：恢复原来的 Layout Visible。

具体代码如下。

```
public void resetScreenSaverListener()
{
  /* 删除现有的 Runnable */
  mHandler01.removeCallbacks(mTasks01);
  mHandler02.removeCallbacks(mTasks02);

  /* 取得单击按键事件时的系统 Time Millis */
  Date timeNow = new Date(System.currentTimeMillis());
  /* 重新计算单击按键距离上一次静止的时间间距 */
  timePeriod =
  (long)timeNow.getTime() - (long)lastUpdateTime.getTime();
  lastUpdateTime.setTime(timeNow.getTime());

  /* 取消屏幕保护 */
  bIfRunScreenSaver = false;

  /* 重置 Runnable1 与 Runnable1 的 Counter */
  intCounter1 = 0;
  intCounter2 = 0;

  /* 恢复原来的 Layout，变为可见状态*/
  recoverOriginalLayout();

  /* 重新 postDelayed()新的 Runnable */
  mHandler01.postDelayed(mTasks01, intervalKeypadeSaver);
}
```

18）定义 onKeyDown(int keyCode, KeyEvent event)方法，用于监听用户的触摸单击事件。具体代码如下。

```
@Override
public boolean onKeyDown(int keyCode, KeyEvent event)
```

```
{
  if(bIfRunScreenSaver==true && keyCode!=4)
  {
    /* 当屏幕保护程序正在运行中，触动解除屏幕保护程序 */
    onUserWakeUpEvent();
  }
  else
  {
    /* 更新用户未触动手机的时间戳记 */
    updateUserActionTime();
  }
  return super.onKeyDown(keyCode, event);
}
@Override
public boolean onTouchEvent(MotionEvent event)
{
  if(bIfRunScreenSaver==true)
  {
    /* 当屏幕保护程序正在运行中，触动解除屏幕保护程序 */
    onUserWakeUpEvent();
  }
  else
  {
    /* 更新用户未触动手机的时间戳记 */
    updateUserActionTime();
  }
  return super.onTouchEvent(event);
}

@Override
protected void onResume()
{
  // TODO Auto-generated method stub
  mHandler01.postDelayed(mTasks01, intervalKeypadeSaver);
  super.onResume();
}
```

19）定义方法 onPause()，用于删除运行中的运行线程 mHandler01、mHandler02、mHandler03 和 mHandler01。具体代码如下。

```
@Override
protected void onPause()
{

  try
  {
    /* 删除运行中的运行线程 */
```

```
        mHandler01.removeCallbacks(mTasks01);
        mHandler02.removeCallbacks(mTasks02);
        mHandler03.removeCallbacks(mTasks03);
        mHandler04.removeCallbacks(mTasks04);
      }
      catch(Exception e)
      {
        e.printStackTrace();
      }
      super.onPause();
    }
  }
```

执行后如果超过 5 秒不动键盘或屏幕，则会进入屏保状态，如图 7-4 所示。可以设置屏保密码，当输入正确的密码后才能解除屏保。如图 7-5 所示。

图 7-4　执行效果

图 7-5　解锁密码

在上述代码中，声明的 4 个 Runnable 接口对象是整个程序的重点。

- mTasks01：设置每 1 秒检查一次 timePeriod，并监视是否超过 5 秒未触发。超过 5 秒则将 blRunScreenSaver 这个 flag 更改为 true，并启动 mTasks02。
- mTasks02：设置每 1 秒运行一次屏保程序，并隐藏原有 Layout 上面的 Widget，并调用 ScreenSaver()加载图片，即轮换显示预设的 5 幅图片。
- mTasks03：是 Fade Out 特效使用的 Runable，每 0.1 秒运行一个 scale（刻度）。
- mTasks04：是 Fade In 特效使用的 Runable，每 0.1 秒运行一个 scale。

7.4　点击移动照片

在触摸屏手机中，点击移动照片的功能十分常见。在本节的内容中，将通过一个具体实例的实现，介绍在 Android 中实现点击移动照片功能的基本流程。本实例的源代码保存在“光盘:\daima\7\example4”，下面开始讲解本实例的具体实现流程。

7.4.1 实现原理

在本实例中，设计了一个 ImageView 对象，并使用了 Drawable 中的照片，在程序开始运行时，将照片放在屏幕中央。通过 onTouchEvent 事件来处理单击、拖动和放开等事件来完成拖动图片的功能。并且设置了 ImageView 的 onClickListener 让用户在单击图片的同时，恢复到图片的初始位置。

7.4.2 具体实现

本实例的主程序文件是 example4.java，下面开始讲解其具体实现代码。

1）先加载相关类然后分别声明 ImageView 变量 mImageView01，声明相关变量存储图片宽高和位置，声明存储屏幕的分辨率变量。具体代码如下。

```
public class example4 extends Activity
{
  /*声明 ImageView 变量*/
  private ImageView mImageView01;
  /*声明相关变量作为存储图片宽高,位置使用*/
  private int intWidth, intHeight, intDefaultX, intDefaultY;
  private float mX, mY;
  /*声明存储屏幕的分辨率变量 */
  private int intScreenX, intScreenY;
```

2）先通过 DisplayMetrics 取得屏幕对象，然后通过 intScreenX 对象和 intScreenY 对象取得屏幕像素，最后分别设置图片的宽高。具体代码如下。

```
/** Called when the activity is first created. */
@Override
public void onCreate(Bundle savedInstanceState)
{
  super.onCreate(savedInstanceState);
  setContentView(R.layout.main);

  /* 取得屏幕对象 */
  DisplayMetrics dm = new DisplayMetrics();
  getWindowManager().getDefaultDisplay().getMetrics(dm);

  /* 取得屏幕像素 */
  intScreenX = dm.widthPixels;
  intScreenY = dm.heightPixels;

  /* 设置图片的宽高 */
  intWidth = 100;
  intHeight = 100;
```

3）通过 findViewById 构造器创建 ImageView 对象，然后将图片赋值给 ImageView 来呈现，并通过 RestoreButton()方法初始化按钮，设置位置居中。具体代码如下。

```
/*通过 findViewById 构造器创建 ImageView 对象*/
mImageView01 =(ImageView) findViewById(R.id.myImageView1);
/*将图片从 Drawable 赋值给 ImageView 来呈现*/
mImageView01.setImageResource(R.drawable.baby);

/* 初始化按钮位置居中 */
RestoreButton();
```

4）定义 setOnClickListener(new Button.OnClickListener())，当单击 ImageView 时，还原初始位置。具体代码如下。

```
  /* 单击 ImageView，还原初始位置 */
  mImageView01.setOnClickListener(new Button.OnClickListener()
  {
    @Override
    public void onClick(View v)
    {
      RestoreButton();
    }
  });
}
```

5）定义 onTouchEvent(MotionEvent event)覆盖触控事件。首先取得手指触控屏幕的位置，然后实现触控事件的处理，即实现下面的动作处理。

- 点击屏幕。
- 移动位置。
- 离开屏幕。

具体代码如下。

```
/*覆盖触控事件*/
@Override
public boolean onTouchEvent(MotionEvent event)
{
  /*取得手指触控屏幕的位置*/
  float x = event.getX();
  float y = event.getY();

  try
  {
    /*触控事件的处理*/
    switch (event.getAction())
    {
```

```
            /*点击屏幕*/
            case MotionEvent.ACTION_DOWN:
              picMove(x, y);
                break;
            /*移动位置*/
            case MotionEvent.ACTION_MOVE:
              picMove(x, y);
                break;
            /*离开屏幕*/
            case MotionEvent.ACTION_UP:
              picMove(x, y);
                break;
          }
        }catch(Exception e)
          {
            e.printStackTrace();
          }
        return true;
      }
```

6）定义 picMove(float x, float y)方法，用于实现图片移动。具体代码如下。

```
  /*移动图片的方法*/
  private void picMove(float x, float y)
  {
    /*默认微调图片与指针的相对位置*/
    mX=x-(intWidth/2);
    mY=y-(intHeight/2);

    /*防图片超过屏幕的相关处理*/
    /*防止屏幕向右超过屏幕*/
    if((mX+intWidth)>intScreenX)
    {
      mX = intScreenX-intWidth;
    }
    /*防止屏幕向左超过屏幕*/
    else if(mX<0)
    {
      mX = 0;
    }
    /*防止屏幕向下超过屏幕*/
    else if ((mY+intHeight)>intScreenY)
    {
      mY=intScreenY-intHeight;
    }
    /*防止屏幕向上超过屏幕*/
    else if (mY<0)
```

```
    {
        mY = 0;
    }
    /*通过 log（导航对象）来查看图片位置*/
    Log.i("jay", Float.toString(mX)+","+Float.toString(mY));
    /* 以 setLayoutParams 方法，重新安排 Layout 上的位置 */
    mImageView01.setLayoutParams
    (
        new AbsoluteLayout.LayoutParams
        (intWidth,intHeight,(int) mX,(int)mY)
    );
}
```

7）定义 RestoreButton()方法，用于还原 ImageView 位置的事件处理。具体代码如下。

```
/* 还原 ImageView 位置的事件处理 */
public void RestoreButton()
{
    intDefaultX = ((intScreenX-intWidth)/2);
    intDefaultY = ((intScreenY-intHeight)/2);
    /*Toast 还原位置坐标*/
    mMakeTextToast
    (
        "("+
        Integer.toString(intDefaultX)+
        ","+
        Integer.toString(intDefaultY)+")",true
    );

    /* 使用 setLayoutParams 方法，重新安排 Layout 上的位置 */
    mImageView01.setLayoutParams
    (
        new AbsoluteLayout.LayoutParams
        (intWidth,intHeight,intDefaultX,intDefaultY)
    );
}
```

8）定义 mMakeTextToast(String str, boolean isLong)方法，是用于自定义发出信息的方法。具体代码如下。

```
/*自定义发出信息的方法*/
public void mMakeTextToast(String str, boolean isLong)
{
    if(isLong==true)
    {
        Toast.makeText(example4.this, str, Toast.LENGTH_LONG).show();
```

```
            }
            else
            {
                Toast.makeText(example4.this, str, Toast.LENGTH_SHORT).show();
            }
        }
    }
```

执行后效果如图 7-6 所示，并能在屏幕中通过鼠标点击来移动指定图片的位置，如图 7-7 所示。

图 7-6　执行效果

图 7-7　移动图片

7.5　显示存储卡中的照片

在手机应用中，为了扩大手机的容量，经常会使用存储卡。在存储卡中，通常会保存一些图片或媒体文件。在本节的内容中，将通过一个具体实例的实现，介绍在 Android 中获取存储卡中图片的基本流程。本实例的源代码保存在“光盘:\daima\7\example5”，下面开始讲解本实例的具体实现流程。

7.5.1　实现原理

在本实例中，定义了获取文件列表方法 getSD()，将内存中扫描过的照片以 File List <String>的方式存储，再利用定义的 ImageAdapter 来初始化 Gallery 对象，最后将存储卡中的图片加载到 Gallery Widget 中。

7.5.2　具体实现

本实例的主程序文件是 example5.java，下面开始讲解其具体实现代码。

1）先加载相关类，然后新增 ImageAdapter 并赋值给 Gallery 对象，最后设置几个 itemclick Listener 事件。具体代码如下。

```
public class example5 extends Activity
{

  /** Called when the activity is first created. */
  @Override
  public void onCreate(Bundle savedInstanceState)
  {
    super.onCreate(savedInstanceState);
    setContentView(R.layout.main);
    Gallery g = (Gallery) findViewById(R.id.mygallery);

    /*新增 ImageAdapter 并赋值给 Gallery 对象*/
    g.setAdapter(new ImageAdapter(this,getSD()));

    /*设置几个 itemclickListener 事件*/
    g.setOnItemClickListener(new OnItemClickListener()
    {
      public void onItemClick(AdapterView<?> parent,
                              View v, int position, long id)
      {

      }
    });
  }
```

2）定义 List<String> getSD()方法，然后访问 SD 卡中数据，并将获取的图片显示在 ArrayList 列表中。具体代码如下。

```
private List<String> getSD()
{
  /* 设定目前所用路径 */
  List<String> it=new ArrayList<String>();
  File f=new File("/sdcard/");
  File[] files=f.listFiles();

  /* 将所有数据返回到 ArrayList 中 */
  for(int i=0;i<files.length;i++)
  {
    File file=files[i];
    if(getImageFile(file.getPath()))
      it.add(file.getPath());
  }
  return it;
}
```

3）定义 getImageFile(String fName)方法，获取存储卡内的图片文件，并根据扩展类型决定数据类型。具体代码如下。

```
private boolean getImageFile(String fName)
{
  boolean re;

  /* 取得扩展名 */
  String end=fName.substring(fName.lastIndexOf(".")+1,
            fName.length()).toLowerCase();

  /* 根据扩展类型决定打开方式*/
  if(end.equals("jpg")||end.equals("gif")||end.equals("png")
          ||end.equals("jpeg")||end.equals("bmp"))
  {
    re=true;
  }
  else
  {
    re=false;
  }
  return re;
}
```

4）定义继承于 ImageAdapter 的子类 BaseAdapter，流程如下。

- 第一步：声明变量。
- 第二步：声明 ImageAdapter 构造器。
- 第三步：获取 Gallery 信息。

具体代码如下。

```
public class ImageAdapter extends BaseAdapter
{
  /*声明变量*/
  int mGalleryItemBackground;
  private Context mContext;
  private List<String> lis;

  /*ImageAdapter 构造器*/
  public ImageAdapter(Context c,List<String> li)
  {
    mContext = c;
    lis=li;
    /* 使用文件 res/values/attrs.xml 中的<declare-styleable>定义
    * Gallery 属性.*/
    TypedArray a = obtainStyledAttributes(R.styleable.Gallery);
    /*取得 Gallery 属性的 Index id*/
```

```
    mGalleryItemBackground = a.getResourceId(
        R.styleable.Gallery_android_galleryItemBackground, 0);
    /*让对象的 styleable 属性能够反复使用*/
    a.recycle();
}
```

5）定义如下几个需要覆盖的方法

- getCount：用于返回图片数目。
- getItem：用于返回位置。
- getView：用于返回 View 对象。

具体代码如下。

```
/*定义要覆盖的方法 getCount,返回图片数目*/
public int getCount()
{
  return lis.size();
}

/*定义要覆盖的方法 getItem, 用于返回位置*/
public Object getItem(int position)
{
  return position;
}

/*定义要覆盖的方法 getItemId, 用于返回 position */
public long getItemId(int position)
{
  return position;
}

/*定义要覆盖的方法 getView,用于返回 View 对象*/
public View getView(int position, View convertView,
            ViewGroup parent)
{
  /*产生 ImageView 对象*/
  ImageView i = new ImageView(mContext);
  /*设定图传给 imageView 对象*/
  Bitmap bm = BitmapFactory.decodeFile(lis.
                get(position).toString());
  i.setImageBitmap(bm);
  /*重新设定图片的宽高*/
  i.setScaleType(ImageView.ScaleType.FIT_XY);
  /*重新设定 Layout 的宽高*/
  i.setLayoutParams(new Gallery.LayoutParams(136, 88));
  /*设定 Gallery 背景图*/
  i.setBackgroundResource(mGalleryItemBackground);
```

```
            /*传输 imageView 对象*/
            return i;
        }
    }
}
```

执行后将会获取存储卡内的图片信息，并以 Gallery 的样式显示出来。具体如图 7-8 所示。鼠标单击后，能够滑动显示图片，如图 7-9 所示。

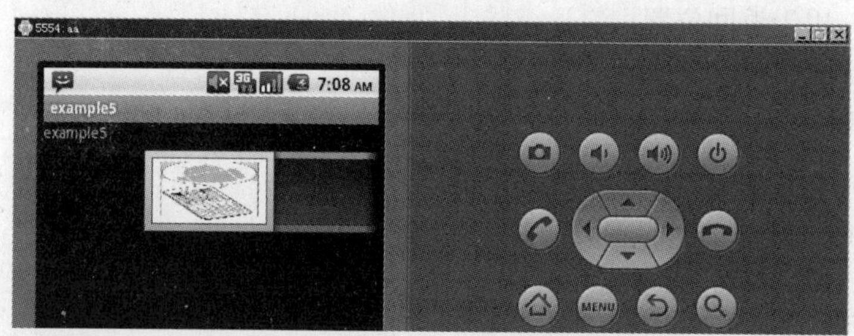

图 7-8　执行效果

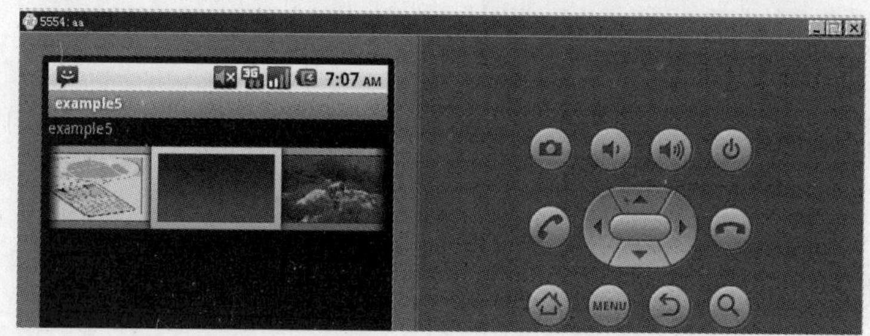

图 7-9　滑动显示

接下来详细讲解下本实例的测试过程。

（1）创建虚拟 SD 卡

使用 cmd 命令进入到 android 的 SDK 的 Tools 目录下，执行 mksdcard 创建。例如，作者的 tools 目录为：

```
E:\skyland\android-sdk-windows\tools>
```

则创建命令为：

```
E:\skyland\android-sdk-windows\tools>mksdcard 128M sdcard.img
```

其中第一个参数为要创建的 sdcard 容量大小（容量大小自己决定），第二个参数为 sdcard 的名字。

（2）启动装有 sdcard 的 Android 模拟器

可以不用 cmd 命令，而直接在 eclipse 中启用。具体方法如下。

第一步：用鼠标右键单击实例工程，在弹出菜单中依次选择“Run As”→“Run Configurations”，命令，如图 7-10 所示。

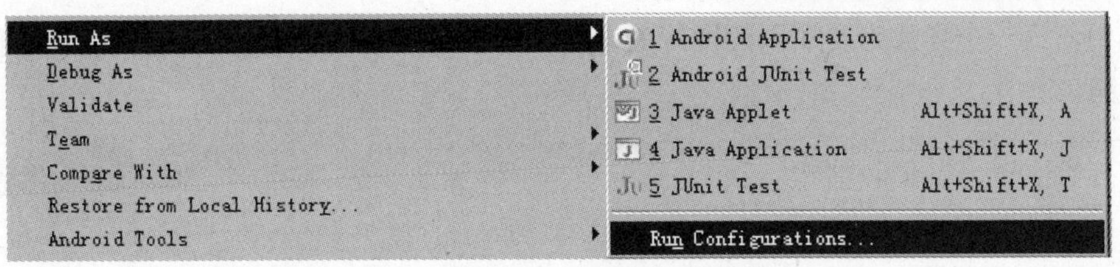

图 7-10　命令

第二步：在弹出的对话框中“Target”选项卡，在最后一行中输入如下字符。

```
-sdcard c:\sdcard.img
```

如图 7-11 所示。

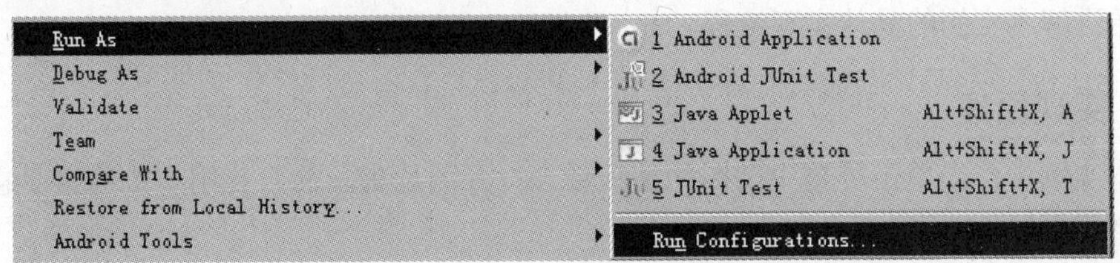

图 7-11　输入-sdcard c:\sdcard.img

（3）添加文件到创建的 SD 卡

经过以上两步操作后，模拟器已经运行了，下面开始添加文件到创建的 SD 卡。具体有如下两种方法。

第一种：eclipse 菜单实现。

在 eclipse 中打开 DDMS 视图，打开“File Explor”选项卡，单击目录下的“sdcard”，如图 7-12 所示。

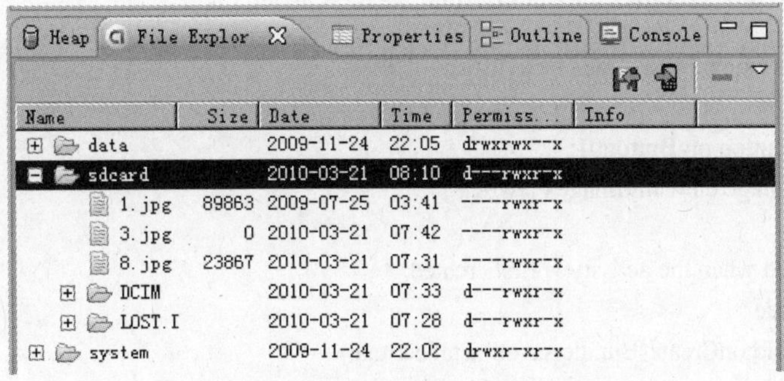

图 7-12　“File Explor”选项卡

通过单击图 7-12 中的按钮，可以上传本地文件到虚拟存储卡。

第二种：用 cmd 命令。

通过 cmd 来到 android 的 SDK 的 Tools 目录下，然后用 adb push 命令添加，格式如下。

```
E:\skyland\android-sdk-windows\tools>adb push new.JPG /sdcard
```

其中，第一个参数为要加入的图片的全名，如果名字中间有空格，要用双引号将其括起来。例如，

```
adb push "c:\1.jpg" /sdcard
```

第二参数就是刚才创建的 sdcard 了。

7.6 获取内置媒体中的图片文件

在移动手机中，内置了多媒体这个 Intent，用户可以直接单击多媒体按钮进入菜单程序。当然也可以通过编程的方式实现上述操作。在本节的内容中，将通过一个具体实例的实现，介绍获取内置媒体中图片文件的基本流程。本实例的源代码保存在“光盘:\daima\7\example6”，下面开始讲解本实例的具体实现流程。

7.6.1 实现原理

在本实例中，当单击屏幕中的图片按钮后会打开图片集画面，让用户挑选自己喜欢的图片。主要包括如下工作。

- ❑ 打开 Intent.ACTION_GET_CONTENT。
- ❑ 获取用户选择的 Image。
- ❑ 通过 ContentResolver 将 Image 转换为 Bitmap。
- ❑ 在 ImageView 中显示图片。

7.6.2 具体实现

本实例的主程序文件是 example6.java，下面开始讲解其具体实现代码。

1）先加载相关类，然后在 onCreate 方法中引入主布局文件 main.xml，最后设置 myButton01 对象，用于激活单击事件 OnClickListener()。具体代码如下。

```
public class example6 extends Activity
{
  private Button myButton01;
  private ImageView myImageView01;

  /** Called when the activity is first created. */
  @Override
  public void onCreate(Bundle savedInstanceState)
  {
    super.onCreate(savedInstanceState);
    setContentView(R.layout.main);

    myImageView01 = (ImageView) findViewById(R.id.myImageView01);
    myButton01 = (Button) findViewById(R.id.myButton01);
```

2）定义 myButton01 的单击事件 OnClickListener()，流程如下。

- ❑ 第一步：打开 Pictures 画面类型设置为 image。
- ❑ 第二步：使用 Intent 中的 ACTION_GET_CONTENT 动作。
- ❑ 第三步：获取照片后返回原来画面。

具体代码如下。

```
myButton01.setOnClickListener(new Button.OnClickListener()
{
  @Override
  public void onClick(View arg0)
  {
    Intent intent = new Intent();
    /* 打开 Pictures 画面类型设置为 image */
    intent.setType("image/*");
    /* 使用 Intent.ACTION_GET_CONTENT 这个 Action */
    intent.setAction(Intent.ACTION_GET_CONTENT);
    /* 取得照片后返回本画面 */
    startActivityForResult(intent, 1);
  }
});
}
```

3）定义 onActivityResult(int requestCode, int resultCode)，用于处理用户挑选的图片。通过 requestCode 和 resultCode 返回标识码以及数据类型为 Intent 的数据参数，调用 Intent 对象的 getData()方法可以获取具体内容。具体代码如下。

```
@Override
protected void onActivityResult(int requestCode, int resultCode,
    Intent data)
{
  if (resultCode == RESULT_OK)
  {
    Uri uri = data.getData();
    ContentResolver cr = this.getContentResolver();
    try
    {
      Bitmap bitmap = BitmapFactory.decodeStream(cr
          .openInputStream(uri));
      /* 将 Bitmap 设置到 ImageView */
      myImageView01.setImageBitmap(bitmap);
    } catch (FileNotFoundException e)
    {
      e.printStackTrace();
    }
  }
  super.onActivityResult(requestCode, resultCode, data);
}
```

```
}
```

执行后的初始效果如图 7-13 所示。单击“图片”按钮后会显示系统内的图片信息，如图 7-14 所示。选中某幅图片并单击后，会突出显示此照片。

图 7-13　初始效果

图 7-14　系统内的图片

注意：如果手机多媒体中没有图片，则可以按照 7.5 中的操作在 SD 存储卡中添加照片，同样可以显示出对应效果。

7.7　调节音量大小

在移动手机中，调节音量大小是一个十分重要的功能。在本节的内容中，将通过一个具体实例的实现，介绍在 Android 手机中调节音量大小的基本流程。本实例的源代码保存在“光盘:\daima\7\example7”，下面开始讲解本实例的具体实现流程。

7.7.1　实现原理

在 Android API 中的 AudioManager 类提供了相关的方法，可以在程序中控制手机音量的大小，也可以切换声音的模式为震动或静音。

本实例使用的素材图片存放在了“res\drawable”目录下。

7.7.2　具体实现

本实例的主程序文件是 example7.java，下面开始讲解其具体实现代码。

1）先加载相关类，然后声明各个需要的变量。具体代码如下。

```
public class example7 extends Activity
{
  /* 变量声明 */
  private ImageView myImage;
  private ImageButton downButton;
  private ImageButton upButton;
```

```
private ImageButton normalButton;
private ImageButton muteButton;
private ImageButton vibrateButton;
private ProgressBar myProgress;
private AudioManager audioMa;
private int volume=0;
```

2）依次对象初始化变量 audioMa、myImage、myProgress、downButton、upButton、normal Button、muteButton、muteButton 和 vibrateButton。具体代码如下。

```
@Override
public void onCreate(Bundle savedInstanceState)
{
  super.onCreate(savedInstanceState);
  setContentView(R.layout.main);

  /* 对象初始化 */
  audioMa = (AudioManager)getSystemService(Context.AUDIO_SERVICE);
  myImage = (ImageView)findViewById(R.id.myImage);
  myProgress = (ProgressBar)findViewById(R.id.myProgress);
  downButton = (ImageButton)findViewById(R.id.downButton);
  upButton = (ImageButton)findViewById(R.id.upButton);
  normalButton = (ImageButton)findViewById(R.id.normalButton);
  muteButton = (ImageButton)findViewById(R.id.muteButton);
  vibrateButton = (ImageButton)findViewById(R.id.vibrateButton);
```

3）分别设置初始的手机音量和初始的声音模式，具体代码如下。

```
/* 设置初始的手机音量 */
volume=audioMa.getStreamVolume(AudioManager.STREAM_RING);
myProgress.setProgress(volume);
/* 设置初始的声音模式 */
int mode=audioMa.getRingerMode();
if(mode==AudioManager.RINGER_MODE_NORMAL)
{
  myImage.setImageDrawable(getResources()
                              .getDrawable(R.drawable.normal));
}
else if(mode==AudioManager.RINGER_MODE_SILENT)
{
  myImage.setImageDrawable(getResources()
                              .getDrawable(R.drawable.mute));
}
else if(mode==AudioManager.RINGER_MODE_VIBRATE)
{
  myImage.setImageDrawable(getResources()
                              .getDrawable(R.drawable.vibrate));
```

```
}
```

4）设置音量调小按钮 downButton 的处理事件 setOnClickListener，流程如下。

- ❑ 第一步：设置音量调小一格。
- ❑ 第二步：设置调整后声音模式。

具体代码如下。

```
/* 音量调小声的按钮 */
downButton.setOnClickListener(new Button.OnClickListener()
{
  @Override
  public void onClick(View arg0)
  {
    /* 设置音量调小一格 */
    audioMa.adjustVolume(AudioManager.ADJUST_LOWER, 0);
    volume=audioMa.getStreamVolume(AudioManager.STREAM_RING);
    myProgress.setProgress(volume);
    /* 设置调整后声音模式 */
    int mode=audioMa.getRingerMode();
    if(mode==AudioManager.RINGER_MODE_NORMAL)
    {
      myImage.setImageDrawable(getResources()
                                    .getDrawable(R.drawable.normal));
    }
    else if(mode==AudioManager.RINGER_MODE_SILENT)
    {
      myImage.setImageDrawable(getResources()
                                    .getDrawable(R.drawable.mute));
    }
    else if(mode==AudioManager.RINGER_MODE_VIBRATE)
    {
      myImage.setImageDrawable(getResources()
                                  .getDrawable(R.drawable.vibrate));
    }
  }
});
```

5）设置音量调大按钮 upButton 的处理事件 setOnClickListener，流程如下。

- ❑ 第一步：设置音量调大一格。
- ❑ 第二步：设置调整后声音模式。

具体代码如下。

```
/* 音量调大声的按钮 */
upButton.setOnClickListener(new Button.OnClickListener()
{
  @Override
```

```
    public void onClick(View arg0)
    {
      /* 设置音量调大一格 */
      audioMa.adjustVolume(AudioManager.ADJUST_RAISE, 0);
      volume=audioMa.getStreamVolume(AudioManager.STREAM_RING);
      myProgress.setProgress(volume);
      /* 设置调整后的声音模式 */
      int mode=audioMa.getRingerMode();
      if(mode==AudioManager.RINGER_MODE_NORMAL)
      {
        myImage.setImageDrawable(getResources()
                                  .getDrawable(R.drawable.normal));
      }
      else if(mode==AudioManager.RINGER_MODE_SILENT)
      {
        myImage.setImageDrawable(getResources()
                                  .getDrawable(R.drawable.mute));
      }
      else if(mode==AudioManager.RINGER_MODE_VIBRATE)
      {
        myImage.setImageDrawable(getResources()
                                  .getDrawable(R.drawable.vibrate));
      }
    }
  });
```

6）设置调整正常铃声模式按钮 normalButton 的处理事件 setOnClickListener，流程如下。

- 第一步：设置铃声模式为 NORMAL。
- 第二步：设置音量与声音模式。

具体代码如下。

```
/* 调整铃声模式为正常模式的按钮*/
normalButton.setOnClickListener(new Button.OnClickListener()
{
  @Override
  public void onClick(View arg0)
  {
    /* 设置铃声模式为正常模式 */
    audioMa.setRingerMode(AudioManager.RINGER_MODE_NORMAL);
    /* 设置音量与声音模式 */
    volume=audioMa.getStreamVolume(AudioManager.STREAM_RING);
    myProgress.setProgress(volume);
    myImage.setImageDrawable(getResources()
                              .getDrawable(R.drawable.normal));
  }
});
```

7）设置调整静音铃声模式按钮 muteButton 的处理事件 setOnClickListener，流程如下。

- 第一步：设置铃声模式为 SILENT（静音）。
- 第二步：设置音量与声音状态。

具体代码如下。

```
/* 调整铃声模式为静音模式的按钮 */
muteButton.setOnClickListener(new Button.OnClickListener()
{
  @Override
  public void onClick(View arg0)
  {
    /* 设置铃声模式为 SILENT */
    audioMa.setRingerMode(AudioManager.RINGER_MODE_SILENT);
    /* 设置音量与声音状态 */
    volume=audioMa.getStreamVolume(AudioManager.STREAM_RING);
    myProgress.setProgress(volume);
    myImage.setImageDrawable(getResources()
                          .getDrawable(R.drawable.mute));
  }
});
```

8）设置调整震动铃声模式按钮 vibrateButton 的处理事件 setOnClickListener，流程如下。

- 第一步：设置铃声模式为 VIBRATE（震动）。
- 第二步：设置音量与声音状态。

具体代码如下。

```
    /* 调整铃声模式为震动模式的按钮 */
    vibrateButton.setOnClickListener(new Button.OnClickListener()
    {
      @Override
      public void onClick(View arg0)
      {
        /* 设置铃声模式为 VIBRATE */
        audioMa.setRingerMode(AudioManager.RINGER_MODE_VIBRATE);
        /* 设置音量与声音状态 */
        volume=audioMa.getStreamVolume(AudioManager.STREAM_RING);
        myProgress.setProgress(volume);
        myImage.setImageDrawable(getResources()
                              .getDrawable(R.drawable.vibrate));
      }
    });
  }
}
```

执行后将会显示一个音量调节界面。具体如图 7-15 所示。

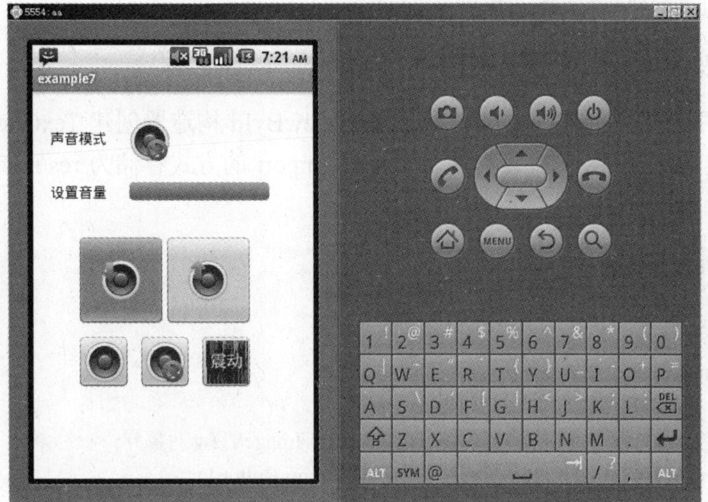

图 7-15　执行效果

7.8　播放 MP3 文件

在移动手机中，播放 MP3 文件是一个十分重要的功能。在本节的内容中，将通过一个具体实例的实现，介绍在 Android 手机中播放 MP3 文件的基本流程。本实例的源代码保存在“光盘:\daima\7\example8”，下面开始讲解本实例的具体实现流程。

7.8.1　实现原理

在本实例中，插入了 3 个按钮，分别用于播放、暂停和停止。当单击“播放”按钮后会从指定的手机资源中获取 MP3 文件，并执行播放处理。在具体实现上，先添加一个 MediaPlayer 对象，并使用 MediaPlayer.create()方法来创建播放器资源，然后通过方法 MediaPlay.create()、MediaPlay.stop()和 MediaPlay.pause()分别实现开始、停止和暂停功能。为了处理按钮所需要的各个事件，需要覆盖各个 ImageButton 的 onClick()方法，通过按钮来控制 MediaPlayer 的状态。

7.8.2　具体实现

本实例的主程序文件是 example8.java，下面开始讲解其具体实现代码。

1）先引入相关类，然后分别声明一个 ImageButton、TextView 和 MediaPlayer 变量，声明一个 Flag 作为确认音乐是否暂停的变量并默认为 false。具体代码如下。

```
public class example8 extends Activity
{
  /*声明一个 ImageButton,TextView,MediaPlayer 变量*/
  private ImageButton mButton01, mButton02, mButton03;
  private TextView mTextView01;
  private MediaPlayer mMediaPlayer01;
```

```
/*声明一个 Flag 作为确认音乐是否暂停的变量并默认为 false*/
private boolean bIsPaused = false;
```

2）先引入主布局文件 main.xml，通过 findViewById 构造器创建 TextView 与 ImageView 对象，然后创建 MediaPlayer 对象，并将音乐以 Import 的方式存储为 res/raw/always.mp3。具体代码如下。

```
public void onCreate(Bundle savedInstanceState)
{
  super.onCreate(savedInstanceState);
  setContentView(R.layout.main);

  /*通过 findViewById 构造器创建 TextView 与 ImageView 对象*/
  mButton01 =(ImageButton) findViewById(R.id.myButton1);
  mButton02 =(ImageButton) findViewById(R.id.myButton2);
  mButton03 =(ImageButton) findViewById(R.id.myButton3);
  mTextView01 = (TextView) findViewById(R.id.myTextView1);

  /* onCreate 时创建 MediaPlayer 对象 */
  mMediaPlayer01 = new MediaPlayer();
  /* 将音乐以 Import 的方式存储在 res/raw/always.mp3 */
  mMediaPlayer01 = MediaPlayer.create(example8.this, R.raw.big);
```

3）运行播放音乐的按钮，流程如下。

❑ 第一步：覆盖 OnClick 事件。

❑ 第二步：在 MediaPlayer 对象中取得播放资源并使用停止播放方法 stop()之后，开始重复播放。

❑ 第三步：改变 TextView 为开始播放状态。

具体代码如下。

```
/* 运行播放音乐的按钮 */
mButton01.setOnClickListener(new ImageButton.OnClickListener()
{
  @Override
  /*覆盖 OnClick 事件*/
  public void onClick(View v)
  {
    try
    {

      if (mMediaPlayer01 != null)
      {
        mMediaPlayer01.stop();
      }
      /*在 MediaPlayer 取得播放资源与 stop()之后
       * 要准备 Playback 的状态前一定要使用 MediaPlayer.prepare()*/
```

```
            mMediaPlayer01.prepare();
            /*开始或恢复播放*/
            mMediaPlayer01.start();
            /*改变 TextView 为开始播放状态*/
            mTextView01.setText(R.string.str_start);
        }
        catch (Exception e)
        {
            mTextView01.setText(e.toString());
            e.printStackTrace();
        }
    }
});
```

4）定义停止播放处理事件，通过 mMediaPlayer01.stop()停止播放 MP3，改变 TextView 对象为停止播放状态。具体代码如下。

```
/* 停止播放 */
mButton02.setOnClickListener(new ImageButton.OnClickListener()
{
    @Override
    public void onClick(View arg0)
    {
        try
        {
            if (mMediaPlayer01 != null)
            {
                /*停止播放*/
                mMediaPlayer01.stop();
                /*改变 TextView 为停止播放状态*/
                mTextView01.setText(R.string.str_close);
            }

        }
        catch (Exception e)
        {
            // TODO Auto-generated catch block
            mTextView01.setText(e.toString());
            e.printStackTrace();
        }
    }
});
```

5）定义暂停播放处理事件，首先判断是否处于暂停状态。具体代码如下。

```
/* 暂停播放 */
mButton03.setOnClickListener(new ImageButton.OnClickListener()
```

```
{
  @Override
  public void onClick(View arg0)
  {
    // TODO Auto-generated method stub
    try
    {
      if (mMediaPlayer01 != null)
      {
        /*是否为暂停状态=否*/
        if(bIsPaused==false)
        {
          /*暂停播放*/
          mMediaPlayer01.pause();
          /*设置 Flag 为 treu 表示播放状态为暂停*/
          bIsPaused = true;
          /*改变 TextView 为暂停播放*/
          mTextView01.setText(R.string.str_pause);
        }
        /*是否为暂停状态=是*/
        else if(bIsPaused==true)
        {
          /*恢复播出状态*/
          mMediaPlayer01.start();
          /*设置 Flag 为 false 表示播放状态为非暂停状态*/
          bIsPaused = false;
          /*改变 TextView 为开始播放*/
          mTextView01.setText(R.string.str_start);
        }
      }
    }
    catch (Exception e)
    {
      // TODO Auto-generated catch block
      mTextView01.setText(e.toString());
      e.printStackTrace();
    }
  }
});
```

6）通过 MediaPlayer.OnCompletionLister 监听是否播放完毕，并覆盖文件播出完毕事件，并改变 TextView 为播放结束。具体代码如下。

```
mMediaPlayer01.setOnCompletionListener(
  new MediaPlayer.OnCompletionListener()
{
  /*覆盖文件播出完毕事件*/
```

```
    public void onCompletion(MediaPlayer arg0)
    {
      try
      {
        /*解除资源与 MediaPlayer 的赋值关系
         * 让资源可以为其他程序利用*/
        mMediaPlayer01.release();
        /*改变 TextView 为播放结束*/
        mTextView01.setText(R.string.str_OnCompletionListener);
      }
      catch (Exception e)
      {
        mTextView01.setText(e.toString());
        e.printStackTrace();
      }
    }
  });
```

7）使用 MediaPlayer.OnErrorListener 监听覆盖错误处理事件，当发生错误时也解除资源与 MediaPlayer 的赋值。具体代码如下。

```
  /* 当 MediaPlayer.OnErrorListener 会运行的 Listener */
  mMediaPlayer01.setOnErrorListener(new MediaPlayer.OnErrorListener()
  {
    @Override
    /*覆盖错误处理事件*/
    public boolean onError(MediaPlayer arg0, int arg1, int arg2)
    {
      try
      {
        /*发生错误时也解除资源与 MediaPlayer 的赋值*/
        mMediaPlayer01.release();
        mTextView01.setText(R.string.str_OnErrorListener);
      }
      catch (Exception e)
      {
        mTextView01.setText(e.toString());
        e.printStackTrace();
      }
      return false;
    }
  });
}
```

8）覆盖主程序暂停状态事件，在主程序暂停时解除资源与媒体播放器的赋值关系，通过 catch 实现异常处理。具体代码如下。

```
    @Override
    /*覆盖主程序暂停状态事件*/
    protected void onPause()
    {
      try
      {
        /*在主程序暂停时解除资源与 MediaPlayer 的赋值关系*/
        mMediaPlayer01.release();
      }
      catch (Exception e)
      {
        mTextView01.setText(e.toString());
        e.printStackTrace();
      }
      super.onPause();
    }
}
```

执行后会在屏幕中通过 3 个播放按钮播放指定的 MP3 文件。如图 7-16 所示。

图 7-16　执行效果

注意：如果手机多媒体中没有 MP3 文件片，则可以按照 7.5 中的操作在 SD 存储卡中添加 MP3 资源，同样可以显示出对应效果。

7.9　录音处理

在移动手机中，录音处理也是一个十分重要的功能。在本节的内容中，将通过一个具体实例的实现，介绍在 Android 手机中实现录音处理的基本流程。本实例的源代码保存在“光盘:\daima\7\example9”，下面开始讲解本实例的具体实现流程。

7.9.1　实现原理

在本实例中，插入了 4 个按钮，分别用于录音、停止录音、播放录音和删除录音。为了能够不限制录音的长度，现将录音暂时保存到存储卡，当录音完毕后，再将录音文件显示在 ListView。单击文件后，可以播放或删除录音文件。

7.9.2　具体实现

本实例的主程序文件是 example9.java，下面开始讲解其具体实现代码。

1）先加载相关类，然后分别声明各个需要的变量。具体代码如下。

```
public class example9 extends Activity
{
  private ImageButton myButton1;
  private ImageButton myButton2;
  private ImageButton myButton3;
  private ImageButton myButton4;
  private ListView myListView1;
  private String strTempFile = "ex07_11_";
  private File myRecAudioFile;
  private File myRecAudioDir;
  private File myPlayFile;
  private MediaRecorder mMediaRecorder01;

  private ArrayList<String> recordFiles;
  private ArrayAdapter<String> adapter;
  private TextView myTextView1;
  private boolean sdCardExit;
  private boolean isStopRecord;
```

2)通过 findViewById 分别构造 4 个按钮对象和 2 个文本对象，然后设置按钮状态不可选。具体代码如下。

```
/** Called when the activity is first created. */
@Override
public void onCreate(Bundle savedInstanceState)
{
  super.onCreate(savedInstanceState);
  setContentView(R.layout.main);
  /*  设置 4 个按钮和 2 个文本  */
  myButton1 = (ImageButton) findViewById(R.id.ImageButton01);
  myButton2 = (ImageButton) findViewById(R.id.ImageButton02);
  myButton3 = (ImageButton) findViewById(R.id.ImageButton03);
  myButton4 = (ImageButton) findViewById(R.id.ImageButton04);
  myListView1 = (ListView) findViewById(R.id.ListView01);
  myTextView1 = (TextView) findViewById(R.id.TextView01);
  /*  设置按钮状态不可选  */
  myButton2.setEnabled(false);
  myButton3.setEnabled(false);
  myButton4.setEnabled(false);
```

3）通过 sdCardExit 对象判断 SD Card 是否插入，然后获取 SD Card 路径作为录音的文件位置，并取得 SD Card 目录里的所有.amr 格式的文件，最后将 ArrayAdapter 添加到 ListView

对象中。具体代码如下。

```
/* 判断 SD Card 是否插入 */
sdCardExit = Environment.getExternalStorageState().equals(
      android.os.Environment.MEDIA_MOUNTED);
/* 取得 SD Card 路径作为录音的文件位置 */
if (sdCardExit)
   myRecAudioDir = Environment.getExternalStorageDirectory();

/* 取得 SD Card 目录里的所有.amr 格式的文件 */
getRecordFiles();

adapter = new ArrayAdapter<String>(this,
     R.layout.my_simple_list_item, recordFiles);
/* 将 ArrayAdapter 添加 ListView 对象中 */
myListView1.setAdapter(adapter);
```

4）设置单击“录音”按钮后的录音处理事件，流程如下。

- 第一步：创建录音文件。
- 第二步：设置录音来源为麦克风。
- 第三步：通过 myTextView1.setText("录音中")方法设置录音过程显示的提示文本。

具体代码如下。

```
/* 录音 */
myButton1.setOnClickListener(new ImageButton.OnClickListener()
{
   @Override
   public void onClick(View arg0)
   {
      try
      {
         if (!sdCardExit)
         {
            Toast.makeText(example9.this, "请插入 SD Card",
                  Toast.LENGTH_LONG).show();
            return;
         }
         /* 创建录音文件 */
         myRecAudioFile = File.createTempFile(strTempFile, ".amr",
               myRecAudioDir);
         mMediaRecorder01 = new MediaRecorder();
         /* 设置录音来源为麦克风 */
         mMediaRecorder01
               .setAudioSource(MediaRecorder.AudioSource.MIC);
         mMediaRecorder01
               .setOutputFormat(MediaRecorder.OutputFormat.DEFAULT);
```

```
            mMediaRecorder01
                .setAudioEncoder(MediaRecorder.AudioEncoder.DEFAULT);
            mMediaRecorder01.setOutputFile(myRecAudioFile
                .getAbsolutePath());
            mMediaRecorder01.prepare();
            mMediaRecorder01.start();
            myTextView1.setText("录音中");
            myButton2.setEnabled(true);
            myButton3.setEnabled(false);
            myButton4.setEnabled(false);
            isStopRecord = false;
          }
          catch (IOException e)
          {
            // TODO Auto-generated catch block
            e.printStackTrace();
          }
        }
      });
```

5）设置单击“停止”按钮后的处理事件，流程如下。

- ❑ 第一步：通过 mMediaRecorder01.stop()方法停止录音。
- ❑ 第二步：将录音频文件名传递给 Adapter。

具体代码如下。

```
    /* 停止播放 */
    myButton2.setOnClickListener(new ImageButton.OnClickListener()
    {
      @Override
      public void onClick(View arg0)
      {
        if (myRecAudioFile != null)
        {
          /* 停止录音 */
          mMediaRecorder01.stop();
          mMediaRecorder01.release();
          mMediaRecorder01 = null;
          /* 将录音频文件名传给 Adapter */
          adapter.add(myRecAudioFile.getName());
          myTextView1.setText("停止：" + myRecAudioFile.getName());
          myButton2.setEnabled(false);
          isStopRecord = true;
        }
      }
    });
```

6）设置单击“播放”按钮后的处理事件，单击后将打开播放的程序。具体代码如下。

```
/* 播放 */
myButton3.setOnClickListener(new ImageButton.OnClickListener()
{
  @Override
  public void onClick(View arg0)
  {
    if (myPlayFile != null && myPlayFile.exists())
    {
      /* 打开播放的程序 */
      openFile(myPlayFile);
    }
  }
});
```

7）设置单击“删除”按钮后的处理事件，流程如下。

- 第一步：将 Adapter 删除文件名。
- 第二步：删除存在的文件。

具体代码如下。

```
/* 删除 */
myButton4.setOnClickListener(new ImageButton.OnClickListener()
{
  @Override
  public void onClick(View arg0)
  {
    if (myPlayFile != null)
    {
      /* 先将 Adapter 删除文件名 */
      adapter.remove(myPlayFile.getName());
      /* 删除文件 */
      if (myPlayFile.exists())
        myPlayFile.delete();
      myTextView1.setText("完成删除");
    }
  }
});
```

8）设置单击列表中文件的处理事件，流程如下。

- 第一步：如果有文件，单击后删除这个文件，并将播放按钮设置为不可用。
- 第二步：输出选择提示语句。

具体代码如下。

```
myListView1.setOnItemClickListener
(new AdapterView.OnItemClickListener()
```

```
{
  @Override
  public void onItemClick(AdapterView<?> arg0, View arg1,
      int arg2, long arg3)
  {
    myButton3.setEnabled(true);
    myButton4.setEnabled(true);
    myPlayFile = new File(myRecAudioDir.getAbsolutePath()
        + File.separator
        + ((CheckedTextView) arg1).getText());
    myTextView1.setText("你选的是："
        + ((CheckedTextView) arg1).getText());
  }
});
}
```

9）定义方法 onStop()，用于停止录音处理。具体代码如下。

```
@Override
protected void onStop()
{
  if (mMediaRecorder01 != null && !isStopRecord)
  {
    /* 停止录音 */
    mMediaRecorder01.stop();
    mMediaRecorder01.release();
    mMediaRecorder01 = null;
  }
  super.onStop();
}
```

10）定义方法 getRecordFiles()，用于获取文件的长度，并设置只获取“.amr”格式的文件。具体代码如下。

```
private void getRecordFiles()
{
  recordFiles = new ArrayList<String>();
  if (sdCardExit)
  {
    File files[] = myRecAudioDir.listFiles();
    if (files != null)
    {
      for (int i = 0; i < files.length; i++)
      {
        if (files[i].getName().indexOf(".") >= 0)
        {
          /* 只获取.amr 文件 */
```

```
                String fileS = files[i].getName().substring(
                    files[i].getName().indexOf("."));
                if (fileS.toLowerCase().equals(".amr"))
                  recordFiles.add(files[i].getName());
              }
            }
          }
        }
      }
```

11）定义方法 openFile(File f)，用于打开指定的录音文件。具体代码如下。

```
/* 打开录音文件的程序 */
private void openFile(File f)
{
  Intent intent = new Intent();
  intent.addFlags(Intent.FLAG_ACTIVITY_NEW_TASK);
  intent.setAction(android.content.Intent.ACTION_VIEW);
  String type = getMIMEType(f);
  intent.setDataAndType(Uri.fromFile(f), type);
  startActivity(intent);
}
```

12）定义方法 getMIMEType(File f)，用于获得文件的类型，在此设置了 audio 类型、image 类型和其他类型，共三大类。具体代码如下。

```
private String getMIMEType(File f)
{
  String end = f.getName().substring(
       f.getName().lastIndexOf(".") + 1, f.getName().length())
       .toLowerCase();
  String type = "";
  if (end.equals("mp3") || end.equals("aac") || end.equals("aac")
       || end.equals("amr") || end.equals("mpeg")
       || end.equals("mp4"))
  {
    type = "audio";
  }
  else if (end.equals("jpg") || end.equals("gif")
     || end.equals("png") || end.equals("jpeg"))
  {
    type = "image";
  }
  else
  {
    type = "*";
  }
```

```
        type += "/*";
        return type;
    }
}
```

最后，需要在文件 AndroidManifest.xml 中打开录音权限。具体代码如下。

```
<uses-permission android:name="android.permission.RECORD_AUDIO">
```

执行后的效果如图 7-17 所示。当单击“录音”按钮时开始录音处理，如图 7-18 所示。当单击“停止”按钮后停止录音处理，并在列表中显示录制的音频文件，如图 7-19 所示；当选中音频文件，单击“删除”按钮后会删除选中音频文件；单击“播放”按钮后会播放选中的音频文件，如图 7-21 所示。

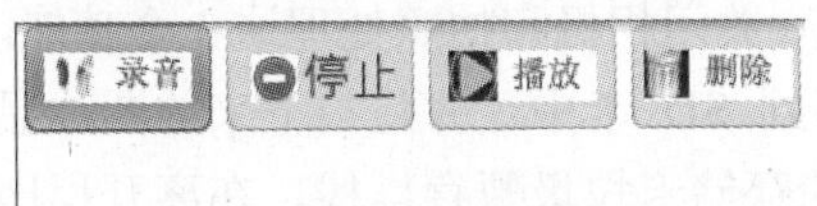

图 7-17　初始效果

图 7-18　正在录音

图 7-19　显示录制的文件

图 7-20　删除音频

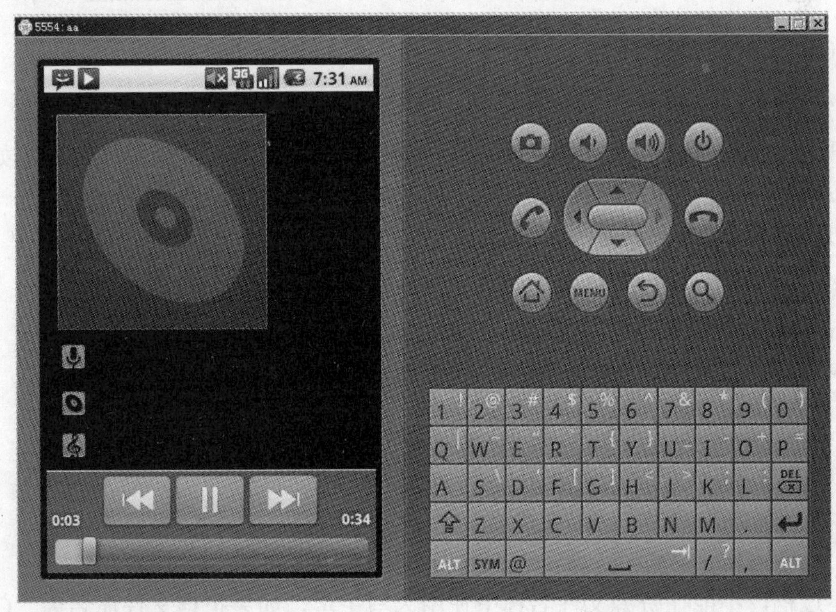

图 7-21　播放音频

7.10 相机预览及拍照

在移动手机中，拍照和录制视频也是十分重要的功能。在本节的内容中，将通过一个具体实例的实现，介绍在 Android 手机中实现相机预览和拍照处理的基本流程。本实例的源代码保存在“光盘:\daima\7\example10”，下面开始讲解本实例的具体实现流程。

7.10.1 实现原理

在相机中，由一个 Preview 功能，能够实现预览处理。在 Preview 的 API Demo 中，没有示范如何在“Activity”中实现 Preview 的讲解，也没有提到如何获取 Preview 中画面的方法。

本实例是一个简单的拍照练习，和 API Demo 的 Preview 程序不同。实例中将以 Activity 为基础，在 Layout 中配置了 3 个按钮，分别实现打开预览、关闭相机和拍照处理这 3 个功能。当单击“拍照”按钮后，需要先将画面截取下来，并存储到 SD 卡中，并将拍下来的图片显示在 Activity 中的 ImageView 中。为避免拍照相片造成的存储卡垃圾暂存堆栈，在离开程序前将临时文件删除。

7.10.2 编程思想

目前的智能手机拥有很多强大的功能，例如摄像头、GPS 和无线上网等，现在是开始充分使用这些功能的时候了，在本篇文章中我们一起学习，如何在 Google Android 编程环境中，以最简单的方式实现 Google Android 摄像头拍照。

注意：从 Android 1.5（代号 cupcake）版本后，在安全方面有诸多改进，其中之一与摄像头权限控制有关。在此之前，能够创建无需用户许可就可实现拍照的应用。现在该问题已被修复，如果想在自己的应用中使用摄像头，需要在文件 AndroidManifest.xml 中增加以下代码。

```
<uses-permission android:name="android.permission.CAMERA"/>
```

1．设定摄像头布局

这是开发工作的基础，也就是我们希望在应用程序中增加多少辅助性元素，如摄像头、各种功能按钮等。本例采取最简方式，除了拍照外，没有多余功能。下面一起看一下示例将要用到的布局文件“camera_surface.xml”。

```
<LinearLayout xmlns:android="http://schemas.android.com/apk/res/android"
android:layout_width="fill_parent" android:layout_height="fill_parent"
android:orientation="vertical">
<SurfaceView android:id="@+id/surface_camera"
android:layout_width="fill_parent" android:layout_height="10dip"
android:layout_weight="1">
</SurfaceView>
</LinearLayout>
```

在上述过程，不能在资源文件名称中使用大写字母，如果把该文件命名为“CameraSurface.xml”，会带来不必要的麻烦。

该布局非常简单，只有一个 LinearLayout 视图组，其中只有一个 SurfaceView（可画图界面）视图，也就是摄像头屏幕。

2．摄像头实现代码

下面再来看一下 Android 代码。创建一个名为“CameraView”的 Activity 类，实现 SurfaceHolder.Callback 接口。

```
public class CamaraView extends Activity implements SurfaceHolder.Callback
```

接口 SurfaceHolder.Callback 被用来接收摄像头预览界面变化的信息。它实现了三个方法。

- ❑ surfaceChanged：当预览界面的格式和大小发生改变时，该方法被调用。
- ❑ surfaceCreated：初次实例化，预览界面被创建时，该方法被调用。
- ❑ surfaceDestroyed：当预览界面被关闭时，该方法被调用。

下面看一下在摄像头应用中如何使用这个接口，首先看一下在 Activity 类中的 onCreate 方法。

```
super.onCreate(icicle);
getWindow().setFormat(PixelFormat.TRANSLUCENT);
requestWindowFeature(Window.FEATURE_NO_TITLE);
getWindow().setFlags(WindowManager.LayoutParams.FLAG_FULLSCREEN,
WindowManager.LayoutParams.FLAG_FULLSCREEN);
setContentView(R.layout.camera);
mSurfaceView = (SurfaceView) findViewById(R.id.surface_camera);
mSurfaceHolder = mSurfaceView.getHolder();
mSurfaceHolder.addCallback(this);
mSurfaceHolder.setType(SurfaceHolder.SURFACE_TYPE_PUSH_BUFFERS);
}
```

下面逐一对代码进行说明。

```
getWindow().setFormat(PixelFormat.TRANSLUCENT);
requestWindowFeature(Window.FEATURE_NO_TITLE);
getWindow().setFlags(WindowManager.LayoutParams.FLAG_FULLSCREEN,
WindowManager.LayoutParams.FLAG_FULLSCREEN);
```

通过上述代码，读者告诉屏幕两点信息。

1）摄像头预览界面将通过全屏显示，没有“标题（title）”。

2）屏幕格式为“半透明”。

```
setContentView(R.layout.camera_surface );
mSurfaceView = (SurfaceView) findViewById(R.id.surface_camera);
```

在以上代码中，读者通过 setContentView 设定 Activity 的布局为前面创建的 camera_surface，并创建一个 SurfaceView 对象，从 XML 文件中获得布局信息。

```
mSurfaceHolder = mSurfaceView.getHolder();
mSurfaceHolder.addCallback(this);
```

```
mSurfaceHolder.setType(SurfaceHolder.SURFACE_TYPE_PUSH_BUFFERS);
```

通过以上代码，读者从 surfaceview 中获得了 holder，并增加 callback（回放）功能到"this"。这意味着我们的操作（Activity）将可以管理这个 surfaceview。

看一下 callback 功能时如何实现的。

```
public void surfaceCreated(SurfaceHolder holder) {
mCamera = Camera.open();
```

mCamera 是"Camera"类的一个对象。在 surfaceCreated 方法中"打开"摄像头，这就是启动它的方式。

```
public void surfaceChanged(SurfaceHolder holder, int format, int w, int h) {
if (mPreviewRunning) {
mCamera.stopPreview();
}
Camera.Parameters p = mCamera.getParameters();
p.setPreviewSize(w, h);
mCamera.setParameters(p);
try {
mCamera.setPreviewDisplay(holder);
} catch (IOException e) {
e.printStackTrace();
}
mCamera.startPreview();
mPreviewRunning = true;
}
```

该方法让摄像头做好拍照准备，设定它的参数，并开始在 Android 屏幕中启动预览画面。本例使用了一个"semaphore"参数来防止冲突：当 mPreviewRunning 为 true 时，意味着摄像头处于激活状态，并未被关闭，因此可以使用它。

```
public void surfaceDestroyed(SurfaceHolder holder) {
mCamera.stopPreview();
mPreviewRunning = false;
mCamera.release();
}
```

通过这个方法，停止摄像头并释放相关的资源。正如大家所看到的，在此设置 mPreviewRunning 为 false，以此来防止在 surfaceChanged 方法中的冲突。原因何在？因为这意味着用户已经关闭了摄像头，而且不能再设置其参数或在摄像头中启动图像预览。

最后看一下本例中最重要的方法。

```
Camera.PictureCallback mPictureCallback = new Camera.PictureCallback() {
public void onPictureTaken(byte[] imageData, Camera c) {
}
};
```

当拍照时，该方法被调用。举例来说，读者可以在界面上创建一个 OnClickListener 方法，单击屏幕时，调用 PictureCallBack 方法。这个方法会提供图像的字节数组，然后你可以使用 Android 提供的 Bitmap 和 BitmapFactory 类，将其从字节数组转换成你想要的图像格式。

7.10.3 具体实现

本实例的主程序文件是 example10.java，下面开始讲解其具体实现代码。

1）先使 Activity 实现 SurfaceHolder.Callback，创建私有 Camera 对象，然后分别创建 mImageView01、mTextView01、TAG、mSurfaceView01、mSurfaceHolder01 和 intScreenY，作为 review 照下来的照片之用。具体代码如下。

```
/* 使 Activity 实现 SurfaceHolder.Callback */
public class example10 extends Activity
implements SurfaceHolder.Callback
{
  /* 创建私有 Camera 对象 */
  private Camera mCamera01;
  private Button mButton01, mButton02, mButton03;

  /* 作为 review 照下来的相片之用 */
  private ImageView mImageView01;
  private TextView mTextView01;
  private String TAG = "HIPPO";
  private SurfaceView mSurfaceView01;
  private SurfaceHolder mSurfaceHolder01;
  //private int intScreenX, intScreenY;
```

2）默认相机预览模式为 false，将照下来的图片存储在“/sdcard/camera_snap.jpg”。具体代码如下。

```
/* 默认相机预览模式为 false */
private boolean bIfPreview = false;

/* 将照下来的图片存储在此 */
private String strCaptureFilePath = "/sdcard/camera_snap.jpg";
```

3）通过 requestWindowFeature(Window.FEATURE_NO_TITLE)，设置程序全屏幕运行，而不使用标题栏。然后判断存储卡是否存在，并提醒用户未安装存储卡。具体代码如下。

```
/** Called when the activity is first created. */
@Override
public void onCreate(Bundle savedInstanceState)
{
  super.onCreate(savedInstanceState);

  /* 应用程序全屏幕运行，不使用标题栏 */
```

```
        requestWindowFeature(Window.FEATURE_NO_TITLE);
        setContentView(R.layout.main);

        /* 判断存储卡是否存在 */
        if(!checkSDCard())
        {
          /* 提醒 User 未安装存储卡 */
          mMakeTextToast
          (
            getResources().getText(R.string.str_err_nosd).toString(),
            true
          );
        }
```

4）通过 DisplayMetrics 对象 dm 取得屏幕解析像素，然后以 SurfaceView 作为相机预览之用，并绑定 SurfaceView，取得 SurfaceHolder 对象，最后通过 setFixedSize 额外设置预览大小。具体代码如下。

```
      /* 取得屏幕解析像素 */
      DisplayMetrics dm = new DisplayMetrics();
      getWindowManager().getDefaultDisplay().getMetrics(dm);
      intScreenX = dm.widthPixels;
      intScreenY = dm.heightPixels;
      Log.i(TAG, Integer.toString(intScreenX));

      mTextView01 = (TextView) findViewById(R.id.myTextView1);
      mImageView01 = (ImageView) findViewById(R.id.myImageView1);

      /* 以 SurfaceView 作为相机 Preview 之用 */
      mSurfaceView01 = (SurfaceView) findViewById(R.id.mSurfaceView1);

      /* 绑定 SurfaceView，取得 SurfaceHolder 对象 */
      mSurfaceHolder01 = mSurfaceView01.getHolder();

      /* Activity 必须实现 SurfaceHolder.Callback */
      mSurfaceHolder01.addCallback(example10.this);

      /* 预览大小设置，在此不使用 */
      mSurfaceHolder01.setFixedSize(320, 240);

      /*
       * 以 SURFACE_TYPE_PUSH_BUFFERS(3)
       * 作为 SurfaceHolder 显示类型
       * */
      mSurfaceHolder01.setType
      (SurfaceHolder.SURFACE_TYPE_PUSH_BUFFERS);
```

```
mButton01 = (Button)findViewById(R.id.myButton1);
mButton02 = (Button)findViewById(R.id.myButton2);
mButton03 = (Button)findViewById(R.id.myButton3);
```

5）设置打开相机及浏览按钮事件，自定义初始化打开相机方法。具体代码如下。

```
/* 打开相机及浏览 */
mButton01.setOnClickListener(new Button.OnClickListener()
{
  @Override
  public void onClick(View arg0)
  {

    /* 自定义初始化打开相机方法 */
    initCamera();
  }
});
```

6）设置停止浏览及相机按钮事件，自定义重置相机，并关闭相机预览方法。具体代码如下。

```
/* 停止及浏览相机 */
mButton02.setOnClickListener(new Button.OnClickListener()
{
  @Override
  public void onClick(View arg0)
  {

    /* 自定义重置相机，并关闭相机预览方法 */
    resetCamera();
  }
});
```

7）设置停止拍照按钮事件，当存储卡存在才允许拍照，存储暂存图像文件，并自定义拍照方法 takePicture()。具体代码如下。

```
/* 拍照 */
mButton03.setOnClickListener(new Button.OnClickListener()
{
  @Override
  public void onClick(View arg0)
  {

    /* 如果存储卡存在才允许拍照，存储暂存图像文件 */
    if(checkSDCard())
    {
```

```
                /* 自定义拍照方法 */
                takePicture();
            }
            else
            {
                /* 存储卡不存在显示提示 */
                mTextView01.setText
                (
                    getResources().getText(R.string.str_err_nosd).toString()
                );
            }
        }
    });
}
```

8）定义 initCamera()方法，若相机在非预览模式，则打开相机。流程如下。

- 第一步：创建 Camera.Parameters 对象。
- 第二步：设置照片格式为 JPEG。
- 第三步：指定 preview 的屏幕大小。
- 第四步：设置图片分辨率大小。
- 第五步：将 Camera.Parameters 参数设置为 Camera。
- 第六步：运行预览模式。

具体代码如下。

```
/* 自定义初始相机方法 */
private void initCamera()
{
    if(!bIfPreview)
    {
        /* 若相机在非预览模式，则打开相机 */
        mCamera01 = Camera.open();
    }

    if (mCamera01 != null && !bIfPreview)
    {
        Log.i(TAG, "inside the camera");

        /* 创建 Camera.Parameters 对象 */
        Camera.Parameters parameters = mCamera01.getParameters();

        /* 设置相片格式为 JPEG */
        parameters.setPictureFormat(PixelFormat.JPEG);

        /* 指定 preview 的屏幕大小 */
        parameters.setPreviewSize(320, 240);
```

```
        /* 设置图片分辨率大小 */
        parameters.setPictureSize(320, 240);

        /* 将 Camera.Parameters 设置给 Camera */
        mCamera01.setParameters(parameters);

        /* setPreviewDisplay 唯一的参数为 SurfaceHolder */
        mCamera01.setPreviewDisplay(mSurfaceHolder01);

        /* 立即运行 Preview */
        mCamera01.startPreview();
        bIfPreview = true;
    }
}
```

9）定义 takePicture()方法，调用 takePicture()方法拍照并截取图像。具体代码如下。

```
/* 拍照截取图像 */
private void takePicture()
{
  if (mCamera01 != null && bIfPreview)
  {
    /* 调用 takePicture()方法拍照 */
    mCamera01.takePicture
    (shutterCallback, rawCallback, jpegCallback);
  }
}
```

10）定义 resetCamera()方法，实现相机重置。具体代码如下。

```
/* 相机重置 */
private void resetCamera()
{
  if (mCamera01 != null && bIfPreview)
  {
    mCamera01.stopPreview();
    /* 扩展学习，释放 Camera 对象 */
    mCamera01.release();
    mCamera01 = null;
    bIfPreview = false;
  }
}

private ShutterCallback shutterCallback = new ShutterCallback()
{
  public void onShutter()
```

```
    {
    }
};

private PictureCallback rawCallback = new PictureCallback()
{
  public void onPictureTaken(byte[] _data, Camera _camera)
  {
  }
};
```

11）通过 onPictureTaken 对象传入的第一个参数表示相片的大小，并用 new File(strCaptureFilePath) 创建一个新文件，并采用压缩转档方法压缩图片，然后调用 flush()方法更新 BufferStream 中的数据，最后将拍照下来且存储完毕的图像显示出来。具体代码如下。

```
private PictureCallback jpegCallback = new PictureCallback()
{
  public void onPictureTaken(byte[] _data, Camera _camera)
  {

    /* onPictureTaken 传入的第一个参数即为相片的 byte */
    Bitmap bm = BitmapFactory.decodeByteArray
              (_data, 0, _data.length);

    /* 创建新文件 */
    File myCaptureFile = new File(strCaptureFilePath);
    try
    {
      BufferedOutputStream bos = new BufferedOutputStream
      (new FileOutputStream(myCaptureFile));

      /* 采用压缩转档方法 */
      bm.compress(Bitmap.CompressFormat.JPEG, 80, bos);

      /* 调用 flush()方法，更新 BufferStream */
      bos.flush();

      /* 结束 OutputStream */
      bos.close();

      /* 将拍照下来且存储完毕的图文件，显示出来 */
      mImageView01.setImageBitmap(bm);

      /* 显示完图文件，立即重置相机，并关闭预览 */
      resetCamera();
```

```
        /* 再重新启动相机继续预览 */
        initCamera();
      }
      catch (Exception e)
      {
        Log.e(TAG, e.getMessage());
      }
    }
  };
```

12）通过自定义方法 delFile(String strFileName)，删除临时文件，具体代码如下。

```
/* 自定义删除文件方法 */
private void delFile(String strFileName)
{
  try
  {
    File myFile = new File(strFileName);
    if(myFile.exists())
    {
      myFile.delete();
    }
  }
  catch (Exception e)
  {
    Log.e(TAG, e.toString());
    e.printStackTrace();
  }
}
```

13）通过方法 mMakeTextToast(String str, boolean isLong)，输出提示语句。具体代码如下。

```
public void mMakeTextToast(String str, boolean isLong)
{
  if(isLong==true)
  {
    Toast.makeText(example10.this, str, Toast.LENGTH_LONG).show();
  }
  else
  {
    Toast.makeText(example10.this, str, Toast.LENGTH_SHORT).show();
  }
}
```

14）通过方法 checkSDCard()，检查是否有存储卡。具体代码如下。

```
private boolean checkSDCard()
{
```

```
    /* 判断存储卡是否存在 */
    if(android.os.Environment.getExternalStorageState().equals
    (android.os.Environment.MEDIA_MOUNTED))
    {
      return true;
    }
    else
    {
      return false;
    }
  }

  @Override
  public void surfaceChanged
  (SurfaceHolder surfaceholder, int format, int w, int h)
  {
    Log.i(TAG, "Surface Changed");
  }

  @Override
  public void surfaceCreated(SurfaceHolder surfaceholder)
  {
    Log.i(TAG, "Surface Changed");
  }
```

15）当 Surface 不存在时需要删除图片，具体代码如下。

```
  @Override
  public void surfaceDestroyed(SurfaceHolder surfaceholder)
  {
    /* 如果 Surface 不存在，需要删除图片 */
    try
    {
      delFile(strCaptureFilePath);
    }
    catch(Exception e)
    {
      e.printStackTrace();
    }
    Log.i(TAG, "Surface Destroyed");
  }
}
```

接下来，需要在文件 AndroidManifest.xml 中设置 android.permission.CAMERA 权限，具体代码如下。

```
  <uses-permission android:name="android.permission.CAMERA">
```

执行后的效果如图 7-22 所示。单击“预览”、“拍照”和“关闭”按钮后，能够实现对应的功能。

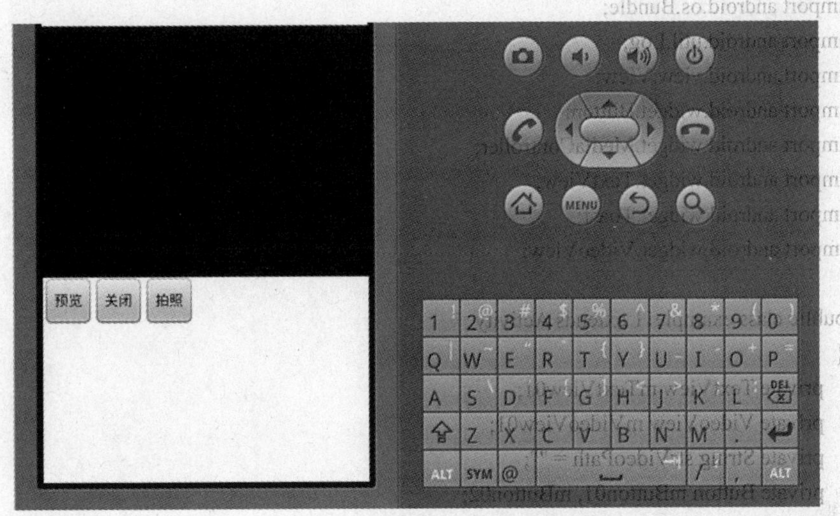

图 7-22 执行效果

7.11 3gp 影片播放器

在移动手机中，录音处理也是十分重要的功能。在本节的内容中，将通过一个具体实例的实现，介绍在 Android 手机中实现录音处理的基本流程。本实例的源代码保存在“光盘:\daima\7\example11”，下面开始讲解本实例的具体实现流程。

7.11.1 实现原理

在 Android 中，内置了 VideoView（视频视图组件）作为多媒体视频播放器，这样可以浏览电影视频。VideoView 和前面介绍的 Widget 使用方法类似，必须先在 Layout XML 中定义 VideoView 属性，在程序中通过 findViewById()方法即可创建 VideoView 对象。

在本实例中，预先准备了 2 个.3gp 格式的视频文件，上传到了虚拟 SD 卡中。然后插入 2 个按钮，当单击按钮后分别实现对这 2 个视频文件的播放。

7.11.2 具体实现

本实例的主程序文件是 example11.java，下面开始讲解其具体实现代码。

1）先加载相关类，然后分别声明变量 mTextView01、mVideoView01、strVideoPath、mButton01、mButton02 和 TAG。具体代码如下。

```
package irdc.example11;

import irdc.example11.R;
import android.app.Activity;
import android.graphics.PixelFormat;
```

```
import android.media.MediaPlayer;
import android.net.Uri;
import android.os.Bundle;
import android.util.Log;
import android.view.View;
import android.widget.Button;
import android.widget.MediaController;
import android.widget.TextView;
import android.widget.Toast;
import android.widget.VideoView;

public class example11 extends Activity
{
   private TextView mTextView01;
   private VideoView mVideoView01;
   private String strVideoPath = "";
   private Button mButton01, mButton02;
   private String TAG = "HIPPO_VIDEOVIEW";
```

2）先设置判别是否安装存储卡变量 flag（标识）为 false，然后设置全屏幕显示。具体代码如下。

```
/* 默认判别是否安装存储卡 flag 为 false */
private boolean bIfSDExist = false;

/** Called when the activity is first created. */
@Override
public void onCreate(Bundle savedInstanceState)
{
  super.onCreate(savedInstanceState);

   /* 全屏幕 */
   getWindow().setFormat(PixelFormat.TRANSLUCENT);
   setContentView(R.layout.main);
```

3）判断存储卡是否存在，不存在则通过 mMakeTextToast 对象输出提示语句。具体代码如下。

```
/* 判断存储卡是否存在 */
if(android.os.Environment.getExternalStorageState().equals
(android.os.Environment.MEDIA_MOUNTED))
{
   bIfSDExist = true;
}
else
{
   bIfSDExist = false;
```

```
        mMakeTextToast
        (
          getResources().getText(R.string.str_err_nosd).toString(),
          true
        );
      }
```

4）分别取得 TextView、EditText，然后通过 findViewById 构造 2 个对象。具体代码如下。

```
mTextView01 = (TextView)findViewById(R.id.myTextView1);
mVideoView01 = (VideoView)findViewById(R.id.myVideoView1);

mButton01 = (Button)findViewById(R.id.myButton1);
mButton02 = (Button)findViewById(R.id.myButton2);
```

5）定义单击处理事件，通过 playVideo(strVideoPath)方法播放第一个影片 A。具体代码如下。

```
mButton01.setOnClickListener(new Button.OnClickListener()
{
  @Override
  public void onClick(View arg0)
  {
    if(bIfSDExist)
    {
      /* 播放影片路径 1 */
      strVideoPath = "file:///sdcard/hello.3gp";
      playVideo(strVideoPath);
    }
  }
});
```

6）定义单击处理事件，通过 playVideo(strVideoPath)方法播放第二个影片 B。具体代码如下。

```
mButton02.setOnClickListener(new Button.OnClickListener()
{
  @Override
  public void onClick(View arg0)
  {
    if(bIfSDExist)
    {
      /* 播放影片路径 2 */
      strVideoPath = "file:///sdcard/test.3gp";
      playVideo(strVideoPath);
    }
  }
});
```

```
    }
```

7）自定义 VideoView()方法，用于播放指定路径的影片。具体代码如下。

```
/* 自定义以 VideoView 播放影片 */
private void playVideo(String strPath)
{
  if(strPath!="")
  {
    /* 调用 VideoURI 方法，指定解析路径 */
    mVideoView01.setVideoURI(Uri.parse(strPath));

    /* 设置控制 Bar 显示于此 Context 中 */
    mVideoView01.setMediaController
    (new MediaController(example11.this));

    mVideoView01.requestFocus();

    /* 调用 VideoView.start()自动播放 */
    mVideoView01.start();
    if(mVideoView01.isPlaying())
    {
      /* 程序不会被运行，因 start()后尚需要 preparing() */
      mTextView01.setText("Now Playing:"+strPath);
      Log.i(TAG, strPath);
    }
  }
}
```

8）定义 mMakeTextToast(String str, boolean isLong)方法，输出提醒语句。具体代码如下。

```
public void mMakeTextToast(String str, boolean isLong)
{
  if(isLong==true)
  {
    Toast.makeText(example11.this, str, Toast.LENGTH_LONG).show();
  }
  else
  {
    Toast.makeText(example11.this, str, Toast.LENGTH_SHORT).show();
  }
}
}
```

执行后的效果如图 7-23 所示。当单击“播放 SD 3gp 影片 1”和“播放 SD 3gp 影片 2”按钮后，分别播放预设的影片。

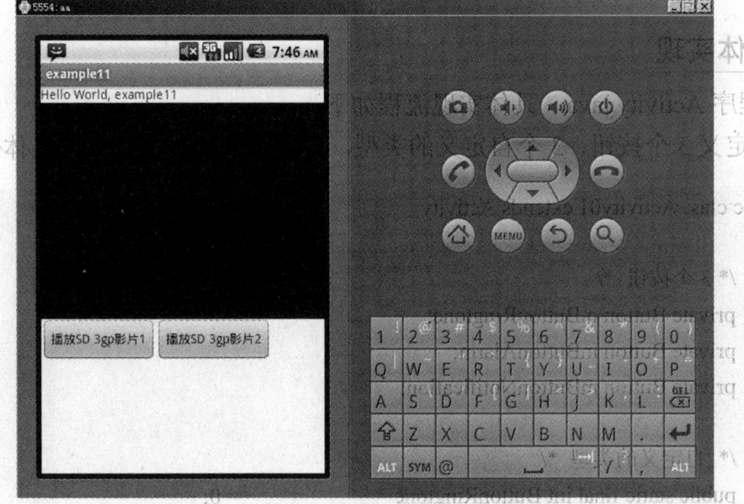

图 7-23 执行效果

7.12 铃声设置

在移动手机中，铃声设置也是十分重要的功能，用户可以去网络中下载自己喜欢的铃声。在本节的内容中，将通过一个具体实例的实现，介绍在 Android 手机中设置指定铃声的基本流程。本实例的源代码保存在“光盘:\daima\7\example12”，下面开始讲解本实例的具体实现流程。

7.12.1 实现原理

在 Android 中，通过 RingtoneManager 类来专门控制各种铃声。例如，常见的来电铃声、闹钟铃声、一些警告和信息通知。RingtoneManager 类的常用方法如下。

- getActualDefaultRingtoneUri：获取指定类型的当前默认铃声。
- getCursor：返回所有可用铃声的游标。
- getDefaultType：获取指定 URL 默认的铃声类型。
- getDefaultUri：返回指定类型默认铃声的 URL。
- getRingtoneUri：返回指定位置铃声的 URL。
- getRingtonePosition：获取指定铃声的位置。
- getValidRingtoneUri：获取一个可用铃声的位置。
- isDefault：获取指定 URL 是否是默认的铃声。
- setActualDefaultRingtoneUri：设置默认的铃声。

在 Android 系统中，默认的铃声存储在“system/medio/audio”目录中，而下载的铃声一般被保存在 SD 卡中，假设下载的铃声分别保存在 SD 卡的下述目录中。

- sdcard/music/ringtone：一般来电铃声。
- sdcard/music/alarm：闹钟铃声。
- sdcard/music/notification：警告、通知铃声。

7.12.2 具体实现

编写成程序 Activity.java，具体实现流程如下：

1）分别定义 3 个按钮、3 个自定义的类型，设置 3 个铃声文件夹。具体代码如下。

```
public class Activity01 extends Activity
{
    /* 3 个按钮 */
    private Button mButtonRingtone;
    private Button mButtonAlarm;
    private Button mButtonNotification;

    /* 自定义的类型 */
    public static final int ButtonRingtone          = 0;
    public static final int ButtonAlarm             = 1;
    public static final int ButtonNotification      = 2;
    /* 铃声文件夹 */
    private String strRingtoneFolder = "/sdcard/music/ringtone";
    private String strAlarmFolder = "/sdcard/music/alarm";
    private String strNotificationFolder = "/sdcard/music/notification";
```

2）设置单击按钮 mButtonRingtone 后的处理事件，打开系统铃声设置，然后进行设置。具体代码如下。

```
/** Called when the activity is first created. */
@Override
public void onCreate(Bundle savedInstanceState)
{
    super.onCreate(savedInstanceState);
    setContentView(R.layout.main);

    mButtonRingtone = (Button) findViewById(R.id.ButtonRingtone);
    mButtonAlarm = (Button) findViewById(R.id.ButtonAlarm);
    mButtonNotification = (Button) findViewById(R.id.ButtonNotification);
    /* 设置来电铃声 */
    mButtonRingtone.setOnClickListener(new Button.OnClickListener()
    {
        @Override
        public void onClick(View arg0)
        {
            if (bFolder(strRingtoneFolder))
            {
                /*打开系统铃声设置*/
                Intent intent = new Intent(RingtoneManager.ACTION_RINGTONE_PICKER);
                /*类型为来电 RINGTONE 模式*/
```

```
                intent.putExtra(RingtoneManager.EXTRA_RINGTONE_TYPE, RingtoneManager.
TYPE_RINGTONE);
                /*设置显示的标题*/
                intent.putExtra(RingtoneManager.EXTRA_RINGTONE_TITLE, "设置来电铃声");
                /*当设置完成之后返回到当前的 Activity*/
                startActivityForResult(intent, ButtonRingtone);
            }
        }
    });
```

3）设置单击按钮 mButtonAlarm 后的处理事件，打开系统铃声设置，然后进行设置。具体代码如下。

```
    /* 设置闹钟铃声 */
    mButtonAlarm.setOnClickListener(new Button.OnClickListener()
    {
        @Override
        public void onClick(View arg0)
        {
            if (bFolder(strAlarmFolder))
            {
                /*打开系统铃声设置*/
                Intent intent = new Intent(RingtoneManager.ACTION_RINGTONE_PICKER);
                /*设置铃声类型和 title*/
                intent.putExtra(RingtoneManager.EXTRA_RINGTONE_TYPE, RingtoneManager.TYPE_
ALARM);
                intent.putExtra(RingtoneManager.EXTRA_RINGTONE_TITLE, "设置闹铃铃声");
                /*当设置完成之后返回到当前的 Activity*/
                startActivityForResult(intent, ButtonAlarm);
            }
        }
    });
```

4）设置单击按钮 mButtonNotification 后的处理事件，打开系统铃声设置，然后进行设置。具体代码如下。

```
    /* 设置通知铃声 */
    mButtonNotification.setOnClickListener(new Button.OnClickListener()
    {
        @Override
        public void onClick(View arg0)
        {
            if (bFolder(strNotificationFolder))
            {
                /*打开系统铃声设置*/
                Intent intent = new Intent(RingtoneManager.ACTION_RINGTONE_PICKER);
```

```
                    /*设置铃声类型和 title*/
                    intent.putExtra(RingtoneManager.EXTRA_RINGTONE_TYPE, RingtoneManager.TYPE_
NOTIFICATION);
                    intent.putExtra(RingtoneManager.EXTRA_RINGTONE_TITLE, "设置通知铃声");
                    /*当设置完成之后返回到当前的 Activity*/
                    startActivityForResult(intent, ButtonNotification);
                }
            }
        });
    }
```

5）定义方法 onActivityResult()，此方法作为设置铃声之后的回调方法。具体代码如下。

```
    /* 当设置铃声之后的回调方法 */
    @Override
    protected void onActivityResult(int requestCode, int resultCode, Intent data)
    {
        if (resultCode != RESULT_OK)
        {
            return;
        }
        switch (requestCode)
        {
            case ButtonRingtone:
                try
                {
                    /*得到选择的铃声*/
                    Uri pickedUri = data.getParcelableExtra(RingtoneManager.EXTRA_RINGTONE_
PICKED_URI);
                    /*将选择的铃声设置成为默认*/
                    if (pickedUri != null)
                    {
                        RingtoneManager.setActualDefaultRingtoneUri(Activity01.this, Ringtone
Manager.TYPE_RINGTONE, pickedUri);
                    }
                }
                catch (Exception e)
                {
                }
                break;
            case ButtonAlarm:
                try
                {
                    /*得到选择的铃声*/
                    Uri pickedUri = data.getParcelableExtra(RingtoneManager.EXTRA_RINGTONE_
PICKED_URI);
```

```
                    /*将选择的铃声设置成为默认*/
                    if (pickedUri != null)
                    {
                        RingtoneManager.setActualDefaultRingtoneUri(Activity01.this, Ringtone
Manager.TYPE_ALARM, pickedUri);
                    }
                }
                catch (Exception e)
                {
                }
                break;
            case ButtonNotification:
                try
                {
                    /*得到选择的铃声*/
                    Uri pickedUri = data.getParcelableExtra(RingtoneManager.EXTRA_RINGTONE_
PICKED_URI);
                    /*将选择的铃声设置成为默认*/
                    if (pickedUri != null)
                    {
                        RingtoneManager.setActualDefaultRingtoneUri(Activity01.this, Ringtone
Manager.TYPE_NOTIFICATION, pickedUri);
                    }
                }
                catch (Exception e)
                {
                }
                break;
        }
        super.onActivityResult(requestCode, resultCode, data);
    }
```

6）定义方法 boolean bFolder，用于检测是否存在指定的文件夹，如果不存在则创建。具体代码如下。

```
    private boolean bFolder(String strFolder)
    {
        boolean btmp = false;
        File f = new File(strFolder);
        if (!f.exists())
        {
            if (f.mkdirs())
            {
                btmp = true;
            }
            else
            {
```

```
                        btmp = false;
                    }
                }
                else
                {
                    btmp = true;
                }
                return btmp;
            }
        }
```

执行后的效果如图 7-24 所示。在此可以分别设置这三种类型的铃声。

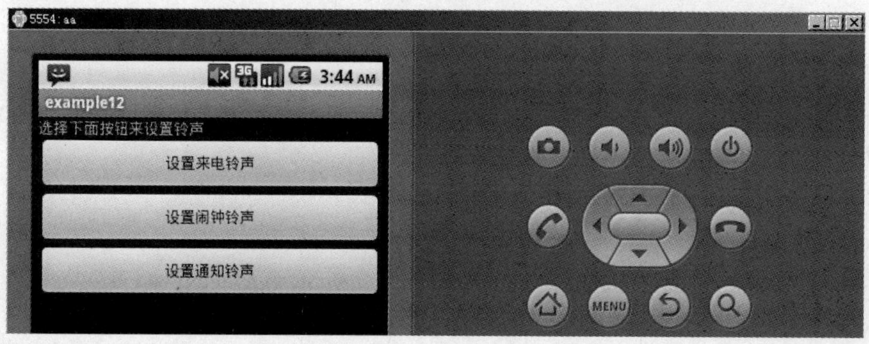

图 7-24　执行效果

第8章 网络应用

在移动手机应用中，网络是一个重要的构成模块，例如电子邮件、QQ聊天、网上冲浪已经充斥了我们眼球。作为智能手机系统，在 Android 平台上可以尽情玩享这些网络应用。在本节的内容中，将通过几个典型实例的实现过程，来详细介绍在 Android 中实现网络编程的基本知识。

8.1 最常见的传递 HTTP 参数

了解 Web 技术的读者，对于 HTTP 应该不会陌生，HTTP 是一种网络传输协议，生活中的大多数网页都是通过“HTTP://WWW.”的形式显示的。在具体应用中，一些需要的数据都是通过其参数传递的。在本节的内容中，将通过一个具体实例的实现，介绍在 Android 中传递 HTTP 参数的基本流程。本实例的源代码保存在“光盘:\daima\8\example1”，下面开始讲解本实例的具体实现流程。

8.1.1 实现原理

和网络 HTTP 有关的协议是 HTTP 协议，在 Android SDK 中，集成了 Apache 的 HttpClient 模块。通过此模块，可以方便地编写出和 HTTP 有关的程序。在 Android SDK 中通常使用 HttpClient 4.0 版。

在本实例中，使用了两个按钮，一个用于以 Post 方式获取网站数据，另外一个用于以 Get 方式获取数据，并以 TextView 对象来显示由服务器端返回的网页内容。当然首先得建立和 HTTP 的连接，连接之后才能获取 Web Server 返回的结果。

8.1.2 具体实现

本实例的主程序文件是 example1.java，下面开始讲解其具体实现代码。

1）分别声明两个 Button 对象和一个 TextView 对象，然后通过 findViewById 构造器创建 TextView 与 Button 对象。具体代码如下。

```
public class example1 extends Activity
{
  /*分别声明两个 Button 对象和一个 TextView 对象*/
  private Button mButton1,mButton2;
  private TextView mTextView1;

  /** Called when the activity is first created. */
  @Override
```

```
public void onCreate(Bundle savedInstanceState)
{
  super.onCreate(savedInstanceState);
  setContentView(R.layout.main);

  /*通过 findViewById 构造器创建 TextView 与 Button 对象*/
  mButton1 =(Button) findViewById(R.id.myButton1);
  mButton2 =(Button) findViewById(R.id.myButton2);
  mTextView1 = (TextView) findViewById(R.id.myTextView1);
```

2）设定 OnClickListener 来监听第一个按钮的 OnClick 事件，首先声明网址字符串，并建立 Post 方式联机，最后通过 mTextView1.setText 对象输出提示字符。具体代码如下。

```
/*设定 OnClickListener 来监听 OnClick 事件*/
mButton1.setOnClickListener(new Button.OnClickListener()
{
  /*覆盖 onClick 事件*/
  @Override
  public void onClick(View v)
  {
    /*声明网址字符串*/
    String uriAPI = "http://www.dubblogs.cc:8751/Android/Test/API/Post/index.php";
    /*建立 HTTP Post 联机方式*/
    HttpPost httpRequest = new HttpPost(uriAPI);
    /*
     * Post 运行传送变量必须用 NameValuePair[]数组存储
    */
    List <NameValuePair> params = new ArrayList <NameValuePair>();
    params.add(new BasicNameValuePair("str", "I am Post String"));
    try
    {
      /*发送 HTTP 请求*/
      httpRequest.setEntity(new UrlEncodedFormEntity(params, HTTP.UTF_8));
      /*取得 HTTP 应答*/
      HttpResponse httpResponse = new DefaultHttpClient().execute(httpRequest);
      /*若状态码为 200*/
      if(httpResponse.getStatusLine().getStatusCode() == 200)
      {
        /*获取应答字符串*/
        String strResult = EntityUtils.toString(httpResponse.getEntity());
        mTextView1.setText(strResult);
      }
      else
      {
        mTextView1.setText("Error Response: "+httpResponse.getStatusLine().toString());
      }
    }
```

```
        catch (ClientProtocolException e)
        {
          mTextView1.setText(e.getMessage().toString());
          e.printStackTrace();
        }
        catch (IOException e)
        {
          mTextView1.setText(e.getMessage().toString());
          e.printStackTrace();
        }
        catch (Exception e)
        {
          mTextView1.setText(e.getMessage().toString());
          e.printStackTrace();
        }

      }
    });
```

3）设定 OnClickListener 来监听第二个按钮的 OnClick 事件，首先声明网址字符串，然后建立 HTTP Get 联机方式，分别实现发出 HTTP 请求、获取应答字符串和删除冗余字符，最后通过 mTextView1.setText 输出提示字符。具体代码如下。

```
mButton2.setOnClickListener(new Button.OnClickListener()
{
  @Override
  public void onClick(View v)
  {
    /*声明网址字符串*/
    String uriAPI = "http://www.XXXX.cc:8751/index.php?str=I+am+Get+String";
    /*建立 HTTP Get 联机方式*/
    HttpGet httpRequest = new HttpGet(uriAPI);
    try
    {
     /*发出 HTTP 请求*/
     HttpResponse httpResponse = new DefaultHttpClient().execute(httpRequest);
      /*若状态码为 200*/
      if(httpResponse.getStatusLine().getStatusCode() == 200)
      {
        /*获取应答字符串*/
        String strResult = EntityUtils.toString(httpResponse.getEntity());
        /*删除冗余字符*/
        strResult = eregi_replace("(\r\n|\r|\n|\n\r)","",strResult);
        mTextView1.setText(strResult);
      }
      else
```

```
            {
                mTextView1.setText("Error Response: "+httpResponse.getStatusLine().toString());
            }
          }
          catch (ClientProtocolException e)
          {
            mTextView1.setText(e.getMessage().toString());
            e.printStackTrace();
          }
          catch (IOException e)
          {
            mTextView1.setText(e.getMessage().toString());
            e.printStackTrace();
          }
          catch (Exception e)
          {
            mTextView1.setText(e.getMessage().toString());
            e.printStackTrace();
          }
        }
      });
    }
```

4）设定字符串替换方法 eregi_replace(String strFrom, String strTo, String strTarget)，替换掉一些非法字符。具体代码如下。

```
      /* 字符串替换方法*/
      public String eregi_replace(String strFrom, String strTo, String strTarget)
      {
        String strPattern = "(?i)"+strFrom;
        Pattern p = Pattern.compile(strPattern);
        Matcher m = p.matcher(strTarget);
        if(m.find())
        {
          return strTarget.replaceAll(strFrom, strTo);
        }
        else
        {
          return strTarget;
        }
      }
    }
```

接下来在文件 AndroidManifest.xml 中添加对网络连接权限，具体代码如下。

```
      <uses-permission android:name="android.permission.INTERNET"></uses-permission>
```

执行后的效果如图 8-1 所示。分别单击图 8-1 中的按钮，能够以不同方式获取 HTTP 参数。

图 8-1　执行效果

8.2　实现网页浏览

网上冲浪功能，对于现在的手机来说已经不是什么难事。在本节的内容中，将通过一个具体实例的实现，介绍在 Android 中编程实现网页浏览的基本流程。本实例的源代码保存在"光盘:\daima\8\example2"，下面开始讲解本实例的具体实现流程。

8.2.1　实现原理

在 Android 中，内置了一个 WebKit 引擎，里面的 WebView 组件能够迅速实现网页浏览。在本实例中，通过 WebView.loadUrl 方法来加载网址，所以从 EditText 中传入要浏览的网址后，就可以在 WebView 中加载网页的内容了。

8.2.2　具体实现

本实例的主程序文件是 example2.java，下面开始讲解其具体实现代码。

通过 setOnClickListener 监听按钮单击事件，单击箭头后先获取 EditText 中的数据，然后打开此网址，并在 WebView 中显示网页内容。具体代码如下。

```
/** Called when the activity is first created. */
@Override
public void onCreate(Bundle savedInstanceState)
{
  super.onCreate(savedInstanceState);
  setContentView(R.layout.main);

  mImageButton1 = (ImageButton)findViewById(R.id.myImageButton1);
  mEditText1 = (EditText)findViewById(R.id.myEditText1);
  mWebView1 = (WebView) findViewById(R.id.myWebView1);
```

```
        /*当单击箭头后*/
        mImageButton1.setOnClickListener(new
                                         ImageButton.OnClickListener()
        {
          @Override
          public void onClick(View arg0)
          {
            {
                mImageButton1.setImageResource(R.drawable.go_2);
                /*获取 EditText 中的数据*/
                String strURI = (mEditText1.getText().toString());
                /*    WebView 显示网页内容*/
                mWebView1.loadUrl(strURI);
                Toast.makeText(
                      example2.this,getString(R.string.load)+strURI,
                                  Toast.LENGTH_LONG)
                      .show();
            }
          }
        });
      }
    }
```

执行后显示一个文本框，在此可以输入网址，如图 8-2 所示。输入网址并单击后面的 ▶ 后，将显示此网页的内容。如图 8-3 所示。

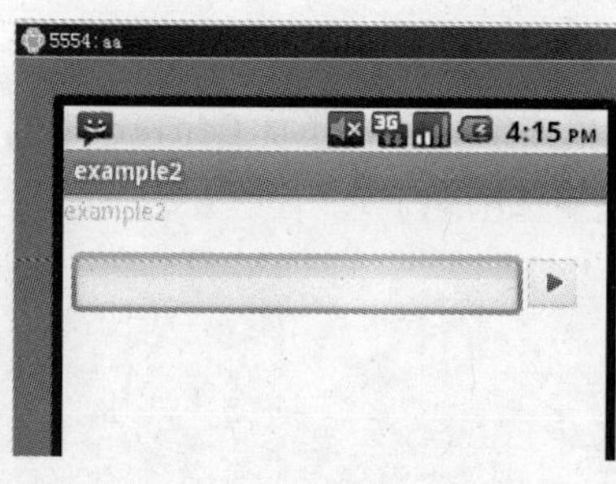

图 8-2 输入网址

图 8-3 打开的网页

8.3 手机使用 HTML 程序

HTML 语言是当前主流的网页技术。在本节的内容中，将通过一个具体实例的实现，介绍

在 Android 中使用 HTML 程序的基本流程。本实例的源代码保存在“光盘:\daima\8\example3”，下面开始讲解本实例的具体实现流程。

8.3.1 实现原理

实际上，WebView 是一个嵌入式的浏览器，可以直接使用 WebView.loadData()方法，将 HTML 标记传递给 WebView 对象，让 Android 手机程序具备 Web 浏览器的功能。这样，网页程序被放在了 WebView 中运行，如同一个 Web 程序。在当前移动程序中，网页下载和动画展示等都利用了 WebView 中的 loadData()方法来载入网页。

8.3.2 具体实现

本实例的主程序文件是 example3.java，在 loadData()方法中插入了指定的 HTML 代码，通过 HTML 代码，显示了一幅图片和文字，并且插入了超级链接功能。具体代码如下。

```
package irdc.example3;

import irdc.example3.R;
import android.app.Activity;
import android.os.Bundle;
import android.webkit.WebView;

public class example3 extends Activity
{
  private WebView mWebView1;
  /** Called when the activity is first created. */
  @Override
  public void onCreate(Bundle savedInstanceState)
  {
    super.onCreate(savedInstanceState);
    setContentView(R.layout.main);

    mWebView1 = (WebView) findViewById(R.id.myWebView1);

    /*自行设置 WebView 要显示的网页内容*/
    mWebView1.
      loadData(
      "<html><body><p>aaaaaaa</p>" +
      "<div class='widget-content'> "+
      "<a href=http://www.sohu.com>" +
      "<img src=http://hiphotos.baidu.com/chaojihedan/pic/item/bbddf5efc260f133fdfa3cd8.jpg />" +
      "<a href=http://www.sohu.com>Link Blog</a>" +
      "</body></html>", "text/html", "utf-8");
  }
}
```

执行后将显示 HTML 产生的页面，如图 8-4 所示。单击超链接后会跳转到指定的目标页面。如图 8-5 所示。

图 8-4 输入网址

图 8-5 打开的网页

8.4 用内置浏览器打开网页

前面几个实例实际上是 Android 浏览器的部分功能，实际上用户可以直接调用它的内置浏览器，来实现上网操作。在本节的内容中，将通过一个具体实例的实现，介绍调用 Android 内置浏览器的基本流程。本实例的源代码保存在“光盘:\daima\8\example4”，下面开始讲解本实例的具体实现流程。

8.4.1 实现原理

在本实例中，定义了一个 ListView，列表显示了 4 个菜单，单击菜单后会连接到指定的页面。当 ListView 的 ItemClick()事件发生时，通过 Intent(Intent.ACTION_VIEW,uri)方法来打开内置的浏览器，并浏览 ListView 中创建的网页 URL。

8.4.2 具体实现

本实例的主程序文件是 example4.java，下面开始介绍其实现流程。

1）分别声明一个 ListView 和 TextView 对象变量，然后声明一个 String array 变量来存储收藏夹列表，最后声明一个 String 变量来存储网址。具体代码如下。

```
package irdc.example4;

public class example4 extends Activity
{
  /*声明一个 ListView,TextView 对象变量
```

```
 * 一个 String array 变量存储收藏夹列表
 * 与 String 变量来存储网址*/
private ListView mListView1;
private TextView mTextView1;
private String[] myFavor;
private String    myUrl;
```

2）先通过 findViewById 构造器创建 ListView 与 TextView 对象，将 string.xml 文件中的信息导入到列表中。具体代码如下。

```
/** Called when the activity is first created. */
@Override
public void onCreate(Bundle savedInstanceState)
{
  super.onCreate(savedInstanceState);
  setContentView(R.layout.main);

  /*通过 findViewById 构造器创建 ListView 与 TextView 对象*/
  mListView1 =(ListView) findViewById(R.id.myListView1);
  mTextView1 = (TextView) findViewById(R.id.myTextView1);
  mTextView1.setText(getResources().getString(R.string.hello));
  /*将 string.xml 文件中信息导入到列表*/
  myFavor = new String[] {
                          getResources().getString
                          (R.string.str_list_url1),
                          getResources().getString
                          (R.string.str_list_url2),
                          getResources().getString
                          (R.string.str_list_url3),
                          getResources().getString
                          (R.string.str_list_url4)
                        };
```

3）自定义 ArrayAdapter 对象，时刻准备传入 ListView 中，并将 myFavor 对象的列表以参数传入。然后自定义完成的 ArrayAdapter 传入自定义的 ListView 中，并将 ListAdapter 的可选(Focusable)菜单选项打开，最后设置 ListView 选项的 nItemClickListener。具体代码如下。

```
/*自定义一 ArrayAdapter 准备传入 ListView 中,并将 myFavor 列表以参数传入*/
ArrayAdapter<String> adapter = new
ArrayAdapter<String>
(example4.this, android.R.layout.simple_list_item_1, myFavor);

/*将自定义完成的 ArrayAdapter 传入自定义的 ListView 中*/
mListView1.setAdapter(adapter);
/*将 ListAdapter 的可选(Focusable)菜单选项打开*/
mListView1.setItemsCanFocus(true);
/*设置 ListView 菜单选项设为每次只能单一选项*/
```

```
mListView1.setChoiceMode
(ListView.CHOICE_MODE_SINGLE);
/*设置 ListView 选项的 nItemClickListener*/
mListView1.setOnItemClickListener
(new ListView.OnItemClickListener()
{
```

4）定义覆盖 onItemClick()方法，当用户单击一个 Item（条目）后，会进行比较，并从文件 string.xml 中取出对应的 URL 网址，并将字符串转换为 URL 对象。具体代码如下。

```
@Override
/*覆盖 OnItemClick()方法*/
public void onItemClick
(AdapterView<?> arg0, View arg1, int arg2,long arg3)
{
  /*如果所选菜单的文字与 myFavor 字符串数组第一个文字相同*/
  if(arg0.getAdapter().getItem(arg2).toString()==
  myFavor[0].toString())
  {
    /*取得网址并调用 goToUrl()方法*/
    myUrl=getResources().getString(R.string.str_url1);
    goToUrl(myUrl);
  }
  /*如果所选菜单的文字与 myFavor 字符串数组第二个文字相同*/
  else if (arg0.getAdapter().getItem(arg2).toString()==
  myFavor[1].toString())
  {
    /*取得网址并调用 goToUrl()方法*/
    myUrl=getResources().getString(R.string.str_url2);
    goToUrl(myUrl);
  }
  /*如果所选菜单的文字与 myFavor 字符串数组第三个文字相同*/
  else if (arg0.getAdapter().getItem(arg2).toString()==
  myFavor[2].toString())
  {
    /*取得网址并调用 goToUrl()方法*/
    myUrl=getResources().getString(R.string.str_url3);
    goToUrl(myUrl);
  }
  /*如果所选菜单的文字与 myFavor 字符串数组第四个文字相同*/
  else if (arg0.getAdapter().getItem(arg2).toString()==
  myFavor[3].toString())
  {
    /*取得网址并调用 goToUrl()方法*/
    myUrl=getResources().getString(R.string.str_url4);
    goToUrl(myUrl);
  }
```

```
        /*以上皆非*/
        else
        {
          /*显示错误信息*/
          mTextView1.setText("Ooops!!出错了");
        }
      }
    });
  }
```

5）定义方法 goToUrl(String url)，用于打开网址为 URL 的网页。具体代码如下。

```
    /*打开网页的方法*/
    private void goToUrl(String url)
    {
      Uri uri = Uri.parse(url);
      Intent intent = new Intent(Intent.ACTION_VIEW, uri);
      startActivity(intent);
    }
  }
```

执行后将列表显示 4 个菜单，如图 8-6 所示。当单击一个菜单后，会跳转到对应的目标页面。如图 8-7 所示。

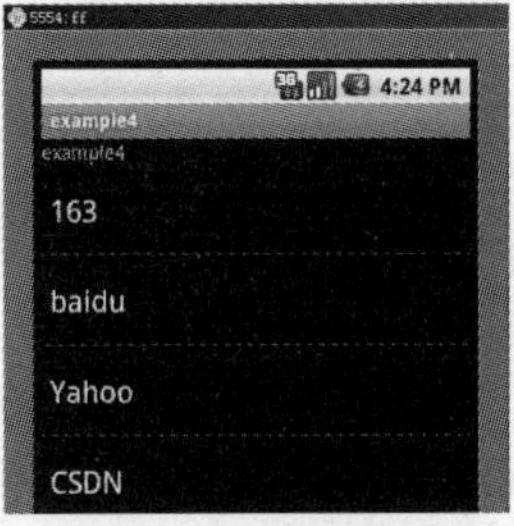

图 8-6　4 个菜单

图 8-7　打开的网页

8.5　Gallery 中显示网络照片

网络真是很神奇，在 QQ 空间中可以存放自己的照片。我们有时并不需要在 Gallery 中存放照片，可以直接从网络中调用照片，并在 Gallery 中显示出来，这样可以节约手机的存储空间。在本节的内容中，将通过一个具体实例的实现，介绍 Gallery 中调用网络照片显示的基本流程。本实例的源代码保存在“光盘:\daima\8\example5”，下面开始讲解本实例的具体实现流程。

8.5.1　实现原理

在本实例中，需要将 URL 网址的照片实时处理下载后，以 InputStream（输入流）转换为 Bitmap 图像，这样才能放入 BaseAdapter 中。在运行实例前，需要预先准备照片并上传到网络空间中，获取照片的连接后，再以 String 数组方式放在程序中，并对 BaseAdapter 稍作修改，增加对 URL 对象的访问以及对 URLConnection 连接的处理。

8.5.2 具体实现

本实例的主程序文件是 example5.java，下面开始介绍其实现流程。

1）分别声明 Gallery 中要显示 5 幅图片的地址栏字符串，具体代码如下。

```
public class example5 extends Activity
{
  private Gallery myGallery01;
  /* 地址栏字符串 */
  private String[] myImageURL = new String[]
  {
      "http://b27.photo.store.qq.com/http_imgload.cgi?/"
          +
"rurl4_b=086a67cbd6a8cfb4389ea2b48efab6f322f755a085107a7aeeaa56fc1358b1bd124186254e021f0655732688e6
9f060725491f8ae82e8e5508dbe9821670e2baf04e92dedc97e3bbf28e5605596aa991c13220f1&a=27&b=27",
      "http://b27.photo.store.qq.com/http_imgload.cgi?/"
          +
"rurl4_b=086a67cbd6a8cfb4389ea2b48efab6f3ea78f5797abbbaa617259f2d2a980a5468f2801897cfcc2b78af92fbb87
565ed7a3a08041daff2dd9ccd26d3cc6198e41f2d205c8a0c445325771e8a179215999afaf9f3&a=27&b=27",
      "http://b27.photo.store.qq.com/http_imgload.cgi?/"
          +
"rurl4_b=2a9dcf1fd909a7ed3ce8951f738608982f26d812b3a5fc96e221b85fc085e7cc3264ee20730f0fd3a1f7aca0674
0db7a6153d9357467ca39f82b866b6fbe3cd94bbdd10ed01841e67c95d8e4af8890b7ced40869&a=30&b=27",
      "http://b27.photo.store.qq.com/http_imgload.cgi?/"
          +
"rurl4_b=2a9dcf1fd909a7ed3ce8951f73860898bb7ff57a8cb7747c9f0eb6a02124850b709c0b86f086a4ba5653eeb71d
d4b01e4a58f407e2eec9433cd8d4bc0b88fda56260c2c8beb34ebab77b610c7131393f82e774ef&a=27&b=27",
      "http://b27.photo.store.qq.com/http_imgload.cgi?/"
          +
"rurl4_b=2a9dcf1fd909a7ed3ce8951f73860898158d252489f84e7d2a83d44c01b7bb12b2c19ca0efdd555dba788407fd
01e9de45524b11a9793f532624197bc8d14c84ae78ddebafe4357e4eedc60e9e510224367490bf&a=27&b=27" };
```

2）引入布局文件 main.xml，定义类成员 myContext Context 对象，然后设置只有一个参数“C”的构造器，即要存储的 Context。获取 Gallery 属性的 Index id，并设置对象的 styleable 属性，使其能够反复使用。具体代码如下。

```
/** Called when the activity is first created. */
@Override
public void onCreate(Bundle savedInstanceState)
{
  super.onCreate(savedInstanceState);
  setContentView(R.layout.main);

  myGallery01 = (Gallery) findViewById(R.id.myGallery01);
```

```
    myGallery01.setAdapter(new myInternetGalleryAdapter(this));
  }

  /* 用 BaseAdapter */
  public class myInternetGalleryAdapter extends BaseAdapter
  {
    /* 类成员 myContext Context 对象 */
    private Context myContext;
    private int mGalleryItemBackground;

    /*构造器只有一个参数，即要存储的 Context */
    public myInternetGalleryAdapter(Context c)
    {
      this.myContext = c;
      TypedArray a = myContext
        .obtainStyledAttributes(R.styleable.Gallery);

      /*     获取 Gallery 属性的 Index id */
      mGalleryItemBackground = a.getResourceId(
          R.styleable.Gallery_android_galleryItemBackground, 0);

      /* 把对象的 styleable 属性能够反复使用 */
      a.recycle();
    }
```

3）定义方法 getCount()，用于返回已定义图片的总量。然后定义方法 getItem(int position)方法，使用 getItem 方法获取当前容器中图像数的数组 ID。具体代码如下。

```
    /*     返回全部已定义图片的总量 */
    public int getCount()
    {
      return myImageURL.length;
    }

    /* 使用 getItem 方法获取当前容器中图像数的数组 ID */
    public Object getItem(int position)
    {
      return position;
    }

    public long getItemId(int position)
    {
      return position;
    }
```

4）定义方法 getScale(boolean focused, int offset)方法，根据中央位移量，利用 getScale()方法返回 views 的大小(0.0f to 1.0f)。具体代码如下。

```
/* 根据中央位移量，利用 getScale 返回 views 的大小(0.0f to 1.0f) */
public float getScale(boolean focused, int offset)
{
  /* Formula: 1 / (2 ^ offset) */
  return Math.max(0, 1.0f / (float) Math.pow(2, Math
       .abs(offset)));
}
```

5）定义 getView()方法，根据中央位移量，获取当前要显示的图像 View，传入数组 ID 值使之读取并成像处理。流程如下。

- 第一步：创建 ImageView 对象。
- 第二步：用 new URL 将对象网址传入。
- 第三步：获取连接。
- 第四步：获取返回的输入流。
- 第五步：将 InputStream 变为 Bitmap。
- 第六步：关闭 InputStream。
- 第七步：设置 Bitmap 到 ImageView 中。
- 第八步：设置 ImageView 的宽和高，单位是 dip。
- 第九步：设置 Gallery 背景图。

具体代码如下。

```
@Override
public View getView(int position, View convertView,
    ViewGroup parent)
{
  /* 创建 ImageView 对象*/

  ImageView imageView = new ImageView(this.myContext);
  try
  {
    /* new URL 将对象网址传入 */
    URL aryURI = new URL(myImageURL[position]);
    /* 获取连接 */
    URLConnection conn = aryURI.openConnection();
    conn.connect();
    /* 获取返回的 InputStream */
    InputStream is = conn.getInputStream();
    /* 将 InputStream 变为 Bitmap */
    Bitmap bm = BitmapFactory.decodeStream(is);
    /* 关闭 InputStream */
    is.close();
    /* 设置 Bitmap 到 ImageView 中 */
    imageView.setImageBitmap(bm);
  } catch (IOException e)
  {
```

```
            e.printStackTrace();
        }

        imageView.setScaleType(ImageView.ScaleType.FIT_XY);
        /* 设置 ImageView 的宽和高，单位是 dip */
        imageView.setLayoutParams(new Gallery.LayoutParams(200, 150));
        /* 设置 Gallery 背景图*/
        imageView.setBackgroundResource(mGalleryItemBackground);
        return imageView;
      }
    }
  }
```

执行后将在 Gallery 中显示网络中的图片。如图 8-8 所示。

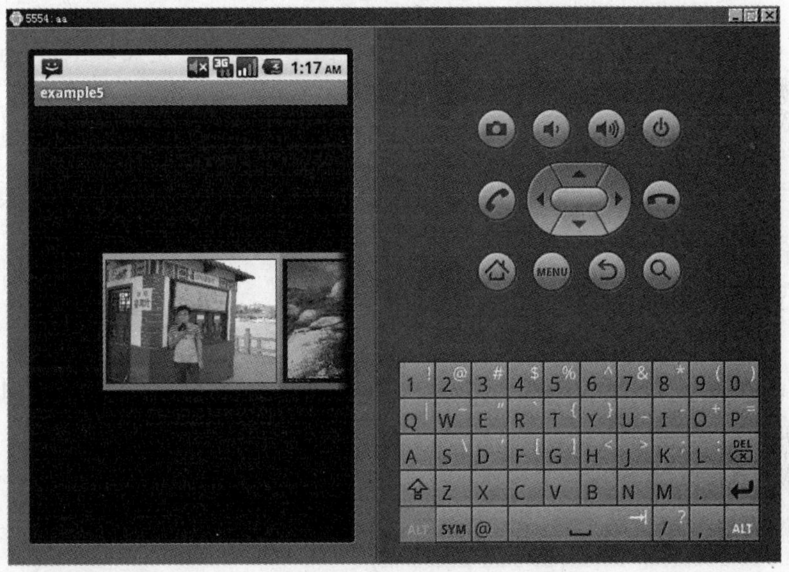

图 8-8 执行效果

8.6 网络播放 MP3

同样是为了节约手机的存储空间，在听音乐时可以从网络中下载播放 MP3。在本节的内容中，将通过一个具体实例的实现，介绍在 Android 中网络下载播放 MP3 的基本流程。本实例的源代码保存在“光盘:\daima\8\example6”，下面开始讲解本实例的具体实现流程。

8.6.1 实现原理

在本实例中，首先插入 4 个按钮，分别用于播放、暂停、重新播放和停止处理。执行后，通过 Runnable 接口发起运行线程，在线程中远程下载指定的 MP3 文件。下载完毕后，临时保存到 SD 卡中，这样可以通过 4 个按钮对其进行控制。当程序关闭后，删除 SD 卡中的临时性文件。

8.6.2 具体实现

本实例的主程序文件是 example6.java，下面开始介绍其实现流程。

1）先加载相关类，然后声明系统中需要的私有变量，具体代码如下。

```
public class example6 extends Activity
{
  private TextView mTextView01;
  private MediaPlayer mMediaPlayer01;
  private ImageButton mPlay, mReset, mPause, mStop;
  private boolean bIsReleased = false;
  private boolean bIsPaused = false;
  private static final String TAG = "Hippo_URL_MP3_Player";
```

2）定义 currentFilePath 用于记录当前正在播放 MP3 的地址，定义 currentTempFilePath 变量表示当前播放 MP3 的路径。具体代码如下。

```
/*  记录当前正在播放 MP3 的地址  */
private String currentFilePath = "";

/*当前播放 MP3 的路径*/
private String currentTempFilePath = "";
private String strVideoURL = "";
```

3）先引入主布局文件 main.xml，然后通过 strVideoURL 对象设置要播放 MP3 文件的网址，并设置透明度。具体代码如下。

```
 /** Called when the activity is first created. */
@Override
public void onCreate(Bundle savedInstanceState)
{
  super.onCreate(savedInstanceState);
  setContentView(R.layout.main);

  /* mp3 文件不会被下载到本地*/
  strVideoURL = "http://www.lrn.cn/zywh/xyyy/yyxs/200805/W020080505536315331317.MP3";

  mTextView01 = (TextView)findViewById(R.id.myTextView1);

  /*设置透明度*/
  getWindow().setFormat(PixelFormat.TRANSPARENT);

  mPlay = (ImageButton)findViewById(R.id.play);
  mReset = (ImageButton)findViewById(R.id.reset);
  mPause = (ImageButton)findViewById(R.id.pause);
```

```
mStop = (ImageButton)findViewById(R.id.stop);
```

4）设置单击“播放”按钮所触发的处理事件，具体代码如下。

```
/* 播放按钮 */
mPlay.setOnClickListener(new ImageButton.OnClickListener()
{
  public void onClick(View view)
  {
    /* 调用播放影片函数 */
    playVideo(strVideoURL);
    mTextView01.setText
    (
      getResources().getText(R.string.str_play).toString()+
      "\n"+ strVideoURL
    );
  }
});
```

5）设置单击“重新播放”按钮所触发的处理事件，具体代码如下。

```
/* 重新播放 */
mReset.setOnClickListener(new ImageButton.OnClickListener()
{
  public void onClick(View view)
  {
    if(bIsReleased == false)
    {
      if (mMediaPlayer01 != null)
      {
        mMediaPlayer01.seekTo(0);
        mTextView01.setText(R.string.str_play);
      }
    }
  }
});
```

6）设置单击“暂停播放”按钮所触发的处理事件，具体代码如下。

```
/* 暂停播放 */
mPause.setOnClickListener(new ImageButton.OnClickListener()
{
  public void onClick(View view)
  {
    if (mMediaPlayer01 != null)
    {
      if(bIsReleased == false)
      {
```

```
                if(bIsPaused==false)
                {
                  mMediaPlayer01.pause();
                  bIsPaused = true;
                  mTextView01.setText(R.string.str_pause);
                }
                else if(bIsPaused==true)
                {
                  mMediaPlayer01.start();
                  bIsPaused = false;
                  mTextView01.setText(R.string.str_play);
                }
              }
            }
          }
        });
```

7）设置单击“停止播放”按钮所触发的处理事件，具体代码如下。

```
/* 终止 */
mStop.setOnClickListener(new ImageButton.OnClickListener()
{
  public void onClick(View view)
  {
    try
    {
      if (mMediaPlayer01 != null)
      {
        if(bIsReleased==false)
        {
          mMediaPlayer01.seekTo(0);
          mMediaPlayer01.pause();
          //mMediaPlayer01.stop();
          //mMediaPlayer01.release();
          //bIsReleased = true;
          mTextView01.setText(R.string.str_stop);
        }
      }
    }
    catch(Exception e)
    {
      mTextView01.setText(e.toString());
      Log.e(TAG, e.toString());
      e.printStackTrace();
    }
  }
});
```

```
    }
```

8）定义方法 playVideo(final String strPath)，用于播放指定的 MP3，其播放的是存储卡中暂时保存的 MP3 文件，具体代码如下。

```
private void playVideo(final String strPath)
{
  try
  {
    if (strPath.equals(currentFilePath)&& mMediaPlayer01 != null)
    {
      mMediaPlayer01.start();
      return;
    }

    currentFilePath = strPath;

    mMediaPlayer01 = new MediaPlayer();
    mMediaPlayer01.setAudioStreamType(2);
```

9）定义 setOnErrorListener 实现错误处理，具体代码如下。

```
/* 错误事件 */
mMediaPlayer01.setOnErrorListener(new MediaPlayer.OnErrorListener()
{
  @Override
  public boolean onError(MediaPlayer mp, int what, int extra)
  {
    //TODO Auto-generated method stub
    Log.i(TAG, "Error on Listener, what: " + what + "extra: " + extra);
    return false;
  }
});
```

10）定义 setOnBufferingUpdateListener，捕捉使用 MediaPlayer 缓冲区的更新事件。具体代码如下。

```
/* 捕捉使用 MediaPlayer 缓冲区的更新事件 */
mMediaPlayer01.setOnBufferingUpdateListener(new MediaPlayer.OnBufferingUpdateListener()
{
  @Override
  public void onBufferingUpdate(MediaPlayer mp, int percent)
  {
    Log.i(TAG, "Update buffer: " + Integer.toString(percent)+ "%");
  }
```

```
});
```

11）定义 setOnCompletionListener，实现播放完毕所触发的事件。具体代码如下。

```
/* 播放完毕所触发的事件 */
mMediaPlayer01.setOnCompletionListener(new MediaPlayer.OnCompletionListener()
{
  @Override
  public void onCompletion(MediaPlayer mp)
  {
    delFile(currentTempFilePath);
    Log.i(TAG,"mMediaPlayer01 Listener Completed");
  }
});
```

12）定义 setOnPreparedListener，用于开始阶段的监听 Listener。具体代码如下。

```
/* 开始阶段的监听 Listener */
mMediaPlayer01.setOnPreparedListener(new MediaPlayer.OnPreparedListener()
{
  @Override
  public void onPrepared(MediaPlayer mp)
  {
    Log.i(TAG,"Prepared Listener");
  }
});
```

13）定义 Runnable 对象 r，用 Runnable 来确保文件在存储完毕后才开始 start()方法。先将文件存到 SD 卡，然后在运行 setDataSource()方法后运行 prepare()方法，最后通过 mMediaPlayer01.start()方法开始播放 MP3。具体代码如下。

```
/* 用 Runnable 来确保文件在存储完毕后才开始 start() */
Runnable r = new Runnable()
{
  public void run()
  {
    try
    {
      /* setDataSource 将文件存到 SD 卡   */
      setDataSource(strPath);
      /*  因为线程顺利进行，所以在 setDataSource 后运行 prepare() */
      mMediaPlayer01.prepare();
      Log.i(TAG, "Duration: " + mMediaPlayer01.getDuration());

      /* 开始播放 mp3 */
      mMediaPlayer01.start();
      bIsReleased = false;
```

```
            }
            catch (Exception e)
            {
                Log.e(TAG, e.getMessage(), e);
            }
        }
    };
    new Thread(r).start();
}
```

14）有异常则输出提示，具体代码如下。

```
    catch(Exception e)
    {
        if (mMediaPlayer01 != null)
        {
            /* 线程发生异常则停止播放    */
            mMediaPlayer01.stop();
            mMediaPlayer01.release();
        }
        e.printStackTrace();
    }
}
```

15）定义方法 setDataSource()，用于存储 URL 的 MP3 文件到存储卡。首先判断传入的地址是否为 URL，然后创建 URL 对象和临时文件，当 fos 存储完毕调用 MediaPlayer.setDataSource()方法。具体代码如下。

```
/*   定义方法用于存储 URL 的 MP3 文件到存储卡    */
private void setDataSource(String strPath) throws Exception
{
    /*    判断传入的地址是否为 URL */
    if (!URLUtil.isNetworkUrl(strPath))
    {
        mMediaPlayer01.setDataSource(strPath);
    }
    else
    {
        if(bIsReleased == false)
        {
            /* 创建 URL 对象 */
            URL myURL = new URL(strPath);
            URLConnection conn = myURL.openConnection();
            conn.connect();

            /*    获取 URL 地址的输入流*/
            InputStream is = conn.getInputStream();
```

```
                if (is == null)
                {
                    throw new RuntimeException("stream is null");
                }

                /* 创建临时文件    */
                File myTempFile = File.createTempFile("yinyue", "."+getFileExtension(strPath));
                currentTempFilePath = myTempFile.getAbsolutePath();

                /* currentTempFilePath = /sdcard/hippoplayertMP39327.MP3 */

                /*
                if(currentTempFilePath!="")
                {
                    Log.i(TAG, currentTempFilePath);
                }
                */

                FileOutputStream fos = new FileOutputStream(myTempFile);
                byte buf[] = new byte[128];
                do
                {
                    int numread = is.read(buf);
                    if (numread <= 0)
                    {
                        break;
                    }
                    fos.write(buf, 0, numread);
                }while (true);

                /* 直到 fos 对象存储完毕，调用 MediaPlayer.setDataSource */
                mMediaPlayer01.setDataSource(currentTempFilePath);
                try
                {
                    is.close();
                }
                catch (Exception ex)
                {
                    Log.e(TAG, "error: " + ex.getMessage(), ex);
                }
            }
        }
    }
```

16）定义方法 getFileExtension(String strFileName)，获取音乐文件扩展名，如果无法顺利获取扩展名，则默认为.dat。具体代码如下。

```
/*    获取音乐文件扩展名自定义方法 */
private String getFileExtension(String strFileName)
{
  File myFile = new File(strFileName);
  String strFileExtension=myFile.getName();
  strFileExtension=(strFileExtension.substring(strFileExtension.lastIndexOf(".")+1)).toLowerCase();
  if(strFileExtension=="")
  {
    /* 如果无法顺利获取扩展名则默认为.dat */
    strFileExtension = "dat";
  }
  return strFileExtension;
}
```

17）定义方法 delFile(String strFileName)，当离开程序时删除临时音乐文件。具体代码如下所示：

```
/* 离开程序时需要调用自定义方法删除临时音乐文件*/
private void delFile(String strFileName)
{
  File myFile = new File(strFileName);
  if(myFile.exists())
  {
    myFile.delete();
  }
}

@Override
protected void onPause()
{

  /*    删除临时文件 */
  try
  {
    delFile(currentTempFilePath);
  }
  catch(Exception e)
  {
    e.printStackTrace();
  }
  super.onPause();
}
}
```

执行后可以通过播放、暂停、重新播放和停止四个按钮，控制指定 MP3 文件的处理。效果如图 8-9 所示。

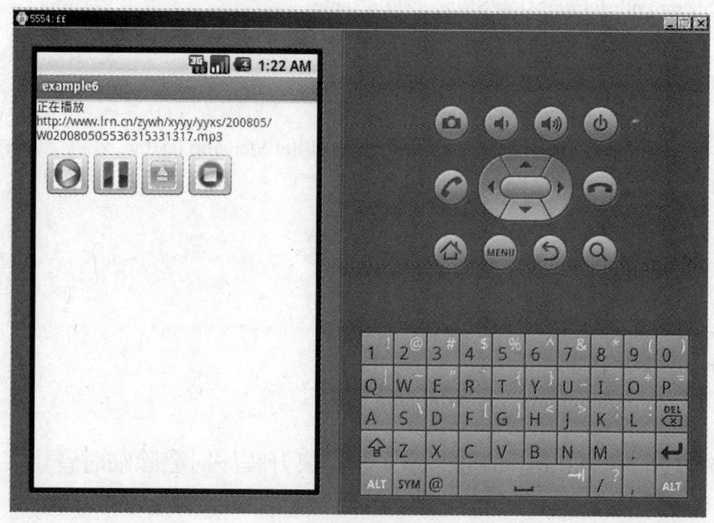

图 8-9 执行效果

8.7 远程下载手机铃声

铃声是移动手机的重要功能之一，同样也可以从网络中直接下载一个 MP3 文件，并设置为手机的铃声。在本节的内容中，将通过一个具体实例的实现，介绍在 Android 中远程下载手机铃声的基本流程。本实例的源代码保存在“光盘:\daima\8\example7”，下面开始讲解本实例的具体实现流程。

8.7.1 实现原理

在本实例中，用户通过 EditText 控件输入一个指定的网址，当下载完成后，打开 RingtoneManager.ACTION_RINGTONE_PICKER 这个 Intent，在打开 Intent 的同时传入一个参数，这个 ACTION_RINGTONE_PICKER 的 Intent 会带入刚才下载文件让用户选择。在实现过程中，会判断下载文件是否完整，并判断用户已设置铃声文件、下载后的铃声文件存储在哪里。在具体实现时，会以 SD 卡中的铃声文件路径作为存储网络下载音乐文件的路径，打开 RingtoneManager 的 ACTION_RINGTONE_PICKER 的 Intent，让用户找到下载的音乐，并作为铃声。

8.7.2 具体实现

本实例的主程序文件是 example7.java，下面开始介绍其实现流程。

1）先引入相关类，然后声明系统中需要的对象，具体代码如下。

```
public class example7 extends Activity
{
  protected static final String APP_TAG = "DOWNLOAD_RINGTONE";
  private Button mButton1;
```

```
private TextView mTextView1;
private EditText mEditText1;
private String strURL = "";
public static final int RINGTONE_PICKED = 0x108;
private String currentFilePath = "";
private String currentTempFilePath = "";
private String fileEx="";
private String fileNa="";
private String strRingtoneFolder = "/sdcard/music/ling";
```

2）判断是否有/sdcard/music/ringtones 文件夹，不存在则输出提示。具体代码如下。

```
/** Called when the activity is first created. */
@Override
public void onCreate(Bundle savedInstanceState)
{
  super.onCreate(savedInstanceState);
  setContentView(R.layout.main);

  mButton1 =(Button) findViewById(R.id.myButton1);
  mTextView1 = (TextView) findViewById(R.id.myTextView1);
  mEditText1 = (EditText) findViewById(R.id.myEditText1);

  /*判断是否有/sdcard/music/ringtones 文件夹*/
  if(bIfExistRingtoneFolder(strRingtoneFolder))
  {
    Log.i(APP_TAG, "Ringtone Folder exists.");
  }
```

3）strURL 是在 String 中设置的，通过 fileEx()方法和 getFile()方法取得文件的名称。具体代码如下。

```
mButton1.setOnClickListener(new Button.OnClickListener()
{
  @Override
  public void onClick(View arg0)
  {
    strURL = mEditText1.getText().toString();

     Toast.makeText(example7.this, getString(R.string.str_msg)
               ,Toast.LENGTH_SHORT).show();

    /*取得文件名称*/
    fileEx = strURL.substring(strURL.lastIndexOf(".")+1,strURL.
                 length()).toLowerCase();
    fileNa = strURL.substring(strURL.lastIndexOf("/")+1,strURL.
                 lastIndexOf("."));
```

```
            getFile(strURL);
        }
    });
}
```

4）定义方法 getMIMEType(File f)，判断文件类型的方法。首先取得扩展名，然后根据扩展名的类型决定 MimeType。具体代码如下。

```
    /* 判断文件 MimeType 的 method */
    private String getMIMEType(File f)
    {
      String type="";
      String fName=f.getName();
      /* 取得扩展名 */
      String end=fName.substring(fName.lastIndexOf(".")+1,
                                          fName.length()).toLowerCase();

      /* 依扩展名的类型决定 MimeType */
      if(end.equals("m4a")||end.equals("MP3")||end.equals("mid")||
           end.equals("xmf")||end.equals("ogg")||end.equals("wav"))
      {
        type = "audio";
      }
      else if(end.equals("3gp")||end.equals("mp4"))
      {
        type = "video";
      }
      else if(end.equals("jpg")||end.equals("gif")||
                  end.equals("png")||end.equals("jpeg")||
                   end.equals("bmp"))
      {
        type = "image";
      }
      else
      {
        type="*";
      }
      /*如果无法直接打开，就跳出软件列表给用户选择 */
      if(end.equals("image"))
      {
      }
      else
      {
        type += "/*";
      }
      return type;
    }
```

5）定义方法 getFile(final String strPath)，用于获取最后的文件。如果地址和当前地址一样，则直接用 getDataSource()方法数据。如果有异常，则输出异常信息。具体代码如下。

```
private void getFile(final String strPath)
{
  try
  {
    if (strPath.equals(currentFilePath) )
    {
     getDataSource(strPath);
    }
    currentFilePath = strPath;
    Runnable r = new Runnable()
    {
      public void run()
      {
        try
        {
          getDataSource(strPath);
        }
        catch (Exception e)
        {
          Log.e(APP_TAG, e.getMessage(), e);
        }
      }
    };
    new Thread(r).start();
  }
  catch(Exception e)
  {
    e.printStackTrace();
  }
}
```

6）定义方法 getDataSource(String strPath)，用于获取远程文件。如果地址错误，则输错误信息。流程如下。

- 第一步：通过 myURL 获取 URL。
- 第二步：创建连接 conn。
- 第三步：通过 InputStream 下载文件。
- 第四步：创建文件地址 myTempFile。
- 第五步：取得在内存中的存储路径。
- 第六步：将文件写入暂存盘。

具体代码如下。

```
/*取得远程文件*/
```

```
private void getDataSource(String strPath) throws Exception
{
  if (!URLUtil.isNetworkUrl(strPath))
  {
    mTextView1.setText("错误的 URL");
  }
  else
  {
    /*取得 URL*/
    URL myURL = new URL(strPath);
    /*创建连接*/
    URLConnection conn = myURL.openConnection();
    conn.connect();
    /*InputStream  下载文件*/
    InputStream is = conn.getInputStream();
    if (is == null)
    {
      throw new RuntimeException("stream is null");
    }

    /*创建文件地址*/
    File myTempFile = new File("/sdcard/music/ling/",
                          fileNa+"."+fileEx);
    /*取得在内存中的存储路径*/
    currentTempFilePath = myTempFile.getAbsolutePath();
    /*将文件写入暂存盘*/
    FileOutputStream fos = new FileOutputStream(myTempFile);
    byte buf[] = new byte[128];
    do
    {
      int numread = is.read(buf);
      if (numread <= 0)
      {
        break;
      }
      fos.write(buf, 0, numread);
    }while (true);
```

7）打开铃声管理对象 RingtonManager 进行铃声选择，通过 Intent 对象 intent 来设置铃声，然后设置显示铃声的文件夹和显示铃声开头。如果有异常则输出异常。具体代码如下。

```
/* 打开 RingtonManager 进行铃声选择 */
String uri = null;
if(bIfExistRingtoneFolder(strRingtoneFolder))
{
  /*设置铃声*/
  Intent intent = new Intent( RingtoneManager.
```

```
                ACTION_RINGTONE_PICKER);
        /*设置显示铃声的文件夹*/
        intent.putExtra( RingtoneManager.EXTRA_RINGTONE_TYPE,
                RingtoneManager.TYPE_RINGTONE);
        /*设置显示铃声开头*/
        intent.putExtra( RingtoneManager.EXTRA_RINGTONE_TITLE,
                "设置铃声");
        if( uri != null)
        {
          intent.putExtra( RingtoneManager.
              EXTRA_RINGTONE_EXISTING_URI, Uri.parse( uri));
        }
        else
        {
          intent.putExtra( RingtoneManager.
              EXTRA_RINGTONE_EXISTING_URI, (Uri)null);
        }
        startActivityForResult(intent, RINGTONE_PICKED);
      }

      try
      {
        is.close();
      }
      catch (Exception ex)
      {
        Log.e(APP_TAG, "error: " + ex.getMessage(), ex);
      }
    }
  }
```

8）定义方法 onActivityResult()，能够根据用户选择的铃声设置保存对应的信息。当选择完毕后，会再次返回选择 Activity。具体代码如下。

```
      @Override
      protected void onActivityResult(int requestCode,
                      int resultCode, Intent data)
      {
        if (resultCode != RESULT_OK)
        {
          return;
        }
        switch (requestCode)
        {
          case (RINGTONE_PICKED):
            try
            {
```

```
                Uri pickedUri = data.getParcelableExtra
                (RingtoneManager.EXTRA_RINGTONE_PICKED_URI);
                if(pickedUri!=null)
                {
                  RingtoneManager.setActualDefaultRingtoneUri
                  (example7.this,RingtoneManager.TYPE_RINGTONE,
                       pickedUri);
                }
              }
              catch(Exception e)
              {
                e.printStackTrace();
              }
              break;
          default:
              break;
        }
        super.onActivityResult(requestCode, resultCode, data);
      }
```

9）定义 bIfExistRingtoneFolder()方法，用于判断是否包含了“/sdcard/music/ringtones”文件夹。具体代码如下。

```
      /*判断是否包含/sdcard/music/ringtones 文件夹*/
      private boolean bIfExistRingtoneFolder(String strFolder)
      {
        boolean bReturn = false;

        File f = new File(strFolder);
        if(!f.exists())
        {
          /*创建/sdcard/music/ringtones 文件夹*/
          if(f.mkdirs())
          {
            bReturn = true;
          }
          else
          {
            bReturn = false;
          }
        }
        else
        {
          bReturn = true;
        }
        return bReturn;
      }
```

```
}
```

执行后会先显示一个下载界面，如图 8-10 所示。单击“下载音乐”按钮后开始下载指定的 MP3 文件，下载完成后会弹出“铃声设置”界面，如图 8-11 所示。选择一种选项，并单击“OK”按钮后，完成铃声设置。

图 8-10　初始界面

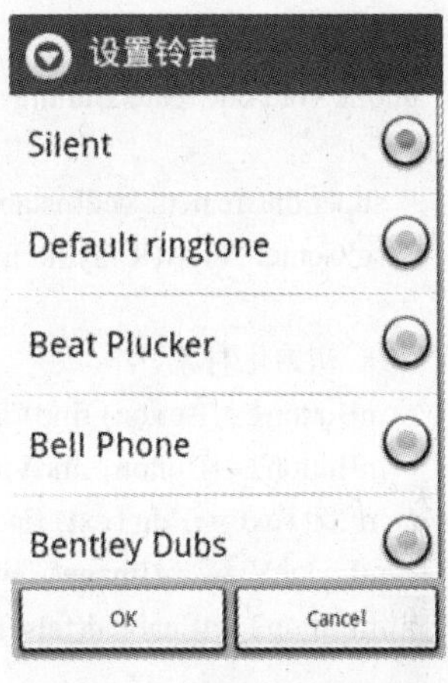

图 8-11　铃声设置界面

8.8　远程下载屏幕背景

可以从网络中直接下载一个图片文件，并设置为手机屏幕背景。在本节的内容中，将通过一个具体实例的实现，介绍在 Android 中远程下载图片作为屏幕背景的基本流程。本实例的源代码保存在“光盘:\daima\8\example8”，下面开始讲解本实例的具体实现流程。

8.8.1　实现原理

在本实例中，可以远程获取网络中的一幅图片，并将这幅图片作为手机屏幕的背景。在具体实现上，将通过 InputStream 传到 ContextWrapper 中重写的 setwallpaper()方法，其中传入的参数是 URLConnection.getInputStream()的 Stream 内容。

8.8.2　具体实现

本实例的主程序文件是 example8.java，下面开始介绍其实现流程。

1）先加载相关类，然后声明系统中需要的私有变量，具体代码如下。

```
public class example8 extends Activity
{
  /* 变量声明 */
  private Button mButton1;
  private Button mButton2;
  private EditText mEditText;
  private ImageView mImageView;
```

```
private Bitmap bm;
```

2）通过 findViewById 初始化各个对象，单击按钮 1，通过 mButton1.setOnClickListener 来实现图片预览。网址为空则输出空白提示，不为空则传入“type=1”表示预览图片。具体代码如下。

```
@Override
public void onCreate(Bundle savedInstanceState)
{
  super.onCreate(savedInstanceState);
  setContentView(R.layout.main);

  /* 初始化对象 */
  mButton1 =(Button) findViewById(R.id.myButton1);
  mButton2 =(Button) findViewById(R.id.myButton2);
  mEditText = (EditText) findViewById(R.id.myEdit);
  mImageView = (ImageView) findViewById(R.id.myImage);
  mButton2.setEnabled(false);

  /* 预览图片的按钮 */
  mButton1.setOnClickListener(new Button.OnClickListener()
  {
    @Override
    public void onClick(View v)
    {
      String path=mEditText.getText().toString();
      if(path.equals(""))
      {
        showDialog("网址不可为空白!");
      }
      else
      {
        /* 传入 type=1 表示预览图片 */
        setImage(path,1);
      }
    }
  });
```

3）单击按钮 2，通过 mButton1.setOnClickListener 将图片设为桌面。网址为空则输出空白提示，否则传入“type=2”表示设置桌面。具体代码如下。

```
/* 将图片设为桌面的 Button */
mButton2.setOnClickListener(new Button.OnClickListener()
{
  @Override
  public void onClick(View v)
```

```
        {
          try
          {
            String path=mEditText.getText().toString();
            if(path.equals(""))
            {
              showDialog("网址不可为空白!");
            }
            else
            {
              /* 传入 type=2 表示设置桌面 */
              setImage(path,2);
            }
          }
          catch (Exception e)
          {
            showDialog("读取错误!网址可能不是图片或网址错误!");
            bm = null;
            mImageView.setImageBitmap(bm);
            mButton2.setEnabled(false);
            e.printStackTrace();
          }
        }
      });
    }
```

4）定义 setImage(String path,int type)方法，用于预览图片并设置为桌面。首先预览图片，然后设置为桌面。有异常则输出对应提示。具体代码如下。

```
    /* 预览图片并设置为桌面的方法 */
    private void setImage(String path,int type)
    {
      try
      {
        URL url = new URL(path);
        URLConnection conn = url.openConnection();
        conn.connect();
        if(type==1)
        {
          /* 预览图片 */
          bm = BitmapFactory.decodeStream(conn.getInputStream());
          mImageView.setImageBitmap(bm);
          mButton2.setEnabled(true);
        }
        else if(type==2)
```

```
            {
                /* 设置为桌面 */
                example8.this.setWallpaper(conn.getInputStream());
                bm = null;
                mImageView.setImageBitmap(bm);
                mButton2.setEnabled(false);
                showDialog("桌面背景设置完成!");
            }
        }
        catch (Exception e)
        {
            showDialog("读取错误!网址可能不是图片或网址错误!");
            bm = null;
            mImageView.setImageBitmap(bm);
            mButton2.setEnabled(false);
            e.printStackTrace();
        }
    }
```

5）定义 showDialog(String mess)方法，用于弹出对话框，单击后完成背景设置。具体代码如下。

```
    /* 弹出对话框的方法 */
    private void showDialog(String mess){
        new AlertDialog.Builder(example8.this).setTitle("Message")
        .setMessage(mess)
        .setNegativeButton("确定", new DialogInterface.OnClickListener()
        {
            public void onClick(DialogInterface dialog, int which)
            {
            }
        })
        .show();
    }
}
```

最后还需要在文件 AndroidManifest.xml 中设置 SET_WALLPAPER 权限和 INTERNET 权限，具体代码如下。

```
<uses-permission android:name="android.permission.SET_WALLPAPER"/>
<uses-permission android:name="android.permission.INTERNET"/>
```

执行后将显示一个输入框和 2 个按钮，如图 8-12 所示。输入图片网址并单击“预览”按钮后，可以查看此图片，如图 8-13 所示。单击“设置”按钮后，将设置此图片为屏幕背景。

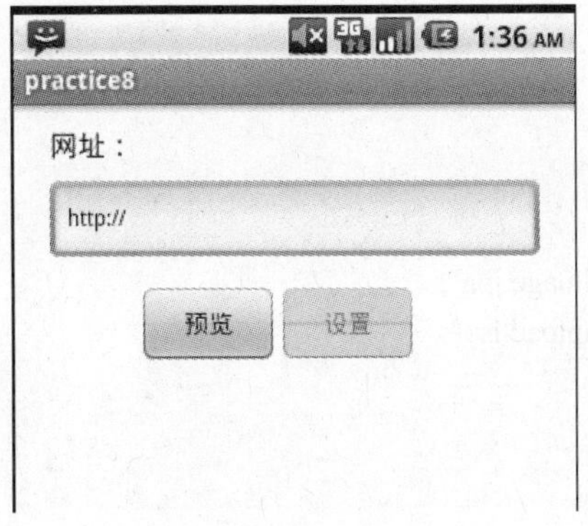

图 8-12　初始效果

图 8-13　设置为背景

8.9　文件上传至服务器

文件上传对于广大读者来说并不陌生，在手机中同样可以实现文件上传功能。在本节的内容中，将通过一个具体实例的实现，介绍在 Android 中实现文件上传的基本流程。本实例的源代码保存在“光盘:\daima\8\example9”，下面开始讲解本实例的具体实现流程。

8.9.1　实现原理

在 Web 应用中，通常通过一个表单和上传程序来实现文件上传。查看下面的代码。

```
<FORM METHOD="POST" ACTION="do_upload.jsp" ENCTYPE="multipart/form-data">
<input type="FILE" name="FILE1" size="30">
<input type="submit" name="Submit" value="上传它！">
</FORM>
```

上述代码是一个上传表单，供用户选择上传文件，并通过文件 do_upload.jsp 实现上传处理。在本实例中，将模拟上述处理过程，以 Post 方式对服务器上的接收程序发出请求，触发该程序运行文件写入服务器的动作。在实现本实例前，需要搭建 Java 的服务器，并在服务器预先编写一个接收文件程序，并在手机目录中准备要上传的资料。在此，设置要上传的文件路径如下。

```
data/data/irdc.example9/image.jpg
```

然后准备上传处理文件 upload.jsp，其代码将不介绍，在网中比比皆是。

8.9.2　具体实现

本实例的主程序文件是 example9.java，下面开始介绍其实现流程。

1）先加载相关类，然后分别声明变量 newName、uploadFile 和 actionUrl，具体代码如下所示。

```
public class example9 extends Activity
{
```

```
/* 变量声明
 * newName：上传后在服务器上的文件名称
 * uploadFile：要上传的文件路径
 * actionUrl：服务器上对应的程序路径 */
private String newName="image.jpg";
private String uploadFile="/data/data/irdc.example9/image.jpg";
private String actionUrl="http://127.127.0.1/upload/upload.jsp";
private TextView mText1;
private TextView mText2;
private Button mButton;
```

2）通过 mText1 获取文件路径，根据 mText2 设置上传网址，单击按钮后调用上传方法 uploadFile()。具体代码如下所示。

```
@Override
public void onCreate(Bundle savedInstanceState)
{
  super.onCreate(savedInstanceState);
  setContentView(R.layout.main);

  mText1 = (TextView) findViewById(R.id.myText2);
  mText1.setText("文件路径：\n"+uploadFile);
  mText2 = (TextView) findViewById(R.id.myText3);
  mText2.setText("上传网址：\n"+actionUrl);
  /* 设置 mButton 的 onClick 事件处理 */
  mButton = (Button) findViewById(R.id.myButton);
  mButton.setOnClickListener(new View.OnClickListener()
  {
    public void onClick(View v)
    {
      uploadFile();
    }
  });
}
```

3）定义方法 uploadFile()，用于将文件上传至远端服务器。流程如下。

- 第一步：设置传送的 method=POST。
- 第二步：设置设置 DataOutputStream。
- 第三步：获取文件的 FileInputStream。
- 第四步：设置每次写入 1024bytes。
- 第五步：从文件中读取数据至缓冲区。
- 第六步：获取 Response 的内容。
- 第七步：把 Response 在 Dialog 中显示。

具体代码如下。

```
/* 上传文件至 Server 的方法 */
private void uploadFile()
```

```
{
  String end = "\r\n";
  String twoHyphens = "--";
  String boundary = "*****";
  try
  {
    URL url =new URL(actionUrl);
    HttpURLConnection con=(HttpURLConnection)url.openConnection();
    /* 允许 Input、Output，不使用 Cache */
    con.setDoInput(true);
    con.setDoOutput(true);
    con.setUseCaches(false);
    /* 设置传送的 method=POST */
    con.setRequestMethod("POST");
    /* setRequestProperty */
    con.setRequestProperty("Connection", "Keep-Alive");
    con.setRequestProperty("Charset", "UTF-8");
    con.setRequestProperty("Content-Type",
                        "multipart/form-data;boundary="+boundary);
    /* 设置 DataOutputStream */
    DataOutputStream ds =
      new DataOutputStream(con.getOutputStream());
    ds.writeBytes(twoHyphens + boundary + end);
    ds.writeBytes("Content-Disposition: form-data; " +
                  "name=\"file1\";filename=\"" +
                  newName +"\"" + end);
    ds.writeBytes(end);

    /* 取得文件的 FileInputStream */
    FileInputStream fStream = new FileInputStream(uploadFile);
    /* 设置每次写入 1024bytes */
    int bufferSize = 1024;
    byte[] buffer = new byte[bufferSize];

    int length = -1;
    /* 从文件读取数据至缓冲区 */
    while((length = fStream.read(buffer)) != -1)
    {
      /* 将资料写入 DataOutputStream 中 */
      ds.write(buffer, 0, length);
    }
    ds.writeBytes(end);
    ds.writeBytes(twoHyphens + boundary + twoHyphens + end);

    /* close streams */
    fStream.close();
    ds.flush();

    /* 取得 Response 内容 */
    InputStream is = con.getInputStream();
    int ch;
```

```
            StringBuffer b =new StringBuffer();
            while( ( ch = is.read() ) != -1 )
            {
              b.append( (char)ch );
            }
            /* 将 Response 显示于 Dialog */
            showDialog(b.toString().trim());
            /* 关闭 DataOutputStream */
            ds.close();
          }
          catch(Exception e)
          {
            showDialog(""+e);
          }
        }
```

4）定义方法 showDialog(String mess)，用于显示提示对话框。具体代码如下。

```
      /* 显示 Dialog 的 method */
      private void showDialog(String mess)
      {
        new AlertDialog.Builder(example9.this).setTitle("Message")
         .setMessage(mess)
         .setNegativeButton("确定",new DialogInterface.OnClickListener()
         {
           public void onClick(DialogInterface dialog, int which)
           {
           }
         })
         .show();
      }
    }
```

执行后的效果如图 8-14 所示。单击“开始上传”按钮后，能够将指定文件上传到服务器。

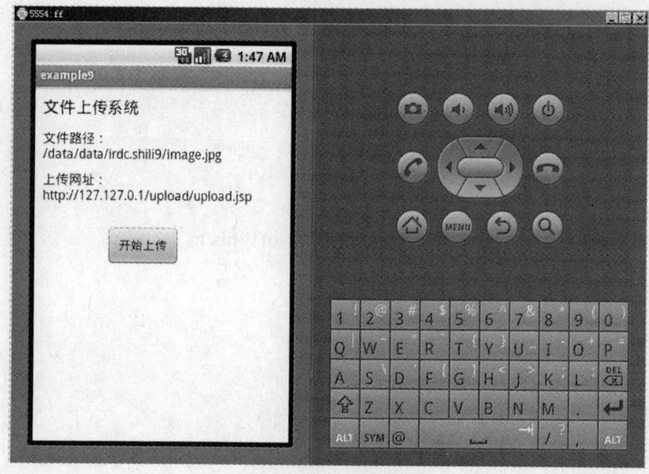

图 8-14　执行效果

8.10　实现一个简单的 RSS 阅读器

RSS 是一个开放的新闻模式，RSS 是在线共享内容的一种简易方式，也叫聚合内容（Really Simple Syndication，RSS）。通常在时效性比较强的内容上使用 RSS 订阅能更快速获取信息，网站提供 RSS 输出，有利于用户获取网站内容的最新更新。在本节的内容中，将通过一个具体实例的实现，介绍在 Android 中查看指定 RSS 新闻的基本流程。本实例的源代码保存在“光盘:\daima\8\example10”，下面开始讲解本实例的具体实现流程。

8.10.1　实现原理

在使用 RSS 订阅时，通常通过网站提供的“订阅 RSS”连接或小图标实现，当单击连接后，会弹出包含 RSS 内容的页面，此页面的网址是网站的 RSS 网址。当连接后，服务器端会返回 RSS 标准规格的 XML 文件，只要按照统一格式来解析 XML 文件，就可以得到 RSS 内的相关信息。

在本实例中，用户只需要输入一个 RSS Feed 网址，通过 SAXParser 解析后就可以直接在手机上浏览在线新闻。

8.10.2　具体实现

本实例的主程序文件包括 example10.java、example10_1.java、example10_2.java、News.java、MyHandler.java 和 MyAdapter.java，下面开始分别介绍其实现流程。

1．主程序 example10.java

主程序 example10.java 中以 EditText 来作为输入 RSS 连接组件。当输入网址后，单击“解析”按钮后，按钮的 onClick（单击事件）会被触发，运行 EditText 的空白检查。当检查无误后，将输入的网址写入 Bundle（绑定）对象中，再将 Bundle 对象指派给 Intent，并通过 startActivityForResult()方法来触发 example10_1 这个 Activity。

主程序 example10.java 的实现代码如下。

```
public class example10 extends Activity
{
  /* 变量声明 */
  private Button mButton;
  private EditText mEditText;

  @Override
  public void onCreate(Bundle savedInstanceState)
  {
    super.onCreate(savedInstanceState);
    setContentView(R.layout.main);
    /* 初始化对象 */
    mEditText=(EditText) findViewById(R.id.myEdit);
    mButton=(Button) findViewById(R.id.myButton);
    /* 设置按钮的单击事件 */
```

```
    mButton.setOnClickListener(new Button.OnClickListener()
    {
      @Override
      public void onClick(View v)
      {
        String path=mEditText.getText().toString();
        if(path.equals(""))
        {
          showDialog("网址不可为空白!");
        }
        else
        {
          /* 新建一个 Intent 对象，并指定类 */
          Intent intent = new Intent();
          intent.setClass(example10.this,example10_1.class);

          /* new 一个 Bundle 对象，并将要传递的数据传入 */
          Bundle bundle = new Bundle();
          bundle.putString("path",path);
          /* 将 Bundle 对象 assign 给 Intent */
          intent.putExtras(bundle);
          /* 调用 Activity EX08_13_1 */
          startActivityForResult(intent,0);
        }
      }
    });
  }

  /* 覆盖 onActivityResult()*/
  @Override
  protected void onActivityResult(int requestCode,int resultCode, Intent data)
  {
    switch (resultCode)
    {
      case 99:
        /* 返回错误时以 Dialog 显示 */
        Bundle bunde = data.getExtras();
        String error = bunde.getString("error");
        showDialog(error);
        break;
      default:
        break;
    }
  }

  /* 显示对话框的方法 */
```

```
    private void showDialog(String mess){
      new AlertDialog.Builder(example10.this).setTitle("Message")
      .setMessage(mess)
      .setNegativeButton("确定", new DialogInterface.OnClickListener()
      {
        public void onClick(DialogInterface dialog, int which)
        {
        }
      })
      .show();
    }
  }
```

2．文件 example10_1.java

文件 example10_1.java 是一个 ListActivity，是通过主程序 example10.java 来调用的，用于显示订阅的 RSS 内容列表。其实现流程如下。

1）加载相关类，分别声明变量 mText、title 和 li。具体代码如下所示。

```
public class example10_1 extends ListActivity
{
  /* 变量声明 */
  private TextView mText;
  private String title="";
  private List<News> li=new ArrayList<News>();
```

2）设置 layout 为 newslist.xml，取得 Intent 中的 Bundle 对象，并取得 Bundle 对象中的数据，然后调用 getRss()取得解析后的 List。具体代码如下。

```
@Override
public void onCreate(Bundle savedInstanceState) {
  super.onCreate(savedInstanceState);
  /* 设置 layout 为 newslist.xml */
  setContentView(R.layout.newslist);

  mText=(TextView) findViewById(R.id.myText);
  /* 取得 Intent 中的 Bundle 对象 */
  Intent intent=this.getIntent();
  Bundle bunde = intent.getExtras();
  /* 取得 Bundle 对象中的数据 */
  String path = bunde.getString("path");
  /* 调用 getRss()取得解析后的 List */
  li=getRss(path);
  mText.setText(title);
  /* 设置自定义的 MyAdapter */
  setListAdapter(new MyAdapter(this,li));
}
```

3）定义 onListItemClick()方法，作为监听 ListItem 被单击时要做的动作。流程如下。

- 第一步：获取 News 对象。
- 第二步：新建一个 Intent 对象，并指定其类名。
- 第三步：新建一个 Bundle 对象，并将要传递的数据传入。
- 第四步：将 Bundle 对象分配给 Intent。
- 第五步：调用 Activity example10_2。

具体代码如下。

```
/* 设置 ListItem 被点击时要做的动作 */
@Override
protected void onListItemClick(ListView l,View v,int position, long id)
{
  /* 取得 News 对象 */
  News ns=(News)li.get(position);
  /* new 一个 Intent 对象，并指定 class */
  Intent intent = new Intent();
  intent.setClass(example10_1.this,example10_2.class);
  /* new 一个 Bundle 对象，并将要传递的数据传入 */
  Bundle bundle = new Bundle();
  bundle.putString("title",ns.getTitle());
  bundle.putString("desc",ns.getDesc());
  bundle.putString("link",ns.getLink());
  /* 将 Bundle 对象分配给 Intent */
  intent.putExtras(bundle);
  /* 调用 Activity example10_2 */
  startActivity(intent);
}
```

4）定义 getRss(String path)方法，用于解析 XML。流程如下。

- 第一步：通过 url 获取地址。
- 第二步：创建 SAXParser 对象和 XMLReader 对象。
- 第三步：设置自定义的 MyHandler 给 XMLReader。
- 第四步：解析 XML。
- 第五步：取得 RSS 标题与内容列表。

具体代码如下。

```
/* 解析 XML 的方法 */
private List<News> getRss(String path)
{
  List<News> data=new ArrayList<News>();
  URL url = null;
  try
  {
    url = new URL(path);
```

```
          /* 产生 SAXParser 对象 */
          SAXParserFactory spf = SAXParserFactory.newInstance();
          SAXParser sp = spf.newSAXParser();
          /* 产生 XMLReader 对象 */
          XMLReader xr = sp.getXMLReader();
          /* 设置自定义的 MyHandler 给 XMLReader */
          MyHandler myExampleHandler = new MyHandler();
          xr.setContentHandler(myExampleHandler);
          /* 解析 XML */
          xr.parse(new InputSource(url.openStream()));
          /* 取得 RSS 标题与内容列表 */
          data =myExampleHandler.getParsedData();
          title=myExampleHandler.getRssTitle();
        }
```

5）有异常则输出错误提示对话框，具体代码如下。

```
          catch (Exception e)
        {
          /* 发生错误时返回结果回上一个 activity */
          Intent intent=new Intent();
          Bundle bundle = new Bundle();
          bundle.putString("error",""+e);
          intent.putExtras(bundle);
          /* 错误的返回值设置为 99 */
          example10_1.this.setResult(99, intent);
          example10_1.this.finish();
        }
      return data;
    }
  }
```

3．文件 example10_2.java

文件 example10_2.java 由 example10_1 唤起，用于显示上一个 Activity 所单击的新闻内容。当程序被唤起后，会首先从 Bundle 对象中获取信息的标题、链接和描述，并显示在画面中。并以 Linkify.addLinks()方法将 link 设置为一个 WEB_URLS 形式的链接。当用户单击链接后，会通过设置的网址直接打开 Web 浏览器来浏览网页。其具体实现代码如下。

```
package irdc.example10;

/* 引入相关类 */
import irdc.example10.R;
import android.app.Activity;
import android.content.Intent;
import android.os.Bundle;
import android.text.util.Linkify;
import android.widget.TextView;
```

```
public class example10_2 extends Activity
{
  /* 变量声明 */
  private TextView mTitle;
  private TextView mDesc;
  private TextView mLink;

  @Override
  public void onCreate(Bundle savedInstanceState)
  {
    super.onCreate(savedInstanceState);
    /* 设置 layout 为 newscontent.xml */
    setContentView(R.layout.newscontent);
    /* 初始化对象 */
    mTitle=(TextView) findViewById(R.id.myTitle);
    mDesc=(TextView) findViewById(R.id.myDesc);
    mLink=(TextView) findViewById(R.id.myLink);

    /* 取得 Intent 中的 Bundle 对象 */
    Intent intent=this.getIntent();
    Bundle bunde = intent.getExtras();
    /* 取得 Bundle 对象中的数据 */
    mTitle.setText(bunde.getString("title"));
    mDesc.setText(bunde.getString("desc")+"....");
    mLink.setText(bunde.getString("link"));
    /* 设置 mLink 为网页连接 */
    Linkify.addLinks(mLink,Linkify.WEB_URLS);
  }
}
```

4．文件 News.java

文件 News.java 是一个 JavaBean 类，用于存放每一篇新闻信息。每一个 News 对象代表了一条新闻，在 News 对象中定义了新闻的标题、描述、网站链接和发布时间这 4 个属性。JavaBean 类中的方法都是以 setAAA()和 getAAA()方式来命名的，所以在 Newsijava 中用 setAAA()来设置属性值，或通过 getAAA()来获取属性值。具体代码如下。

```
package irdc.example10;

public class News
{
  private String _title="";
  private String _link="";
  private String _desc="";
  private String _date="";

  public String getTitle()
  {
```

```
    return _title;
  }
  public String getLink()
  {
    return _link;
  }
  public String getDesc()
  {
    return _desc;
  }
  public String getDate()
  {
    return _date;
  }
  public void setTitle(String title)
  {
    _title=title;
  }
  public void setLink(String link)
  {
    _link=link;
  }
  public void setDesc(String desc)
  {
    _desc=desc;
  }
  public void setDate(String date)
  {
    _date=date;
  }
}
```

5. 文件 MyAdapter.java

在文件 MyAdapter.java 中定义了 Adapter 对象，它继承自 android.widget.BaseAdapter，用于设置 ListView 中要显示的信息，用文件 news_row.xml 作为布局。具体代码如下。

```
/* 自定义的 Adapter，继承 android.widget.BaseAdapter */
public class MyAdapter extends BaseAdapter
{
  /* 变量声明 */
  private LayoutInflater mInflater;
  private List<News> items;

  /* MyAdapter 的构造器，传递两个参数 */
  public MyAdapter(Context context,List<News> it)
  {
    /* 参数初始化 */
    mInflater = LayoutInflater.from(context);
```

```
        items = it;
    }

    /* 因继承 BaseAdapter，需重写以下方法 */
    @Override
    public int getCount()
    {
        return items.size();
    }

    @Override
    public Object getItem(int position)
    {
        return items.get(position);
    }

    @Override
    public long getItemId(int position)
    {
        return position;
    }

    @Override
    public View getView(int position,View convertView,ViewGroup par)
    {
        ViewHolder holder;

        if(convertView == null)
        {
            /* 使用自定义的 news_row 作为 Layout */
            convertView = mInflater.inflate(R.layout.news_row, null);
            /* 初始化 holder 的 text 与 icon */
            holder = new ViewHolder();
            holder.text = (TextView) convertView.findViewById(R.id.text);
            convertView.setTag(holder);
        }
        else
        {
            holder = (ViewHolder) convertView.getTag();
        }
        News tmpN=(News)items.get(position);
        holder.text.setText(tmpN.getTitle());

        return convertView;
    }

    /* class ViewHolder */
    private class ViewHolder
    {
        TextView text;
    }
}
```

6．文件 MyHandler.java

在文件 MyHandler.java 中定义了 MyHandler 对象，它继承于 org.xml.sax.helpers.DefaultHandler 类，用于解析 XML 文件，并获取对应的信息。

下面开始讲解文件 MyHandler.java 的具体实现流程。

1）加载相关类，然后分别声明各个变量。具体代码如下。

```
public class MyHandler extends DefaultHandler
{
  /* 变量声明 */
  private boolean in_item = false;
  private boolean in_title = false;
  private boolean in_link = false;
  private boolean in_desc = false;
  private boolean in_date = false;
  private boolean in_mainTitle = false;
  private List<News> li;
  private News news;
  private String title="";
  private StringBuffer buf=new StringBuffer();
```

2）返回将转换成 List<News>的 XML 数据，通过 getRssTitle()方法返回解析出的 RSS 标题。然后调用 startDocument()方法开始解析操作。当解析结束时，调用 endDocument()方法。当解析到 Element 开头时，调用 startElement()方法。具体代码如下。

```
  /* 将转换成 List<News>的 XML 数据返回 */
  public List<News> getParsedData()
  {
    return li;
  }
  /* 返回解析出的 RSS 标题*/
  public String getRssTitle()
  {
    return title;
  }
  /* XML 文件开始解析时调用此方法 */
  @Override


  public void startDocument() throws SAXException
  {
    li = new ArrayList<News>();
  }
  /* XML 文件结束解析时调用此方法 */
  @Override
  public void endDocument() throws SAXException
  {
  }
  /* 解析到 Element 的开头时调用此方法 */
```

```
@Override
public void startElement(String namespaceURI, String localName,
                String qName, Attributes atts) throws SAXException
{
  if (localName.equals("item"))
  {
    this.in_item = true;
    /* 解析到 item 的开头时 new 一个 News 对象 */
    news=new News();
  }
  else if (localName.equals("title"))
  {
    if(this.in_item)
    {
      this.in_title = true;
    }
    else
    {
      this.in_mainTitle = true;
    }
  }
  else if (localName.equals("link"))
  {
    if(this.in_item)
    {
      this.in_link = true;
    }
  }
  else if (localName.equals("description"))
  {
    if(this.in_item)
    {
      this.in_desc = true;
    }
  }
  else if (localName.equals("pubDate"))
  {
    if(this.in_item)
    {
      this.in_date = true;
    }
  }
}
```

3）当解析到 Element（元素）的结尾时调用 endElement()方法，流程如下。

- 第一步：将解析到 Item 的结尾时将 News 对象写入 List 中。
- 第二步：根据 Item 选项分别设置 News 对象的标题，设置 RSS 的标题，设置 News 对象的链接，设置 News 对象的描述，设置 News 对象的 pubDate 时间。

具体代码如下。

```
/* 解析到 Element 的结尾时调用此方法 */
@Override
public void endElement(String namespaceURI, String localName,
                      String qName) throws SAXException
{
  if (localName.equals("item"))
  {
    this.in_item = false;
    /* 解析到 item 的结尾时将 News 对象写入 List 中 */
    li.add(news);
  }
  else if (localName.equals("title"))
  {
    if(this.in_item)
    {
      /* 设置 News 对象的标题*/
      news.setTitle(buf.toString().trim());
      buf.setLength(0);
      this.in_title = false;
    }
    else
    {
      /* 设置 RSS 的标题*/
      title=buf.toString().trim();
      buf.setLength(0);
      this.in_mainTitle = false;
    }
  }
  else if (localName.equals("link"))
  {
    if(this.in_item)
    {


      /* 设置 News 对象的链接*/
      news.setLink(buf.toString().trim());
      buf.setLength(0);
      this.in_link = false;
    }
  }
  else if (localName.equals("description"))
  {
    if(in_item)
    {
      /* 设置 News 对象的描述*/
      news.setDesc(buf.toString().trim());
      buf.setLength(0);
      this.in_desc = false;
    }
```

```
            }
            else if (localName.equals("pubDate"))
            {
              if(in_item)
              {
                /*  设置 News 对象的时间*/
                news.setDate(buf.toString().trim());
                buf.setLength(0);
                this.in_date = false;
              }
            }
          }
```

4）定义方法 characters()，用于获取 Element 开头和结尾中间的字符串。具体代码如下。

```
          /*  取得 Element 的开头结尾中间夹的字符串  */
          @Override
          public void characters(char ch[], int start, int length)
          {
            if(this.in_item||this.in_mainTitle)
            {
              /*  将 char[]添加 StringBuffer */
              buf.append(ch,start,length);
            }
          }
        }
```

执行后的效果如图 8-15 所示。在文本框中输入 RSS 网址 http://rss.sina.com.cn/news /marquee/ ddt.xml，然后单击“开始解析”按钮后，会在屏幕中列表显示 RSS 新闻，如图 8-16 所示。单击某条新闻后，会显示此新闻的详情，如图 8-17 所示。

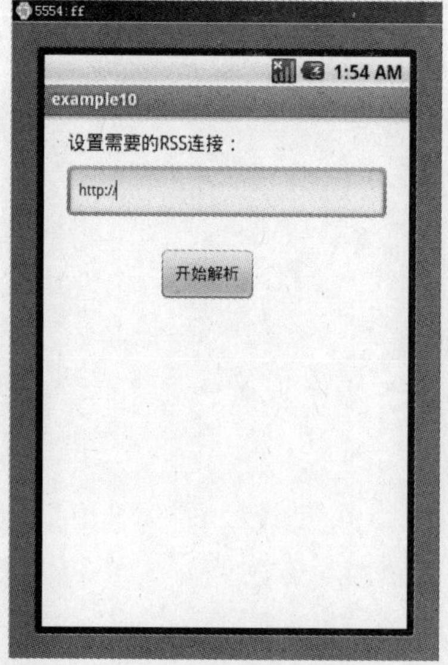

图 8-15　初始效果

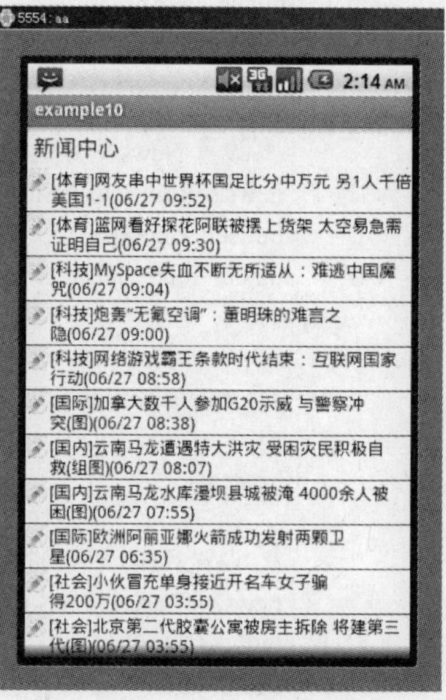

图 8-16　RSS 列表

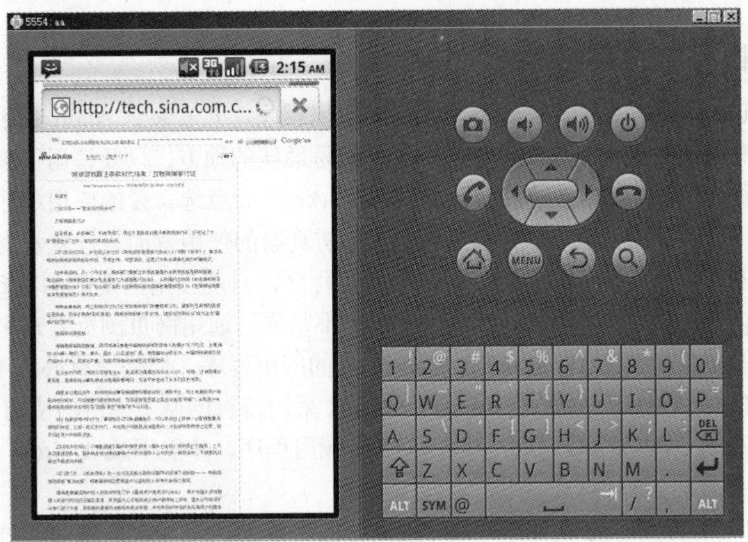

图 8-17　新闻详情

8.11　远程下载安装 Android 程序

在使用智能手机时，经常需要直接利用手机来下载网络中的软件程序，然后安装这个下载的软件。在本节的内容中，将通过一个具体实例的实现过程，介绍在 Android 中远程下载 Android 软件的基本流程。本实例的源代码保存在“光盘:\daima\8\example11”，下面开始讲解本实例的具体实现流程。

8.11.1　APK 简介

APK 是 Android Package 的缩写，即 Android 安装包。APK 是类似 Symbian Sis 或 Sisx 的文件格式。通过将 APK 文件直接传到 Android 模拟器或 Android 手机中执行即可安装。

一个 APK 文件结构为：META-INF\Jar，此文件结构的具体说明如下所示。

- “res\”：存放资源文件的目录；
- AndroidManifest.xml：程序全局配置文件；
- classes.dex：Dalvik 字节码；
- resources.arsc：编译后的二进制资源文件。

Android 在运行程序时，首先需要解压缩，这一点和 Symbian 相似，而和 Windows Mobile 中的 PE 文件有所区别。这样做程序的保密性和可靠性不是很高，通过 dexdump 命令可以反编译，但这样做符合发展规律，微软的 Windows Gadgets 或者说 WPF 也采用了这种构架方式。在 Android 平台中 dalvik vm 的执行文件被打包为 APK 格式，最终运行时加载器会解压然后获取编译后的 androidmanifest.xml 文件中的 permission 分支相关的安全访问，但仍然存在很多安全限制，如果你将 APK 文件传到“/system/app”文件夹下会发现执行是不受限制的。最终安装的文件可能不放这个文件夹中，系统的 APK 文件默认会放入这个文件夹，它们拥有着 ROOT 权限。

8.11.2 下载 APK 应用程序

读者可以从哪里获得好用的 Android APK 应用程序，并安装到 Android 手机上呢？对拥有 G1 实体手机的使用者而言，Android Market 就是最佳的地方，只要使用手机内应用程序列表的 Market 程序，就可以直接连接到 Android Market，而点选喜爱的应用程序后，就会直接下载并安装到 G1 手机上。不过对使用 Android 仿真器的使用者而言，就没有如此方便了，Android 仿真器并没有 Android Market 这个应用程序，只能使用内附的浏览器浏览 Android Market，为何说是浏览呢？因为 Android Market 不是采用通用网页浏览方式来下载文件，虽然可以使用常见的浏览器看到 Android Market 上的应用程序，但是没有办法下载到 Android 仿真器或一般的计算机上，原因是 Android Market 采用特有的网页 API，使用 native UI 的方式来访问，唯有通过内建在 G1 手机内的 Market 应用程序，才能下载 Android Market 网页中的应用程序，并自动安装到 G1 手机上。

所以 Android 仿真器的使用者，只好浏览该网页上的应用程序，然后通过搜索引擎去找找看有没有开发人员将应用程序放到 Android Market 之后，另外还将 APK 文件放置在一般网页上了。到此为止，使用 Android 仿真器，也不需要这么灰心，因为有太多的人遇到同样的问题，也就生成非常多的 Android 应用程序网页，您可以浏览这些网页并把上面的 APK 文件下载到一般计算机上，再进行安装。

下面列出了常用的 APK 应用程序下载网站。

http://andappstore.com/

http://www.getjar.com/software

http://www.phoload.com/android

http://slideme.org/

http://androidforums.com/market/

http://www.cyrket.com/

http://www.androidfreeware.org/

http://androidsoftwaredownload.com/

http://www.freeandroidsoft.com/

http://code.google.com/p/apps-for-android/

http://code.google.com/p/openintents/downloads/list

8.11.3 安装 APK 应用程序

所有的 APK 应用程序要安装到 Android 仿真器上，就只有一个指令，就是在 3.3 节安装影片播放软件时就曾经使用过的 adb install 指令，请开启一个命令字符的终端机窗口，并运行 APK 安装指令。

```
adb install filename.apk
```

这样 adb 指令就会自动将 filename.apk 应用程序安装到 Android 仿真器上，而仿真器上的应用程序列表也会立即出现刚刚安装的应用程序图标，如果应用程序没有安装成功，或安装不完善，也可以重复运行 adb install -r filename.apk 指令，这样会保留已经设置的信息，而仅

是重新安装应用程序本身。

不过在安装 APK 应用程序组件时，不可以同时运行多个 Android 仿真器，因为 adb 不知要将 APK 应用程序安装到哪一个仿真器，最好的方法就是仅运行一个 Android 仿真器。如果需要同时运行多个仿真器，就要在安装 APK 组件时，先使用 adb 指定某一个仿真器。您可以从 Android 仿真器的窗口上，看到类似 Android Emulator（5554）的字样，而 5554 就是仿真器的运行序号，每一个仿真器有其独特的运行序号，只要将 adb 加上-s <serialNumber> 参数，就可以将 APK 应用程序安装到指定的仿真器上。

```
adb -s emulator-5554 install filename.apk
(指定安装 APK 组件在 5554 的 Android 仿真器中)
```

8.11.4 移除 APK 应用程序

如果已经安装了很多 Android 应用程序，如果想要删除一些应用程序图标，也非常的简单，同样是一行指令就搞定了。adb uninstall 指令可以将 APK 应用程序移除。

```
adb uninstall package
```

例如，

```
adb uninstall com.android.email          (把 email 程序移除)
```

Android 使用的 package 名称类似我们浏览网页时常用的域名方式，所以上面的示例是将 email 包移除，请记住，包名称不是您安装 APK 组件时的文件名或是显示在 Android 仿真器中的应用程序名称。另外包名称也并不一定都是 com.android 这样的形式，它可以是各式各样的域名方式。例如 org.iii.ro.iiivpa 或 com.deafcode.android.Cinema。APK 文件的包名称完全是由当初的开发人员所制定的，所以并没有统一的命名方式，唯一相同的就是它一定是类似域名的命名格式。

另外，在移除该 APK 应用程序时，如果想要保留信息与缓存目录，则加上–k 参数即可。

```
adb uninstall -k package                    (移除程序时，保留信息)
```

不过麻烦的是，可能不知道这个想要移除的应用程序包名称，所以必须先运行 adb shell 进入 Android 操作系统的指令列模式，然后到 /data/data 或 /data/app 目录下，得知欲移除的包名称，然后使用 adb uninstall 指令删除 APK 应用程序，这样就可以简易地从 Android 仿真器将不想使用的 APK 应用程序移除了。

```
adb shell
  ls /data/data  或  /data/app                (查询包名称)
  exit
   adb uninstall package                      (移除查询到的包)
```

幸运的是，从 Android SDK 1.5 版起，已经内建应用程序管理系统，不需要再辛苦的使用 adb uninstall 指令移除 APK 应用程序组件，只要在 Android 手机主画面点选 MENU 按键，然后依序选择“Settings”→“Applications”→“Manage applications”，就可以启动应用程序管理系统。当前 Android 系统已经安装的所有应用程序都会列出来，您只要选择想要移除的应用程序，然后选择卸载就可以移除该程序了，这样就不需要使用 adb uninstall 指令。

8.11.5 实现原理

本实例运行后，能够远程下载指定网址的 Android 应用程序，下载到手机后打开 Application installer 软件来安装这个软件。在具体实现上，先设置一个 EditText 来获取远程程序的 URL，按后通过自定义按钮打开下载程序（使用 java.net 的 URLConnection 对象来创建连接，通过 InputStream 将下载文件写入到存储卡的缓存。)，下载后通过自定义方法 openFile()打开文件，并根据文件扩展名，判断是否为 APK 格式，是则启动内置的安装程序，开始安装。安装完成后，在退出安装时通过方法 delFile()将存储卡中的临时文件删除。

8.11.6 具体实现

本实例的主程序文件是 example11.java，下面开始分别介绍其实现流程。

1）先引用 java.io 与 java.net，然后定义各个变量。具体代码如下。

```
public class example11 extends Activity
{
  private TextView mTextView01;
  private EditText mEditText01;
  private Button mButton01;
  private static final String TAG = "DOWNLOADAPK";
  private String currentFilePath = "";
  private String currentTempFilePath = "";
  private String strURL="";
  private String fileEx="";
  private String fileNa="";
  /** Called when the activity is first created. */
  @Override
  public void onCreate(Bundle savedInstanceState)
  {
    super.onCreate(savedInstanceState);
    setContentView(R.layout.main);

    mTextView01 = (TextView)findViewById(R.id.myTextView1);
    mButton01 = (Button)findViewById(R.id.myButton1);
    mEditText01 =(EditText)findViewById(R.id.myEditText1);
```

2）定义 setOnClickListener 用于监听按钮单击事件，设置文件下载到本地端，取得要安装程序的文件名称。具体代码如下。

```
mButton01.setOnClickListener(new Button.OnClickListener()
  {
    public void onClick(View v)
    {
```

```
            /* 文件会下载至本地端 */
            mTextView01.setText("下载中...");
            strURL = mEditText01.getText().toString();
            /*取得欲安装程序之文件名称*/
            fileEx = strURL.substring(strURL.lastIndexOf(".")
            +1,strURL.length()).toLowerCase();
            fileNa = strURL.substring(strURL.lastIndexOf("/")
            +1,strURL.lastIndexOf("."));
            getFile(strURL);
          }
        }
      );
```

3）如果文本框中的远程地址为空，则输出“请输入 URL”的提示。具体代码如下。

```
mEditText01.setOnClickListener(new EditText.OnClickListener()
{
  @Override
  public void onClick(View arg0)
  {
    mEditText01.setText("");
    mTextView01.setText("远程安装程序(请输入 URL)");
  }
});
}
```

4）定义方法 getFile(final String strPath)，用于获取下载的 URL 文件，有异常则输出提示。具体代码如下。

```
/* 处理下载 URL 文件自定义方法*/
private void getFile(final String strPath)   {
  try
  {
    if (strPath.equals(currentFilePath) )
    {
      getDataSource(strPath);
    }
    currentFilePath = strPath;
    Runnable r = new Runnable()
    {
      public void run()
      {
       try
        {
          getDataSource(strPath);
        }
       catch (Exception e)
```

```
            {
              Log.e(TAG, e.getMessage(), e);
            }
          }
        };
        new Thread(r).start();
      }
      catch(Exception e)
      {
        e.printStackTrace();
      }
    }
```

5）定义方法 getDataSource(String strPath)，用于获取远程文件。流程如下。

- 第一步：如果 URL 错误则输出提示。
- 第二步：通过 myURL 对象取得 URL。
- 第三步：创建连接对象 conn。
- 第四步：通过 File 创建临时文件。
- 第五步：取得暂存路径，并将文件写入暂存 SD 存储卡。
- 第六步：通过 openFile(myTempFile)方法打开文件进行安装。

具体代码如下。

```
    /*取得远程文件*/
    private void getDataSource(String strPath) throws Exception
    {
      if (!URLUtil.isNetworkUrl(strPath))
      {
        mTextView01.setText("错误的 URL");
      }
      else
      {
        /*取得 URL*/
        URL myURL = new URL(strPath);
        /*创建连接*/
        URLConnection conn = myURL.openConnection();
        conn.connect();
        /*InputStream  下载文件*/
        InputStream is = conn.getInputStream();
        if (is == null)
        {
          throw new RuntimeException("stream is null");
        }
        /*创建临时文件*/
        File myTempFile = File.createTempFile(fileNa, "."+fileEx);
        /*取得暂存盘路径*/
```

```
            currentTempFilePath = myTempFile.getAbsolutePath();
            /*将文件写入暂存盘*/
            FileOutputStream fos = new FileOutputStream(myTempFile);
            byte buf[] = new byte[128];
            do
            {
              int numread = is.read(buf);
              if (numread <= 0)
              {
                break;
              }
              fos.write(buf, 0, numread);
            }while (true);

            /*打开文件进行安装*/
            openFile(myTempFile);
            try
            {
              is.close();
            }
            catch (Exception ex)
            {
              Log.e(TAG, "error: " + ex.getMessage(), ex);
            }
          }
        }
```

6）定义方法 openFile(File f)，设置在手机上打开文件。流程如下。

- 第一步：调用 getMIMEType()方法来取得文件类型。
- 第二步：设置 intent 的文件位置与文件类型。
- 第三步：判断文件类型的方法，并取得扩展名。
- 第四步：根据扩展名的类型决定文件类型。
- 第五步：如果无法直接打开则弹出软件列表，供用户选择。

具体代码如下。

```
        /* 在手机上打开文件的方法*/
        private void openFile(File f)
        {
          Intent intent = new Intent();
          intent.addFlags(Intent.FLAG_ACTIVITY_NEW_TASK);
          intent.setAction(android.content.Intent.ACTION_VIEW);

          /* 调用 getMIMEType()来取得 MimeType */
          String type = getMIMEType(f);
          /* 设置 intent 的 file 与 MimeType */
```

```
    intent.setDataAndType(Uri.fromFile(f),type);

    startActivity(intent);
  }

  /* 判断文件 MimeType 的方法*/
  private String getMIMEType(File f)
  {
    String type="";
    String fName=f.getName();
    /* 取得扩展名 */
    String end=fName.substring(fName.lastIndexOf(".")
    +1,fName.length()).toLowerCase();

    /* 依扩展名的类型决定 MimeType */
    if(end.equals("m4a")||end.equals("MP3")||end.equals("mid")||
    end.equals("xmf")||end.equals("ogg")||end.equals("wav"))
    {
      type = "audio";
    }
    else if(end.equals("3gp")||end.equals("mp4"))
    {
      type = "video";
    }
    else if(end.equals("jpg")||end.equals("gif")||end.equals("png")||
    end.equals("jpeg")||end.equals("bmp"))
    {
      type = "image";
    }
    else if(end.equals("apk"))
    {
      /* android.permission.INSTALL_PACKAGES */
      type = "application/vnd.android.package-archive";
    }
    else
    {
      type="*";
    }
    /*如果无法直接打开，就跳出软件列表给用户选择 */
    if(end.equals("apk"))
    {
    }
    else
```

```
    {
        type += "/*";

    }
    return type;
}
```

7）定义方法 delFile(String strFileName)，用于删除 SD 卡上的临时文件。具体代码如下。

```
/*自定义删除文件方法*/
private void delFile(String strFileName)
{
    File myFile = new File(strFileName);
    if(myFile.exists())
    {
        myFile.delete();
    }
}
```

8）定义方法 onPause()和 onResume()，分别设置 onPause（暂停）和 onResume（重播）状态。具体代码如下。

```
/*当 Activity 处于 onPause 状态时,更改 TextView 文字状态*/
@Override
protected void onPause()
{
    mTextView01 = (TextView)findViewById(R.id.myTextView1);
    mTextView01.setText("下载成功");
    super.onPause();
}

/*当 Activity 处于 onResume 状态时，删除临时文件*/
@Override
protected void onResume()
{
    /* 删除临时文件 */
    delFile(currentTempFilePath);
    super.onResume();
}
}
```

执行后将在文本框中显示目标安装程序的路径，如图 8-18 所示。实例中的默认路径是 http://mz.ruan8.com/soft/2/sougoushoujishurufa_7786.apk，是一个 sogou 输入法程序。单击“按下开始安装”按钮后，开始下载目标文件，如图 8-19 所示。

图 8-18　下载目标文件

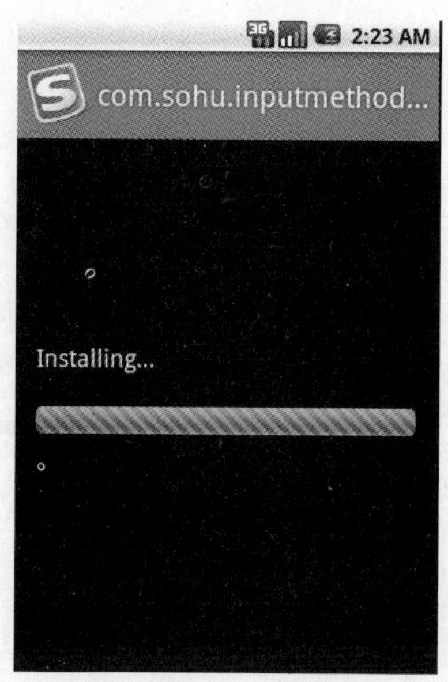

图 8-19　下载界面

下载完成后弹出安装界面，如图 8-20 所示。单击图 8-20 中的“Install”按钮后开始安装，安装完成后输出提示，如图 8-21 所示。

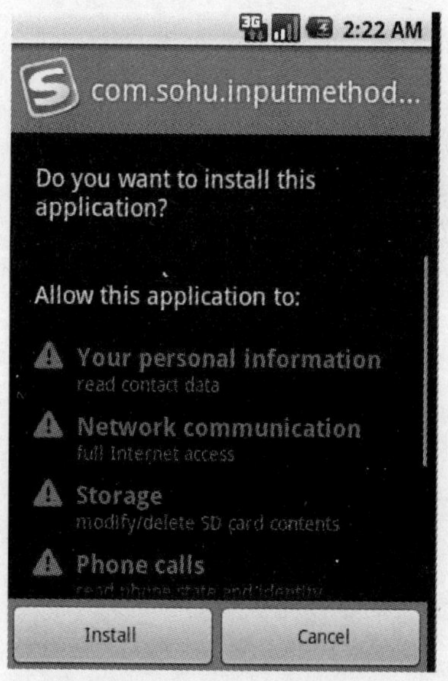

图 8-20　安装界面

图 8-21　安装完成界面

8.12　下载观看 3gp 视频

观看在线视频，是当前智能手机的主要功能之一。在本节的内容中，将通过一个具体实例的实现，介绍在 Android 中下载观看 3gp 视频的基本流程。本实例的源代码保存在“光

盘:\daima\8\example12”，下面开始讲解本实例的具体实现流程。

8.12.1　实现原理

视频一般比较大，必须保证手机有足够空间能够存储，还要确保下载的视频能够被 MediaPlayer 所支持。在本实例中，通过 EditText 来获取远程视频的 URL，然后将此网址的视频下载到手机的存储卡中，然后通过控制按钮来控制对视频的处理。在播放完毕并终止程序后，将暂存到 SD 中的临时视频删除。

8.12.2　具体实现

本实例的主程序文件是 example12.java，下面开始介绍其实现流程。

1）加载相关类，然后通过 Activity 实现 SurfaceHolder.Callback，分别创建 MediaPlayer 对象和 SurfaceHolder 对象。具体代码如下。

```
/* Activity 实现 SurfaceHolder.Callback */
public class example12 extends Activity
implements SurfaceHolder.Callback
{
  private TextView mTextView01;
  private EditText mEditText01;
  /* 创建 MediaPlayer 对象*/
  private MediaPlayer mMediaPlayer01;
  /* 用以配置 MediaPlayer 的 SurfaceView */
  private SurfaceView mSurfaceView01;
  /* SurfaceHolder 对象*/
  private SurfaceHolder mSurfaceHolder01;
  private ImageButton mPlay, mReset, mPause, mStop;
```

2）识别 MediaPlayer 对象是否已被释放，识别 MediaPlayer 对象是否正处于暂停状态，并用 LogCat 输出状态提示标识。具体代码如下。

```
/* 识别 MediaPlayer 是否已被释放*/
private boolean bIsReleased = false;

/* 识别 MediaPlayer 是否正处于暂停*/
private boolean bIsPaused = false;

/* LogCat 输出状态提示标识*/
private static final String TAG = "HippoMediaPlayer";
private String currentFilePath = "";
private String currentTempFilePath = "";
private String strVideoURL = "";
```

3）设置播放视频的 URL 地址，用 mSurfaceView01 对象绑定 Layout 上的 Surface View。然后设置 PixnelFormat（视频流格式），并设置 SurfaceHolder 为界面布局。具体代码如下。

```
/** Called when the activity is first created. */
@Override
public void onCreate(Bundle savedInstanceState)
{
    super.onCreate(savedInstanceState);
    setContentView(R.layout.main);

    /* 将.3gp 图像文件存放 URL 网址*/
    strVideoURL =
    "http://new4.sz.3gp2.com//20100205xyy/喜羊羊与灰太狼%20 踩高跷(www.3gp2.com).3gp";
    //http://www.dubblogs.cc:8751/Android/Test/Media/3gp/test2.3gp

    mTextView01 = (TextView)findViewById(R.id.myTextView1);
    mEditText01 = (EditText)findViewById(R.id.myEditText1);
    mEditText01.setText(strVideoURL);

    /* 绑定 Layout 上的 SurfaceView */
    mSurfaceView01 = (SurfaceView) findViewById(R.id.mSurfaceView1);

    /* 设置 PixnelFormat */
    getWindow().setFormat(PixelFormat.TRANSPARENT);

    /* 设置 SurfaceHolder 为 Layout SurfaceView */
    mSurfaceHolder01 = mSurfaceView01.getHolder();

    mSurfaceHolder01.addCallback(this);
```

4）为影片设置大小比例，设置 4 个控制按钮 mPlay、mReset、mPause 和 mStop。具体代码如下。

```
/* 由于原有的影片尺寸较小，故指定其为固定比例*/
mSurfaceHolder01.setFixedSize(160, 128);
mSurfaceHolder01.setType(SurfaceHolder.SURFACE_TYPE_PUSH_BUFFERS);

mPlay = (ImageButton) findViewById(R.id.play);
mReset = (ImageButton) findViewById(R.id.reset);
mPause = (ImageButton) findViewById(R.id.pause);
mStop = (ImageButton) findViewById(R.id.stop);
```

5）定义 mPlay.setOnClickListener()方法，用于播放监听处理，具体代码如下。

```
/* 播放按钮*/
mPlay.setOnClickListener(new ImageButton.OnClickListener()
{
  public void onClick(View view)
  {
    if(checkSDCard())
```

```
        {
            strVideoURL = mEditText01.getText().toString();
            playVideo(strVideoURL);
            mTextView01.setText(R.string.str_play);
        }
        else
        {
            mTextView01.setText(R.string.str_err_nosd);
        }
    }
});
```

6）定义 mReset.setOnClickListener()方法，用于重新播放监听处理。具体代码如下。

```
/* 重新播放按钮*/
mReset.setOnClickListener(new ImageButton.OnClickListener()
{
  public void onClick(View view)
  {
    if(checkSDCard())
    {
      if(bIsReleased == false)
      {
        if (mMediaPlayer01 != null)
        {

          mMediaPlayer01.seekTo(0);
          mTextView01.setText(R.string.str_play);
        }
      }
    }
    else
    {
      mTextView01.setText(R.string.str_err_nosd);
    }
  }
});
```

7）定义 mReset.setOnClickListener()方法，用于暂停播放监听处理。具体代码如下。

```
/* 暂停按钮*/
mPause.setOnClickListener(new ImageButton.OnClickListener()
{
  public void onClick(View view)
  {
    if(checkSDCard())
    {
```

```
            if (mMediaPlayer01 != null)
            {
              if(bIsReleased == false)
              {
                if(bIsPaused==false)
                {
                  mMediaPlayer01.pause();
                  bIsPaused = true;
                  mTextView01.setText(R.string.str_pause);
                }
                else if(bIsPaused==true)
                {
                  mMediaPlayer01.start();
                  bIsPaused = false;
                  mTextView01.setText(R.string.str_play);
                }
              }
            }
          }
          else
          {
            mTextView01.setText(R.string.str_err_nosd);
          }
        }
      });
```

8）定义 mStop.setOnClickListener()方法，用于停止播放监听处理。具体代码如下。

```
/* 终止按钮*/
mStop.setOnClickListener(new ImageButton.OnClickListener()
{
  public void onClick(View view)
  {
    if(checkSDCard())
    {
      try
      {
        if (mMediaPlayer01 != null)
        {
          if(bIsReleased==false)
          {
            mMediaPlayer01.seekTo(0);
            mMediaPlayer01.pause();
            mTextView01.setText(R.string.str_stop);
          }
        }
      }
```

```
            catch(Exception e)
            {
              mTextView01.setText(e.toString());
              Log.e(TAG, e.toString());
              e.printStackTrace();
            }
          }
          else
          {
            mTextView01.setText(R.string.str_err_nosd);
          }
        }
      });
    }
```

9）定义方法 playVideo()，用于下载指定 URL 影片并实现播放处理。流程如下。

- 第一步：判断传入的 strPath（地址）是否是现有播放的连接。
- 第二步：重新构建 MediaPlayer 对象。
- 第三步：设置播放音量。

具体代码如下。

```
    /* 自定义下载 URL 影片并播放*/
    private void playVideo(final String strPath)
    {
      try

      {
        /* 若传入的 strPath 为现有播放的连接，则直接播放*/
        if (strPath.equals(currentFilePath) && mMediaPlayer01 != null)
        {
          mMediaPlayer01.start();
          return;
        }
        else if(mMediaPlayer01 != null)
        {
          mMediaPlayer01.stop();
        }

        currentFilePath = strPath;

        /* 重新构建 MediaPlayer 对象*/
        mMediaPlayer01 = new MediaPlayer();
        /* 设置播放音量*/
        mMediaPlayer01.setAudioStreamType(2);

        /* 设置显示于 SurfaceHolder */
```

```
mMediaPlayer01.setDisplay(mSurfaceHolder01);

mMediaPlayer01.setOnErrorListener
(new MediaPlayer.OnErrorListener()
{
  @Override
  public boolean onError(MediaPlayer mp, int what, int extra)
  {
    // TODO Auto-generated method stub
    Log.i
    (
      TAG,
      "Error on Listener, what: " + what + "extra: " + extra
    );
    return false;
  }
});
```

10）定义 onBufferingUpdate()方法，用于监听缓冲进度。具体代码如下。

```
mMediaPlayer01.setOnBufferingUpdateListener
(new MediaPlayer.OnBufferingUpdateListener()
{
  @Override
  public void onBufferingUpdate(MediaPlayer mp, int percent)
  {
    // TODO Auto-generated method stub
    Log.i
    (
      TAG, "Update buffer: " +
      Integer.toString(percent) + "%"
    );
  }
});
```

11）分别用 setOnPreparedListener()方法和 setOnCompletionListener()方法监听播放处理，具体代码如下。

```
mMediaPlayer01.setOnCompletionListener
(new MediaPlayer.OnCompletionListener()
{
  @Override
  public void onCompletion(MediaPlayer mp)
  {
    // TODO Auto-generated method stub
    Log.i(TAG,"mMediaPlayer01 Listener Completed");
    mTextView01.setText(R.string.str_done);
```

```
    }
  });
  mMediaPlayer01.setOnPreparedListener
  (new MediaPlayer.OnPreparedListener()
  {
    @Override
    public void onPrepared(MediaPlayer mp)
    {
      // TODO Auto-generated method stub
      Log.i(TAG,"Prepared Listener");
    }
  });
```

12）定义方法 run()，用于接受连接并记录线程信息。首先在线程运行中，调用自定义方法下载指定的文件，当下载完后调用 prepare()方法。当有异常时，则输出错误信息。具体代码如下。

```
    Runnable r = new Runnable()
    {
      public void run()
      {
        try
        {
          /* 在线程运行中，调用自定义方法下载指定文件*/
          setDataSource(strPath);
          /* 下载完后才会调用 prepare */
          mMediaPlayer01.prepare();
          Log.i

          (
            TAG, "Duration: " + mMediaPlayer01.getDuration()
          );
          mMediaPlayer01.start();
          bIsReleased = false;
        }
        catch (Exception e)
        {
          Log.e(TAG, e.getMessage(), e);
        }
      }
    };
    new Thread(r).start();
  }
  catch(Exception e)
  {
    if (mMediaPlayer01 != null)
    {
```

```
        mMediaPlayer01.stop();
        mMediaPlayer01.release();
      }
    }
  }
```

13）定义方法 setDataSource()，用于线程启动的方式播放视频。具体代码如下。

```
/* 自定义 setDataSource，由线程启动*/
private void setDataSource(String strPath) throws Exception
{
  if (!URLUtil.isNetworkUrl(strPath))
  {
    mMediaPlayer01.setDataSource(strPath);
  }
  else
  {
    if(bIsReleased == false)
    {
      URL myURL = new URL(strPath);
      URLConnection conn = myURL.openConnection();
      conn.connect();
      InputStream is = conn.getInputStream();
      if (is == null)
      {
        throw new RuntimeException("stream is null");
      }
      File myFileTemp = File.createTempFile
      ("hippoplayertmp", "."+getFileExtension(strPath));


      currentTempFilePath = myFileTemp.getAbsolutePath();

      /*currentTempFilePath = /sdcard/mediaplayertMP39327.dat */

      FileOutputStream fos = new FileOutputStream(myFileTemp);
      byte buf[] = new byte[128];
      do
      {
        int numread = is.read(buf);
        if (numread <= 0)
        {
          break;
        }
        fos.write(buf, 0, numread);
      }while (true);
      mMediaPlayer01.setDataSource(currentTempFilePath);
```

```
            try
            {
                is.close();
            }
            catch (Exception ex)
            {
                Log.e(TAG, "error: " + ex.getMessage(), ex);
            }
        }
    }
}
```

14）定义方法 getFileExtension()，用于获取视频的扩展名。具体代码如下。

```
private String getFileExtension(String strFileName)
{
    File myFile = new File(strFileName);
    String strFileExtension=myFile.getName();
    strFileExtension=(strFileExtension.substring
    (strFileExtension.lastIndexOf(".")+1)).toLowerCase();

    if(strFileExtension=="")
    {
        /* 若无法顺利取得扩展名，默认为.dat */
        strFileExtension = "dat";
    }
    return strFileExtension;
}
```

15）定义方法 checkSDCard()，用于判断存储卡是否存在。具体代码如下。

```
private boolean checkSDCard()
{
    /* 判断存储卡是否存在*/
    if(android.os.Environment.getExternalStorageState().equals
    (android.os.Environment.MEDIA_MOUNTED))
    {
        return true;
    }
    else
    {
        return false;
    }
}

@Override
```

```
    public void surfaceChanged
    (SurfaceHolder surfaceholder, int format, int w, int h)
    {
      Log.i(TAG, "Surface Changed");
    }

    @Override
    public void surfaceCreated(SurfaceHolder surfaceholder)
    {
      Log.i(TAG, "Surface Changed");
    }

    @Override
    public void surfaceDestroyed(SurfaceHolder surfaceholder)
    {
      Log.i(TAG, "Surface Changed");
    }
  }
```

执行后在文本框中显示指定播放视频的 URL，下载完毕，能实现播放处理。如图 8-22 所示。

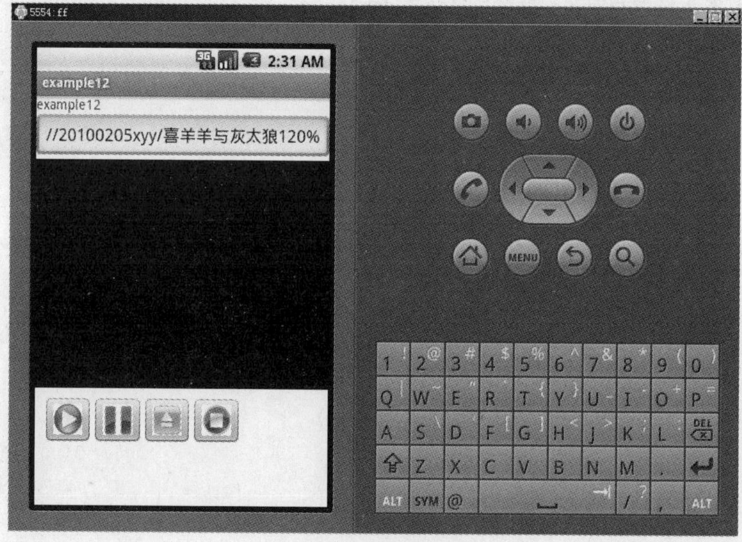

图 8-22　执行效果

第 9 章　绑定官方的服务

Android 作为 Google 旗下的产品，它能够使用官方的很多强大的服务功能。例如 Google API、日历、相册和文件等。在本章的内容中，将通过几个典型实例的实现，来详细介绍 Android 和其官方服务相结合的基本知识和具体使用流程。

9.1　模拟验证官方账号

在现实应用中，基于 Google 的大多数服务都需要通过官方进行的账号验证。在本节的内容中，将通过一个具体实例的实现，介绍在 Android 中通过账号验证来获取“Google Account Authentication Service”所发出的凭据的过程。本实例的源代码保存在“光盘:\daima\9\example1”，下面开始讲解本实例的具体实现流程。

9.1.1　Google Account Authentication Service 介绍

当前的 Google Account 阵容强大，由单一的 AuthSub 发展到 OAuth Federated Hybird 一共四种。具体说明如下。

- AuthSub：提供单独的 google service access，比如 gmail contact。如果你要一次使用多个 google service，比如同时需要读取 picasaweb 和 youtube 的数据，那么用 authsub 就不那么合适了，因为你需要让用户傻傻的登录 google 两次。
- OAuth：和 authsub 相比，提供了 sign-on，一次登录可以取得的 Authtoken 可以 access 好几个 Google services。
- Federated：本质上是 openid，他是唯一可以让你在使用 Google Account 登录自己网站的同事，拿到用户电子邮件的的方法。
- Hybird：集成了 oauth 和 openid，可以让你的网站同时拿到 Google service 访问权限，以及用户的电子邮件。

由此可见。AuthSub 即将过时，现在的好处是，使用 Authsub 不需要在 Google 注册自己的网站。

9.1.2　具体实现

本实例的主文件是 example1.java、example1_01_02.java 和 GoogleAuthSub.java，下面分别介绍其实现流程。

1．文件 example1.java

example1.java 文件的功能是登录 UI 和获取用户账号的密码。首先以 TextView 的 onClick()事件为起点，调用自定义的 showLoginForm()方法显示登录表单。example1.java 文件的具体实现代码如下。

```
public class example1 extends Activity
{
  /*声明变量*/
  private TextView mTextView01;
  private LayoutInflater mInflater01;
  private View mView01;
  private EditText mEditText01,mEditText02;
  private String TAG = "HIPPO_DEBUG";
  /* 中文字的间距 */
  private int intShiftPadding = 14;

  /** Called when the activity is first created. */
  @Override
  public void onCreate(Bundle savedInstanceState)
  {
    super.onCreate(savedInstanceState);
    setContentView(R.layout.main);

    /* 创建 DisplayMetrics 对象，取得屏幕分辨率 */
    DisplayMetrics dm = new DisplayMetrics();
    getWindowManager().getDefaultDisplay().getMetrics(dm);

    mTextView01 = (TextView)findViewById(R.id.myTextView1);

    /* 将文字 Label 放在屏幕右上方 */
    mTextView01.setLayoutParams
    (

      new  AbsoluteLayout.LayoutParams(intShiftPadding*mTextView01.getText().toString().length(),
18,(dm.widthPixels-(intShiftPadding*mTextView01.getText().toString().length()))-10,0)
    );

    /* 处理用户单击 TextView 文字的事件，显示登录对话框*/
    mTextView01.setOnClickListener(new TextView.OnClickListener()
    {
      @Override
      public void onClick(View v)
      {
        // TODO Auto-generated method stub

        /* 显示登录对话框 */
        showLoginForm();
      }
    });
  }
```

```
/* 自定义登录对话框方法 */
private void showLoginForm()
{
  try
  { /* 以 LayoutInflater 取得主 Activity 的 context */
    mInflater01 = LayoutInflater.from(example1.this);
    /* 设置创建的 View 所要使用的 Layout Resource */
    mView01 = mInflater01.inflate(R.layout.login, null);

    /* 账号 EditText */
    mEditText01=(EditText)mView01.findViewById(R.id.myEditText1);

    /* 密码 EditText */
    mEditText02=(EditText)mView01.findViewById(R.id.myEditText2);

    /*创建 AlertDialog 窗口来取得用户账号密码*/
    new AlertDialog.Builder(this)
    .setView(mView01)
    .setPositiveButton("OK",
    new DialogInterface.OnClickListener()
    {
      /*覆盖 onClick()来触发取得 Token 事件与完成登录事件*/
      public void onClick(DialogInterface dialog, int whichButton)
      {
        if(processGoogleLogin(mEditText01.getText().toString(),
mEditText02.getText().toString()))
        {
          Intent i = new Intent();

          /*登录后调用注销程序(example1_01_02.java)*/
          i.setClass(example1.this, example1_01_02.class);
          startActivity(i);
          finish();
        }
      }
    }).show();
  }
  catch(Exception e)
  {
    e.printStackTrace();
  }
}
/*调用 GoogleAuthSub 来取的 Google 账号的 Authentication Token*/
private boolean processGoogleLogin(String strUID, String strUPW)
{
  try
```

```
        {
          /*建构自定义的 GoogtleAuthSub 对象*/
          GoogleAuthSub gas = new GoogleAuthSub(strUID, strUPW);
          /*取得 Google Token*/
          String strAuth =    gas.getAuthSubToken();
          /*将取回的 Google Token 写入 log 中*/
          Log.i(TAG, strAuth);

        }
        catch (Exception e)
        {
          e.printStackTrace();
        }
        return true;
      }
    }
```

showLoginForm()方法是一个使用 LayoutInflater 获取主 Activity（活动程序）的关联程序，搭配 AlertDialog（提示框）所构建的登录表单。当用户输入账号和密码后，开始重写 DialogInterface.OnClickListener()的 onClick()事件来调用自定义的 processGoogleLogin()方法处理和 Google 账号验证的连接事件。当通过 Google 验证后取得 Google Authentication Token（令牌），通过 Intent 打开 example1_01_02.java 以改变 UI 的状态。

2．文件 example1_01_02.java

example1_01_02.java 文件的功能是注销并返回登录界面。即将原来的登录状态，改为注销状态，并实现 TextView 的 onClick()方法。当用户单击 TextView，则通过自定义的 Intent 来调用 example.java，返回到程序的等待状态。文件 example1_01_02.java 的具体实现代码如下。

```
public class example1_01_02 extends Activity
{
  private TextView mTextView03;
  /*  中文字的间距  */
  private int intShiftPadding = 14;

  @Override
  protected void onCreate(Bundle savedInstanceState)
  {
    super.onCreate(savedInstanceState);
    setContentView(R.layout.loginok);

    /*  创建 DisplayMetrics 对象，取得屏幕分辨率  */
    DisplayMetrics dm = new DisplayMetrics();
    getWindowManager().getDefaultDisplay().getMetrics(dm);

    /*通过  findViewById()来取得 TextView 对象*/
    mTextView03 = (TextView)findViewById(R.id.myTextView3);
```

```
        /* 将文字 Label 放在屏幕右上方 */
        mTextView03.setLayoutParams
        (
          new
AbsoluteLayout.LayoutParams(intShiftPadding*mTextView03.getText().toString().length(),18,(dm.widthPixels-(int
ShiftPadding*mTextView03.getText().toString().length()))-10,0)
        );

        /* 处理用户单击 TextView 文字的事件处理 */
        mTextView03.setOnClickListener(new TextView.OnClickListener()
        {
          /*覆盖 onClick()事件*/
          @Override
          public void onClick(View v)
          {
            Intent i = new Intent();
            /*注销后调用登录程序(example1_01_02.java)*/
            i.setClass(example1_01_02.this, example1.class);
            startActivity(i);
            finish();
          }
        });
      }
    }
```

3．文件 GoogleAuthSub.java

GoogleAuthSub.java 的功能是通过 HttpGet 传输网络数据，并获取 Google 的日历服务 Token。此文件是整个实例的核心，通过 Google 提供的 ClientLogin 机制，使用 HttpPost 连接到 https://www.google.com/accounts/ClientLogin，并同时将用户账号和密码及其相关参数以 Name Value Pair 字符串带入，通过自定义的 getAuth()方法获取 Google 认证的 Authentication Token，然后模拟 Google 网络服务的的 AuthSub 的方法，将自定义的 Header()和 HttpGet 方法传入 Token 来获取用户 Google Calendar 服务中的所有日历数据，并以 XML 文件存储在临时文件中，作为使用 Google 服务的规范。文件 GoogleAuthSub.java 的具体实现代码如下。

```
public class GoogleAuthSub
{
  /*声明变量*/
  private DefaultHttpClient httpclient;
  private HttpPost httpost;
  private HttpResponse response;
  private String strGoogleAccount;
  private String strGooglePassword;
  private String TAG = "IRDC_DEBUG";
```

```
/*GoogleAuthSub 对象的构造器*/
public GoogleAuthSub(String strUID, String strPWD)
{
  this.strGoogleAccount = strUID;
  this.strGooglePassword = strPWD;
  httpclient = new DefaultHttpClient();
  httpost = new HttpPost("https://www.google.com/accounts/ClientLogin");
}

/*定义取得 Google Token 方法*/
public String getAuthSubToken()
{
  /*创建 Name Value Pair 格式的字符串*/
  List <NameValuePair> nvps = new ArrayList <NameValuePair>();
  nvps.add(new BasicNameValuePair("Email", this.strGoogleAccount));
  nvps.add(new BasicNameValuePair("Passwd", this.strGooglePassword));
  nvps.add(new BasicNameValuePair("source", "MyApiV1"));
  nvps.add(new BasicNameValuePair("service", "cl"));
  String GoogleLoginAuth="";
  try
  {
    /*创建 Http Post 连接*/
    httpost.setEntity(new UrlEncodedFormEntity(nvps, HTTP.DEFAULT_CONTENT_ CHARSET));
    response = httpclient.execute(httpost);
    if( response.getStatusLine().getStatusCode()!=200 )
    {
      return "";
    }
    /*取回 Google Token*/
    InputStream is = response.getEntity().getContent();
    GoogleLoginAuth = getAuth(is);



    /*模拟 HTTP Header*/
    Header[] headers = new BasicHeader[6];

    headers[0] = new BasicHeader("Content-type", "application/x-www-form-urlencoded");
    headers[1] = new BasicHeader("Authorization", "GoogleLogin auth=\""+GoogleLoginAuth+"\"");
    headers[2] = new BasicHeader("User-Agent", "Java/1.5.0_06");
    headers[3] = new BasicHeader("Accept", "text/html, image/gif, image/jpeg, *; q=.2, */*; q=.2");
    headers[4] = new BasicHeader("Connection", "keep-alive");
    /*发出 Http Get 请求登录 Google Calendar 服务作范例*/
    HttpGet httpget;
    String feedUrl2 = "http://www.google.com/calendar/feeds/default/allcalendars/full";
    httpget = new HttpGet(feedUrl2);
    httpget.addHeader(headers[0]);
```

```
        httpget.addHeader(headers[1]);
        httpget.addHeader(headers[2]);
        httpget.addHeader(headers[3]);
        httpget.addHeader(headers[4]);
        /*取得 Google Calendar 服务应答*/
        response = httpclient.execute(httpget);
        String strTemp01 = convertStreamToString(response.getEntity().getContent());
        Log.i(TAG, strTemp01);
        /*指定暂存盘位置*/
        String strEarthLog = "/sdcard/googleauth.log";
        BufferedWriter bw;
        bw = new BufferedWriter (new FileWriter(strEarthLog));
        /*将取回文件写入暂存盘中*/
        bw.write( strTemp01, 0, strTemp01.length() );
        bw.flush();

    }
    catch (UnsupportedEncodingException e)
    {
        e.printStackTrace();
    }
    catch (ClientProtocolException e)
    {
        e.printStackTrace();
    }
    catch (IOException e)
    {
        e.printStackTrace();
    }
    catch(Exception e)
    {
        e.printStackTrace();
    }
    return GoogleLoginAuth;
}
/*自定义读取 token 内容的方法*/

public String getAuth(InputStream is)
{
    BufferedReader reader = new BufferedReader(new InputStreamReader(is));
    String line = null;
    String strAuth="";
    try
    {
        while ((line = reader.readLine()) != null)
        {
```

```
                Log.d(TAG, ": "+line);
                if( line.startsWith("Auth="))
                {
                    strAuth=line.substring(5);
                    Log.i("auth",": "+strAuth);
                }
            }
        }
        catch (IOException e)
        {
            e.printStackTrace();
        }
        finally
        {
            try
            {
                is.close();
            }
            catch (IOException e)
            {
                e.printStackTrace();
            }
        }
        return strAuth;
    }
    /*将数据转为字符串方法*/
    public String convertStreamToString(InputStream is)
    {
        BufferedReader reader = new BufferedReader(new InputStreamReader(is));
        StringBuilder sb = new StringBuilder();
        String line = null;
        try
        {
            while ((line = reader.readLine()) != null)
            {
                sb.append(line);
            }
        }
        catch (IOException e)
        {
            e.printStackTrace();
        }
        finally
        {
            try
```

```
                {
                    is.close();
                }
                catch (IOException e)
                {
                    e.printStackTrace();
                }
            }
            return sb.toString();
        }
    }
```

执行后的效果如图 9-1 所示；单击“登录”链接后显示登录表单界面，如图 9-2 所示；输入账号和密码并单击“OK”按钮后弹出成功获取 Token 提示。如图 9-3 所示。

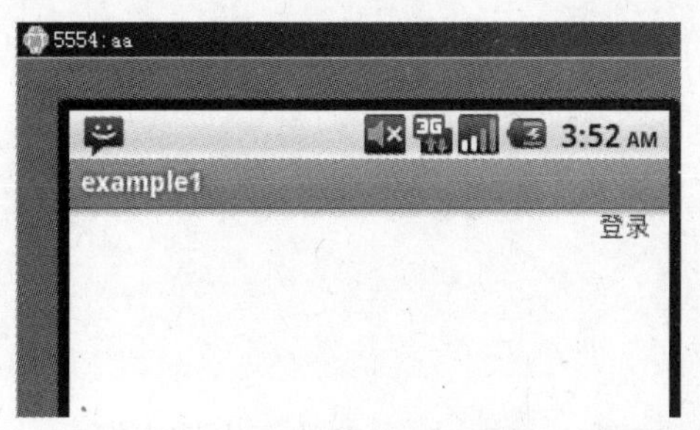

图 9-1　执行效果

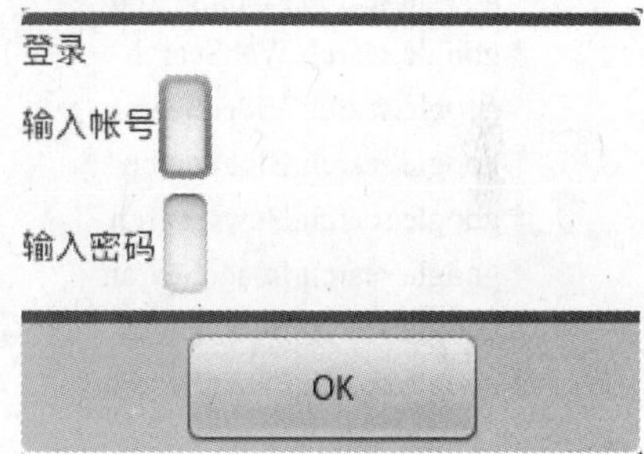

图 9-2　登录表单

图 9-3　成功获取 Token 提示

9.2　模拟实现 Google 搜索

当今社会，信息已经成为第一生产力。在巨大的网络资源中，检索站点蓬勃发展，百度、雅虎和 Google，都已经站在了时代的最前沿。在 Android 官方服务中，提供了 Google Search API 实现检索处理。在本节的内容中，将通过一个具体实例的实现，介绍在 Android 中通过 Google Search API 实现检索处理的流程。本实例的源代码保存在“光盘:\daima\9\example2”，下面开始讲解本实例的具体实现流程。

9.2.1 Google Search API 的使用流程

使用 Google Search API 的基本流程如下。

第一步：构造一个搜索服务“容器”。google.search.SearchControl 的实例代表页面上的一个搜索控件，这个控件是多种“搜索服务”的“容器”。

第二步：构造一个搜索“服务”对象。构造方法 google.search.LocalSearch()可以构造一个“搜索服务”对象。尽管默认的是有一个搜索框和结果列表，但这些都是可以改变的，甚至搜索结果的内容都是可以改变的。

第三步：向容器中添加“搜索服务”。此功能通过方法 searchControl.addSearcher(searcher, options)实现，其中参数 options 的类型为 google.search.Search- Options。在此可以添加的搜索服务有多种，到目前为止 Google 公司提供了以下类型的搜索器。

```
* google.search.LocalSearch
* google.search.WebSearch
* google.search.VideoSearch
* google.search.BlogSearch
* google.search.NewsSearch
* google.search.ImageSearch
* google.search.BookSearch
* google.search.PatentSearch
```

第四步：SearchControl 对象调用 draw 方法，按照 drawOptions 参数画出搜索框。

searchControl.draw(document.getElementById("from"), drawOptions)

上述步骤是一般的流程，但 Google 提供了很多选项来定制这些服务。主要的选项有两种，一种是 SearcherOptions，另一种是 DrawOptions。

9.2.2 具体实现

本实例的主文件是 example2.java 和 MyAdapter.java，下面分别介绍其实现流程。

1. 文件 example2.java

文件 example2.java 功能是创建 MyAdapter 对象，此对象是自己实现的 BaseMyAdapter 类。文件 example2.java 的具体实现代码如下。

```
package irdc.example2;

import irdc.example2.R;
import android.app.Activity;
import android.os.Bundle;
import android.widget.AutoCompleteTextView;

public class example2 extends Activity
{
  private AutoCompleteTextView myAutoCompleteTextView1;
```

```
        /** Called when the activity is first created. */
        @Override
        public void onCreate(Bundle savedInstanceState)
        {
          super.onCreate(savedInstanceState);
          setContentView(R.layout.main);
          myAutoCompleteTextView1 = (AutoCompleteTextView) findViewById(R.id.myAutoComplete-
TextView1);

          /* new 一个自己实现的 BaseAdapter */
          MyAdapter adapter = new MyAdapter(this);
          myAutoCompleteTextView1.setAdapter(adapter);
        }
      }
```

2．文件 MyAdapter.java

MyAdapter 继承于 BaseAdapter，可以通过覆盖 Filterable 对象中的 getFilter()方法来对输入的关键字进行动态处理。当用户输入关键字时，performFiltering()方法所返回的 FilterResults（过滤结果）就是查询后的结果。文件 MyAdapter.java 的具体实现代码如下。

```
public class MyAdapter extends BaseAdapter implements Filterable
{
  ArrayList<String> keyWordValue = new ArrayList<String>();
  ArrayList<String> resultValue = new ArrayList<String>();
  private Context mContext;
  LinearLayout.LayoutParams param1;

  public MyAdapter(Context context)
  {
    mContext = context;

    param1 = new LinearLayout.LayoutParams(
        LinearLayout.LayoutParams.WRAP_CONTENT,
        LinearLayout.LayoutParams.WRAP_CONTENT);
  }

  @Override
  public int getCount()
  {
    return keyWordValue.size();
  }

  @Override
  public Object getItem(int position)
  {
    return keyWordValue.get(position);
```

```
}

@Override
public long getItemId(int position)
{
  return position;
}

@Override
public View getView(int position, View view, ViewGroup viewGroup)
{

  LinearLayout myLinearLayout = new LinearLayout(mContext);
  myLinearLayout.setOrientation(LinearLayout.HORIZONTAL);

  if (position >= keyWordValue.size())
    return myLinearLayout;
  /* 第一个 TextView 存放关键字 */
  TextView keyWordTextView = new TextView(this.mContext);
  keyWordTextView.setTextColor(mContext.getResources().getColor(
      R.drawable.blue));
  keyWordTextView.setTextSize(18);
  keyWordTextView.setWidth(180);
  try
  {
    keyWordTextView
        .setText(keyWordValue.get(position).toString());
  } catch (java.lang.IndexOutOfBoundsException i)
  {
    keyWordTextView.setText("");
  }
  /* 第二个 TextView 存放关键字结果数量 */
  TextView resultTextView = new TextView(this.mContext);
  resultTextView.setTextColor(mContext.getResources().getColor(
      R.drawable.red));
  resultTextView.setTextSize(18);
  try
  {
    resultTextView.setText(resultValue.get(position).toString());
  } catch (java.lang.IndexOutOfBoundsException i)
  {
    resultTextView.setText("");
  }
  myLinearLayout.addView(keyWordTextView, param1);
  myLinearLayout.addView(resultTextView, param1);

  return myLinearLayout;
```

```
}

@Override
public Filter getFilter()
{
  // TODO Auto-generated method stub
  Filter myFilter = new Filter()
  {

    @Override
    protected FilterResults performFiltering(CharSequence text)

    {

      FilterResults fr = new FilterResults();
      keyWordValue = new java.util.ArrayList<String>();
      resultValue = new java.util.ArrayList<String>();
      if (text == null || text.length() == 0)
      {
        fr.count = keyWordValue.size();
        fr.values = keyWordValue;
        return fr;
      }

      /* 输入关键字后调用 Google  关键字 API */
      changeResult(getGoogleAPI(text.toString()));

      fr.count = keyWordValue.size();
      fr.values = keyWordValue;
      return fr;
    }

    @Override
    protected void publishResults(CharSequence text,
        FilterResults filterResults)
    {
      if (filterResults != null && filterResults.count > 0)
        notifyDataSetChanged();
      else
        notifyDataSetInvalidated();

    }
  };
  return myFilter;
}

/* 访问 Google-API 取得返回的结果字符串 */
```

```
private String getGoogleAPI(String text)
{
  String uri = "";
  try
  {
    /* 输入的字要 encode */
    uri = "http://www.google.com/complete/"
        + "search?hl=en&js=true&qu="
        + URLEncoder.encode(text, "utf-8");
  } catch (UnsupportedEncodingException e1)
  {

    // TODO Auto-generated catch block
    e1.printStackTrace();
  }

  URL googleUrl = null;
  HttpURLConnection conn = null;
  InputStream is = null;
  BufferedReader in = null;
  String resultStr = "";
  /* 取得连接 */
  try
  {
    googleUrl = new URL(uri);
    /* 打开连接 */
    conn = (HttpURLConnection) googleUrl.openConnection();
    int code = conn.getResponseCode();
    /* 连接 OK 时 */
    if (code == HttpURLConnection.HTTP_OK)
    {
      /* 取得返回的 InputStream */
      is = conn.getInputStream();

      in = new BufferedReader(new InputStreamReader(is));
      String inputLine;

      /* 一行一行读取 */
      while ((inputLine = in.readLine()) != null)
      {
        resultStr += inputLine;
      }

    }
  } catch (IOException e)
  {
    // TODO Auto-generated catch block
```

```
                e.printStackTrace();
            } finally
            {
                try
                {
                    if (is != null)
                        is.close();
                    if (conn != null)
                        conn.disconnect();
                } catch (Exception e)
                {

                }
            }

            return resultStr;
        }

        /* 处理返回的字符串变成 ArrayList */
        private void changeResult(String text)
        {

            String resultStr = "";
            String startSub = "new Array(2, ";
            String endSub = "), new Array";
            int start = text.indexOf(startSub);
            int end = text.indexOf(endSub);
            if (start != -1 && end != -1)
            {
                resultStr = text.substring(start + startSub.length(), end);
                /* 去掉前后的" */
                resultStr = resultStr.substring(1, resultStr.length() - 1);
                /* 以 "," 来分隔字符串变成字符串数组 */
                String total[] = resultStr.split("\\\", \\\"");
                for (int i = 0; i < total.length / 2; i++)
                {
                    keyWordValue.add(total[i * 2]);
                    /* 将 results 字符串去掉 */
                    resultValue
                        .add(total[i * 2 + 1].replaceAll(" results", ""));
                }
            }

        }

    }
```

在上述代码中，在 MyAdapter 类中重写了 getView()方法，在其中放了 2 个 TextView，一个显示关键字，另一个显示结果数量。因为 AutoCompleteTextView 组件绑定了 MyAdapter，所以当 Adapter 动态向 Google 获取查询结果时，也顺便更新了 AutoCompleteTextView 下拉菜单里的结果。

执行后的效果如图 9-4 所示。

图 9-4　执行效果

9.3　Google Chart API 生成二维条码

通过 Google 公司提供的 Google Chart API，可以方便的生成动态二维条码。这样开发人员无需掌握 GD Libray 等知识，降低了开发门槛。在本节的内容中，将通过一个具体实例的实现，介绍在 Android 中通过 Google Chart API 动态生成二维条码的具体流程。本实例的源代码保存在“光盘:\daima\9\example3”，下面开始讲解本实例的具体实现流程。

9.3.1　Google Chart API 基础

Google Chart API 为每个请求返回一个 PNG 格式图片。目前提供如下类型图表：折线图、柱状图、饼图、维恩图和散点图。可以设定图表尺寸、颜色和图例。可以在网页中使用<img>元素插入图表，当浏览器打开该网页时，Chart API 提供即时图表。

所有 Chart API URL 都应使用如下格式。

```
http://chart.apis.google.com/chart?<parameter 1>&<parameter 2>&<parameter n>
```

注意：每个 URL 所有字符必须在同一行内。

多个参数间使用&作为分隔符，可以使用任意多个参数，例如图 9-5 所示。

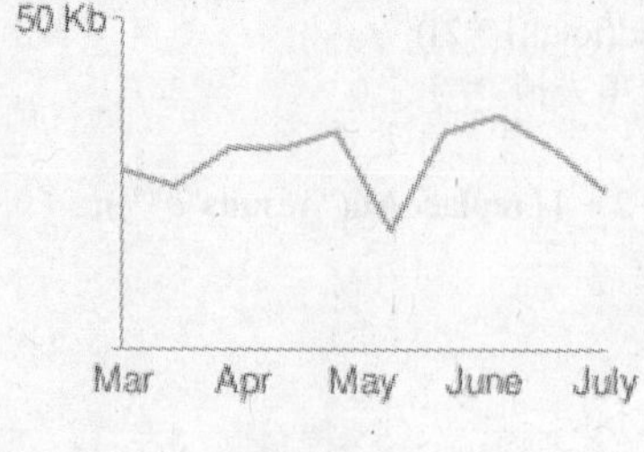

图 9-5

http://chart.apis.google.com/chart?cht=lc&chs=200x125&chd=s:helloWorld&chxt=x,y&chxl=0:|Mar|Apr|May|June|July|1:||50+Kb

各个参数的具体说明如下。

- ❑ http://chart.apis.google.com/chart?：Chart API 的调用地址。
- ❑ &：参数分隔符。
- ❑ chs=200x125：图表尺寸。
- ❑ chd=s:helloWorld：图表数据值。
- ❑ cht=lc：图表类型。
- ❑ chxt=x,y：显示 x、y 轴坐标。
- ❑ chxl=0:|Mar|Apr|May|June|July|1:||50+Kb：x、y 轴坐标值。

可以在网页中使用 img 元素插入图表，例如，

```
<img src="http://chart.apis.google.com/chart?chs=200x125&amp;chd=s:helloWorld&amp;cht=lc&amp;
chxt=x,y&amp;chxl=0:|Mar|Apr|May|June|July|1:||50+Kb" alt="Sample chart" />
```

注意：在 HTMLimg 元素中，URL 属性中&字符应书写为转义字符&。

所有请求必须包含以下参数。

- ❑ 图表尺寸。
- ❑ 图表数据。
- ❑ 图表类型。

其他参数均为可选参数，各类型图表有效参数如表 9-1 所示。

表 9-1　有效参数

参　　数	柱状图	折线图 谱线图	雷达图	散点图	维恩图	饼图	仪表图	地图
数据颜色	√	√	√	√	√	√	√	√
区域、背景填充	√	√	√	√	√	√	仅背景	仅背景
数值缩放	√	√	√	√	√	√	√	
线性过渡填充	√	√	√	√	√	仅背景		
线性条纹填充	√	√	√	√	√	仅背景		
图表标题	√	√	√	√	√	√		
图表图例	√	√	√	√	√			
多轴标注	√	√	√	√				
网格线	√	√	√	√				
形状标记	√	√	√	√				
水平区间填充	√	√	√	√				
垂直区间填充	√	√	√	√				
折线样式	√	√	√					
数据区块填充	√	√	√					
柱形、间隔宽度	√							
柱状图基准线	√							
饼图、仪表图标注						√	√	

9.3.2 具体实现

本实例的主文件是 example3.java 和 main.xml，下面分别介绍其实现流程。

在文件 example3.java 中，通过自定义方法 genGoogleQRChart()获取要显示远程图像的网址，然后以“<img src="">”的方式来组成 HTML 标记。在具体成像时，通过了 WebView 组件来显示 HTML 的内容。文件 example3.java 的具体实现代码如下。

```
public class example3 extends Activity
{
  private Button mButton01;
  private EditText mEditText01;
  private WebView mWebView01;
  private boolean bInternetConnectivity=false;

  /** Called when the activity is first created. */
  @Override
  public void onCreate(Bundle savedInstanceState)
  {
    super.onCreate(savedInstanceState);
    setContentView(R.layout.main);

    /* 测试手机是否具有连接 Google API 的连接能力 */
    if(checkInternetConnection
       ("http://code.google.com/intl/zh-TW/apis/chart/","utf-8")
      )
    {
      bInternetConnectivity = true;
    }

    mWebView01 = (WebView)findViewById(R.id.myWebView1);
    mButton01 = (Button)findViewById(R.id.myButton1);
    mButton01.setOnClickListener(new Button.OnClickListener()
    {
      @Override
      public void onClick(View v)
      {
        // TODO Auto-generated method stub
        if(mEditText01.getText().toString()!="" &&
           bInternetConnectivity==true)
        {
          mWebView01.loadData
          (
              /* 调用自定义云端生成条码的方法 */
            genGoogleQRChart
            (
```

```
                mEditText01.getText().toString(),120
              ),"text/html", "utf-8"
            );
          }
        }
      });

      mEditText01 = (EditText)findViewById(R.id.myEditText1);
      mEditText01.setText(R.string.str_text);

      mEditText01.setOnKeyListener(new EditText.OnKeyListener()
      {
        @Override
        public boolean onKey(View v, int keyCode, KeyEvent event)
        {

          if(mEditText01.getText().toString()!="" &&
             bInternetConnectivity==true)
          {
            mWebView01.loadData
            (
              genGoogleQRChart
              (
                mEditText01.getText().toString(),120
              ),"text/html", "utf-8"
            );
          }
          return false;
        }
      });
    }

    /* 调用 Google API，产生二维条形码 */
    public String genGoogleQRChart(String strToQRCode, int strWidth)
    {
      String strReturn="";
      try
      {
        strReturn = new String(strToQRCode.getBytes("utf-8"));

        /* 组成 Google API 需要的传输参数字符串 */
        strReturn = "<html><body>"+
        "<img src=http://chart.apis.google.com/chart?chs="+
        strWidth+"x"+strWidth+"&chl="+
        URLEncoder.encode(strReturn, "utf-8")+


        "&choe=UTF-8&cht=qr></body></html>";
      }
```

```
        catch (Exception e)
        {
            e.printStackTrace();
        }
        return strReturn;
    }

    /* 检查网络连接是否正常 */
    public boolean checkInternetConnection
    (String strURL, String strEncoding)
    {
        /* 最多延时 n 秒，若无应答则表示无法连接 */
        int intTimeout = 5;
        try
        {
            HttpURLConnection urlConnection= null;
            URL url = new URL(strURL);
            urlConnection=(HttpURLConnection)url.openConnection();
            urlConnection.setRequestMethod("GET");
            urlConnection.setDoOutput(true);
            urlConnection.setDoInput(true);
            urlConnection.setRequestProperty
            (
                "User-Agent","Mozilla/4.0"+
                " (compatible; MSIE 6.0; Windows 2000)"
            );

            urlConnection.setRequestProperty
            ("Content-type","text/html; charset="+strEncoding);
            urlConnection.setConnectTimeout(1000*intTimeout);
            urlConnection.connect();
            if (urlConnection.getResponseCode() == 200)
            {
                return true;
            }
            else
            {
                return false;
            }
        }
        catch (Exception e)
        {
            e.printStackTrace();
            return false;


        }
    }
```

```
    /*  自定义 BIG5 转 UTF-8 */
    public String big52unicode(String strBIG5)
    {
      String strReturn="";
      try
      {
        strReturn = new String(strBIG5.getBytes("big5"), "UTF-8");
      }
      catch (Exception e)
      {
        e.printStackTrace();
      }
      return strReturn;
    }

    /*  自定义 UTF-8 转 BIG5 */
    public String unicode2big5(String strUTF8)
    {
      String strReturn="";
      try
      {
        strReturn = new String(strUTF8.getBytes("UTF-8"), "big5");
      }
      catch (Exception e)
      {
        e.printStackTrace();
      }
      return strReturn;
    }
}
```

执行后的效果如图 9-6 所示。

图 9-6　执行效果

9.4 Google 地图的典型运用

Google 公司在很早的时候就推出了地图应用。在本节的内容中，将通过一个具体实例的实现，介绍在 Android 中通过 Google MapView（谷歌地图）和 GeoPoint（表示经度和纬度的类）实现地图经纬应用的流程。本实例的源代码保存在“光盘:\daima\9\example4”，下面开始讲解本实例的具体实现流程。

9.4.1 Google MapView 基础

Android 支持 GPS 和网络地图，通常将各种不同的定位技术成为 LBS。LBS 是基于位置的服务（Location Based Service）的简称，它是通过电信移动运营商的无线电通讯网络（如 GSM 网、CDMA 网）或外部定位方式（如 GPS）获取移动终端用户的位置信息（地理坐标或大地坐标），在地理信息系统（Geographic Information System，GIS）平台的支持下，为用户提供相应服务的一种增值业务。

Android 支持地理定位服务的 API。该地理定位服务可以用来获取当前设备的地理位置。应用程序可以定时请求更新设备当前的地理定位信息。应用程序也可以借助一个 Intent 接收器来实现如下功能：以经纬度和半径划定的一个区域，当设备出入该区域时，可以发出提醒信息。在下面的内容中，开始讲解 android.location 中和定位有关的功能类。

1. android.location 的功能类

（1）Android Location API

以下是包中关于定位功能的比较重要的类。

- LocationManager：本类提供访问定位服务的功能，也提供获取最佳定位提供者的功能。另外，临近警报功能也可以借助该类来实现。
- LocationProvider：该类是定位提供者的抽象类。定位提供者具备周期性报告设备地理位置的功能。
- LocationListener：提供定位信息发生改变时的回调功能。必须事先在定位管理器中注册监听器对象。
- Criteria：该类使得应用能够通过在 LocationProvider 中设置的属性来选择合适的定位提供者。

（2）Map API

Android 也提供了一组访问地图的 API，借助 Map 及定位 API，就能在地图上显示用户当前的地理位置。在 Android 中定义了一个名为 com.google.android.maps 的包，其中包含了一系列用于在 Map 上显示、控制和层叠信息的功能类，以下是该包中最重要的几个类。

- MapActivity：这个类是用于显示 MAP 的 Activity 类，它需要连接底层网络。
- MapView：MapView 是用于显示地图的 View 组件，它必须和 MapActivity 配合使用。
- MapController：MapController 用于控制地图的移动。
- Overlay：这是一个可显示于地图之上的可绘制的对象。
- GeoPoint：一个包含经纬度位置的对象。

2．Android 定位的基本流程

了解了 Android 中和定位处理相关的类后，下面将简要介绍在 Android 中实现定位处理的基本流程。

（1）准备 Activity 类

目标是使用 MAP API 来显示地图，然后使用定位 API 来获取设备的当前定位信息以在 MAP 上设置设备的当前位置，用户定位会随着用户的位置移动而发生改变。

首先需要一个继承了 MapActivity 的 Activity 类，例如下面的代码。

```
class MyGPSActivity extends MapActivity {
        …
        }
```

要成功引用 MAP API，还必须先在 AndroidManifest.xml 中定义如下信息。

```
<uses-library android:name="com.google.android.maps" />
```

（2）使用 MapView

要让地图显示，需要将 MapView 加入到应用中来。例如，在布局文件（main.xml）中加入如下代码。

```
<com.google.android.maps.MapView
                android:id="@+id/myGMap"
                android:layout_width="fill_parent"
                android:layout_height="fill_parent"
                android:enabled="true"
                android:clickable="true"
                android:apiKey="API_Key_String"
/>
```

另外，要使用 MAP 服务的话，还需要一个 API key。可以通过如下方式获取 API key。

1）找到“USER_HOME\Local Settings\Application Data\Android”目录下的“debug.keystore”文件。

2）使用 keytool 工具来生成认证信息（MD5），使用如下命令行。

```
keytool -list -alias androiddebugkey -keystore <path_to_debug_keystore>.keystore -storepass
android -keypass android
```

3）打开“Sign Up for the Android Maps API”页面，输入之前生成的认证信息（MD5）后将获取到你的 API key。

4）替换上面 AndroidManifest.xml 配置文件中“API_Key_String”为刚才获取的 API key。

注意：上面获取 API key 的介绍比较简单，在本章后面的内容中，将通过一个具体实例的实现过程来演示获取 API key 的方法。

接下来继续补全 MyGPSActivity 类的代码。例如下面的代码。

```
class MyGPSActivity extends MapActivity {
    @Override


    public void onCreate(Bundle savedInstanceState) {
    //创建并初始化地图
      gMapView = (MapView) findViewById(R.id.myGMap);
        GeoPoint p = new GeoPoint((int) (lat * 1000000), (int) (long * 1000000));
        gMapView.setSatellite(true);
        mc = gMapView.getController();
        mc.setCenter(p);
        mc.setZoom(14);
    }
    …
}
```

另外，如果要使用定位信息的话，必须设置一些权限，在 AndroidManifest.xml 中的具体配置如下。

```
<uses-permission android:name="android.permission.INTERNET"></uses-permission>
    <uses-permission
android:name="android.permission.ACCESS_COARSE_LOCATION"></uses-permission>
    <uses-permission
android:name="android.permission.ACCESS_FINE_LOCATION"></uses-permission>
```

（3）使用定位管理器

可以通过使用 Context.getSystemService()方法，传入 Context.LOCATION_SERVICE 参数获取定位管理器的实例。例如下面的代码。

```
LocationManager lm = (LocationManager) getSystemService(Context.LOCATION_SERVICE);
```

之后，需要将原来的 MyGPSActivity 对象做一些修改，让它实现一个 LocationListener 接口，使其能够监听定位信息的改变。

```
class MyGPSActivity extends MapActivity implements LocationListener {
…
ublic void onLocationChanged(Location location) {}
public void onProviderDisabled(String provider) {}
public void onProviderEnabled(String provider) {}
public void onStatusChanged(String provider, int status, Bundle extras) {}
protected boolean isRouteDisplayed() {
return false;
}
}
```

添加一些代码，对 LocationManager 进行一些初始化工作，并在它的 onCreate()方法中注册定位监听器。

```
@Override
public void onCreate(Bundle savedInstanceState) {
LocationManager lm = (LocationManager)getSystemService(Context.LOCATION_SERVICE);
lm.requestLocationUpdates(LocationManager.GPS_PROVIDER, 1000L, 500.0f, this);
        }
```

此时代码中的 onLocationChanged()方法就会在用户的位置发生 500 米距离的改变之后进行调用。这里默认使用的 LocationProvider 是"gps"（GSP_PROVIDER），但是可以根据你的需要，使用特定的 Criteria 对象调用 LocationManger 类的 getBestProvider()方法获取其他的 LocationProvider。以下代码是 onLocationChanged 方法的参考实现。

```
public void onLocationChanged(Location location) {
  if (location != null) {
  double lat = location.getLatitude();
  double lng = location.getLongitude();
    p = new GeoPoint((int) lat * 1000000, (int) lng * 1000000);
    mc.animateTo(p);
  }
}
```

通过上面的代码，获取了当前的新位置并更新地图上的位置显示。还可以为应用程序添加一些诸如缩放效果、地图标注和文本等功能。

（4）添加缩放控件

```
/*将缩放控件添加到地图上*/
ZoomControls zoomControls =   (ZoomControls) gMapView.getZoomControls();
zoomControls.setLayoutParams(new ViewGroup.LayoutParams(LayoutParams.WRAP_CONTENT,
LayoutParams.WRAP_CONTENT));
gMapView.addView(zoomControls);
gMapView.displayZoomControls(true);
```

（5）添加 Map Overlay

来到最后一步，添加 Map Overlay（层）。通过下面的代码可以定义一个 Overlay。

```
class MyLocationOverlay extends com.google.android.maps.Overlay {
@Override
  public boolean draw(Canvas canvas, MapView mapView, boolean shadow, long when) {
super.draw(canvas, mapView, shadow);
Paint paint = new Paint();
/* 将经纬度转换成实际屏幕坐标 */
Point myScreenCoords = new Point();
mapView.getProjection().toPixels(p, myScreenCoords);
paint.setStrokeWidth(1);
paint.setARGB(255, 255, 255, 255);
paint.setStyle(Paint.Style.STROKE);
Bitmap bmp = BitmapFactory.decodeResource(getResources(), R.drawable.marker);
```

```
        canvas.drawBitmap(bmp, myScreenCoords.x, myScreenCoords.y, paint);
        canvas.drawText("how are you…", myScreenCoords.x, myScreenCoords.y, paint);
        return true;

    }
}
```

上面的 Overlay 会在地图上显示一个“Here I am”的文本，然后把这个 Overlay 添加到地图上去。

```
MyLocationOverlay myLocationOverlay = new MyLocationOverlay();
List<Overlay> list = gMapView.getOverlays();
list.add(myLocationOverlay);
```

3. 使用前的设置

Google 地图给人们的生活带来了极大的方便。例如可以通过 Google 地图查找商户信息、查看地图和获取行车路线等。Android 平台也提供了一个 map 包（com.google.android.maps），通过其中的 MapView 就能够方便的利用 Google 地图的资源来进行编程。在使用前需要预先进行如下必要的设置。

（1）添加 maps.jar 到项目

在 Android SDK 中，以 JAR 库的形式提供了和 MAP 有关的 API，此 JAR 库位与“android-sdk-windows\add-ons\google_apis-4”目录下。要把 maps.jar 添加到项目中，可以在项目属性中的“Android”栏中指定使用包含 Google API 的 Target 作为项目的构建目标。如图 9-7 所示。

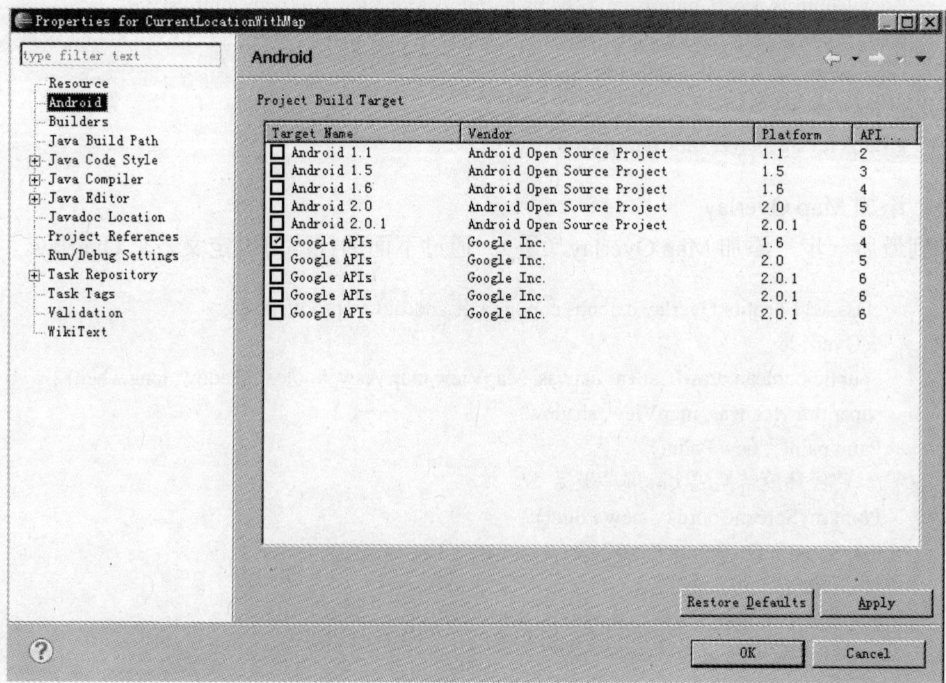

图 9-7　在项目中包含 Google API

（2）将地图嵌入到应用

通过使用 MapActivity 和 MapView 控件，可以轻松地将地图嵌入到应用程序当中。在此步骤中，需要将 Google API 添加到构建路径中。方法是在图 9-7 所示界面中选择“Java Build Path”，然后在 Target 中勾选 Google API，设置项目中包含 Google API。如图 9-8 所示。

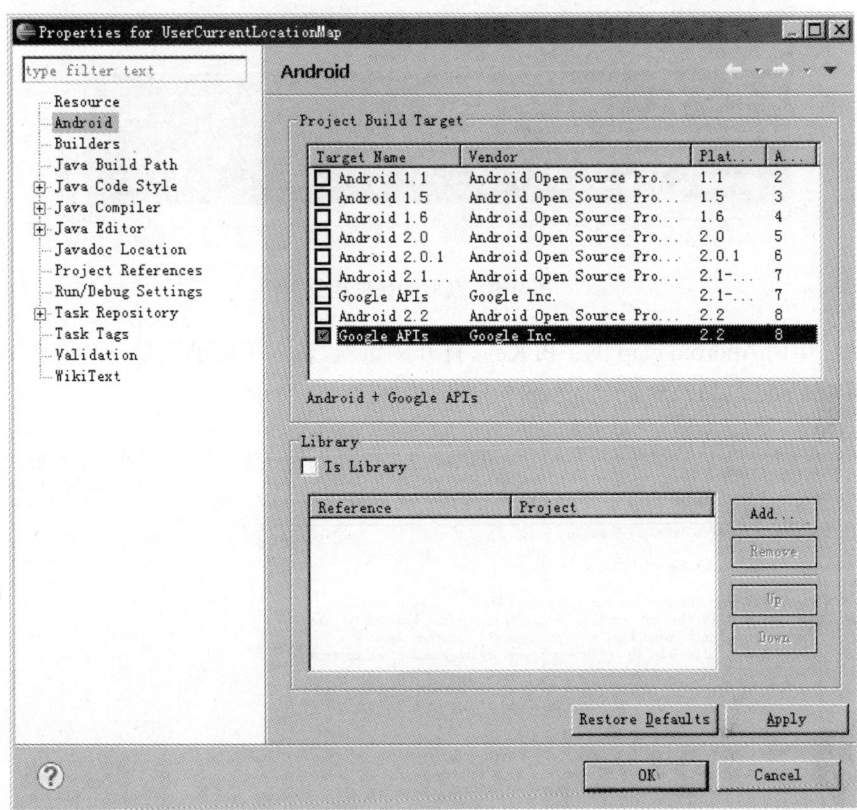

图 9-8　将 Google API 添加到构建路径

（3）获取 Map API 密钥

在利用 MapView 之前，必须要先申请一个 Android Map API Key。具体步骤如下。

第 1 步：找到你的 debug.keystore 文件，通常位于如下目录。

C:\Documents and Settings\你的当前用户\Local Settings\Application Data\Android

第 2 步：获取 MD5 指纹。运行 cmd.exe，执行如下命令获取 MD5 指纹。

```
>keytool -list -alias androiddebugkey -keystore "debug.keystore 的路径" -storepass android -keypass android
```

例如，作者机器输入如下命令。

```
keytool -list -alias androiddebugkey -keystore "C:\Documents and Settings\Administrator\. android\debug.keystore" -storepass android -keypass android
```

此时系统会提示输入 keystore 密码，这时候输入 android，系统就会输出我们申请到的 MD5 认证指纹。如图 9-9 所示。

注意：因为在 CMD 中不能直接复制、粘贴使用 CMD 命令，这样很影响编程效率，所以笔者使用了第三方软件 PowerCmd 来代替机器中自带的 CMD 工具。

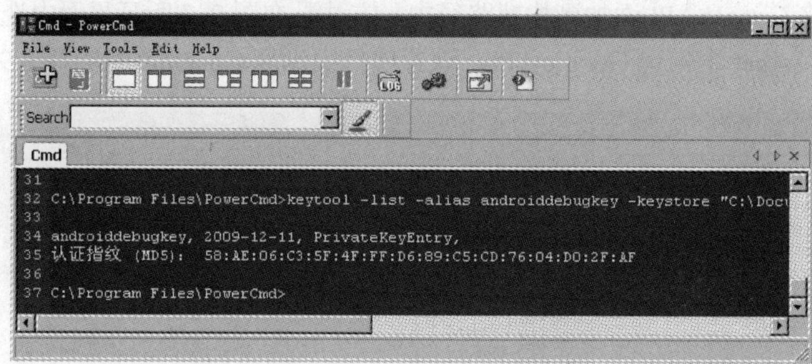

图 9-9　获取的认证指纹

第 3 步：申请 Android map 的 API Key。打开浏览器，输入下面的网址："http://code.google.com/intl/zh-CN/android/maps-api-signup. html"，如图 9-10 所示。

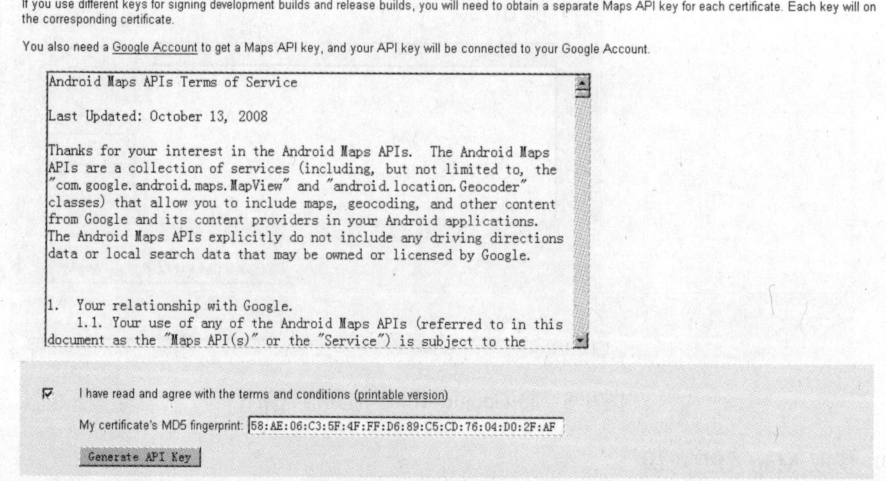
If you use different keys for signing development builds and release builds, you will need to obtain a separate Maps API key for each certificate. Each key will on the corresponding certificate.

You also need a Google Account to get a Maps API key, and your API key will be connected to your Google Account.

Android Maps APIs Terms of Service

Last Updated: October 13, 2008

Thanks for your interest in the Android Maps APIs. The Android Maps APIs are a collection of services (including, but not limited to, the "com.google.android.maps.MapView" and "android.location.Geocoder" classes) that allow you to include maps, geocoding, and other content from Google and its content providers in your Android applications. The Android Maps APIs explicitly do not include any driving directions data or local search data that may be owned or licensed by Google.

1. Your relationship with Google.
1.1. Your use of any of the Android Maps APIs (referred to in this document as the "Maps API(s)" or the "Service") is subject to the

☑ I have read and agree with the terms and conditions (printable version)

My certificate's MD5 fingerprint: 58:AE:06:C3:5F:4F:FF:D6:89:C5:CD:76:04:D0:2F:AF

Generate API Key

图 9-10　申请主页

在 Google 的 Android Map API Key 申请页面上输入图 9-10 中得到的 MD5 认证指纹，按下"Generate API Key"按钮后即可转到下面的这个画面，得到我们申请到的 API Key。如图 9-11 所示。

Google 地图 API

感谢您注册 Android 地图 API 密钥！

您的密钥是：

0by7ffx8jX0A_LWXeKCMTWAh8CqHA1qvzetFqjQ

此密钥适用于所有使用以下指纹所对应证书进行验证的应用程序：

58:AE:06:C3:5F:4F:FF:D6:89:C5:CD:76:04:D0:2F:AF

下面是一个 xml 格式的示例，帮助您了解地图功能：

```
<com.google.android.maps.MapView
    android:layout_width="fill_parent"
    android:layout_height="fill_parent"
    android:apiKey="0by7ffx8jX0A_LWXeKCMTWAh8CqHA1qvzetFqjQ"
    />
```

图 9-11　得到的 API Key

至此，成功获取了一个 API Key，编程前的整个准备工作也完成了。

4．使用 Map API 密钥的基本流程

通过上面的讲解，我们已经申请到了一个 Android Map API Key，下面开始讲解使用 Map API 密钥实现编程的基本流程。

第 1 步：在文件 AndroidManifest.xml 中声明权限。在 Anroid 系统中，如果程序执行需要读取到安全敏感的项目，那么必须在 AndroidManifest.xml 中声明相关权限请求，比如这个地图程序需要从网络读取相关数据。所以必须声明 android.permission.INTERNET 权限。具体方法是在文件 AndroidManifest.xml 中添加如下代码。

```
<uses-permission android:name="android.permission.INTERNET" />
```

另外，因为 maps 类不是 Android 启动的缺省类，所以还需要在文件 AndroidManifest.xml 的 application 标签中声明要用 maps 类。

```
<uses-library android:name="com.google.android.maps" />
```

下面是基本的 AndroidManifest.xml 文件代码。

```
<manifest xmlns:android="http://schemas.android.com/apk/res/android"
    <application android:icon="@drawable/icon" android:label="@string/app_name">
     <uses-library android:name="com.google.android.maps" />
    </application>
  <uses-permission android:name="android.permission.INTERNET" />
</manifest>
```

第 2 步：在 main.xml 主文件中完成 Layout。下面开始着手来完成界面，假设设置要显示杭州的卫星地图，并在地图上方有 5 个按钮，分别可以放大地图、缩小地图或者切换显示模式（卫星、交通和街景）。即整个界面主要由 2 个部分组成，上面是一排 5 个按钮，下面是 Map View。

在 Android 中，LinearLayout 是可以互相嵌套的，在此可以把上面 5 个按钮放在一个子 LinearLayout 里边（子 LinearLayout 的指定可以由 android:addStatesFromChildren="true"实现），然后再把这个子 LinearLayout 加到外面的父 LinearLayout 里边。

第 3 步：完成主 Java 程序代码。

9.4.2　具体实现

本实例的主文件是 example4.java 和 main.xml，下面分别介绍其实现流程。

1．文件 example4.java

在文件 example4.java 中，通过 EditText 来输入坐标的经度和维度，将坐标转换为 GeoPoint 对象后，再利用 MapController 对象的 animateTo()方法将地图的中心点移到 GeoPoint 坐标上。文件 example4.java 的具体实现代码如下。

```
public class example4 extends MapActivity
{
  private MapController mMapController01;
```

```
private MapView mMapView01;
private Button mButton01,mButton02,mButton03;
private EditText mEditText01;
private EditText mEditText02;
private int intZoomLevel=0;
/* Map 启动时的默认坐标*/
private double dLat=120.391177;
private double dLng=39.9067452;

@Override
protected void onCreate(Bundle icicle)
{
  super.onCreate(icicle);
  setContentView(R.layout.main);

  /* 创建 MapView 对象 */
  mMapView01 = (MapView)findViewById(R.id.myMapView1);
  mMapController01 = mMapView01.getController();
  /* 设置 MapView 的显示选项（卫星、街道） */
  mMapView01.setSatellite(false);
  mMapView01.setStreetView(true);
  /* 默认放大的层级 */
  intZoomLevel = 17;
  mMapController01.setZoom(intZoomLevel);
  /* 设置 Map 的中点为默认经纬度 */
  refreshMapView();

  mEditText01 = (EditText)findViewById(R.id.myEdit1);
  mEditText02 = (EditText)findViewById(R.id.myEdit2);


  /* 送出查询的 Button */
  mButton01 = (Button)findViewById(R.id.myButton1);
  mButton01.setOnClickListener(new Button.OnClickListener()
  {
    @Override
    public void onClick(View v)
    {
      /* 经纬度空白检查 */
      if(mEditText01.getText().toString().equals("")||
          mEditText02.getText().toString().equals(""))
      {
        showDialog("经度或纬度填写不正确!");
      }
      else
      {
        /* 取得输入的经纬度 */
```

```
          dLng=Double.parseDouble(mEditText01.getText().toString());
          dLat=Double.parseDouble(mEditText02.getText().toString());
          /* 根据输入的经纬度重置地图*/
          refreshMapView();
        }
      }
    });

    /* 放大地图的按钮*/
    mButton02 = (Button)findViewById(R.id.myButton2);
    mButton02.setOnClickListener(new Button.OnClickListener()
    {
      @Override
      public void onClick(View v)
      {
        intZoomLevel++;
        if(intZoomLevel>mMapView01.getMaxZoomLevel())
        {
          intZoomLevel = mMapView01.getMaxZoomLevel();
        }
        mMapController01.setZoom(intZoomLevel);
      }
    });

    /* 缩小地图的按钮*/
    mButton03 = (Button)findViewById(R.id.myButton3);
    mButton03.setOnClickListener(new Button.OnClickListener()
    {
      @Override
      public void onClick(View v)

      {
        intZoomLevel--;
        if(intZoomLevel<1)
        {
          intZoomLevel = 1;
        }
        mMapController01.setZoom(intZoomLevel);
      }
    });
  }

  /* 重置地图的方法*/
  public void refreshMapView()
  {
    GeoPoint p = new GeoPoint((int)(dLat* 1E6), (int)(dLng* 1E6));
    mMapView01.displayZoomControls(true);
```

```
        /* 将 Map 的中点移至 GeoPoint */
        mMapController01.animateTo(p);
        mMapController01.setZoom(intZoomLevel);
    }

    @Override
    protected boolean isRouteDisplayed()
    {
        return false;
    }

    /* 显示 Dialog 的方法*/
    private void showDialog(String mess){
        new AlertDialog.Builder(example4.this).setTitle("Message")
        .setMessage(mess)
        .setNegativeButton("确定", new DialogInterface.OnClickListener()
        {
            public void onClick(DialogInterface dialog, int which)
            {
            }
        })
        .show();
    }
}
```

2．文件 main.xml

文件 main.xml 是一个布局文件，主要代码如下。

```
    </Button>
    <!-- Google MapView Widget -->
    <com.google.android.maps.MapView
        android:id="@+id/myMapView1"
        android:layout_width="fill_parent"
        android:layout_height="fill_parent"
        android:layout_x="0px"
        android:layout_y="102px"
        android:enabled="true"
        android:clickable="true"
        android:apiKey="0by7ffx8jX0A_LWXeKCMTWAh8CqHAlqvzetFqjQ"
    >
    </com.google.android.maps.MapView>
</AbsoluteLayout>
```

上述代码的核心是 android:apiKey="0by7ffx8jX0A_LWXeKCMTWAh8CqHAlqvzetFqjQ"，即调用了 Android Maps API key。

9.5 Geocoder 实现地址查询

Google 提供了一个 Geocoder 服务，在非商用情况下，Geocoder 能够反查 Address（地址）对象。在本节的内容中，将通过一个具体实例的实现，介绍在 Android 中通过 Geocoder 实现地址反查功能的流程。本实例的源代码保存在“光盘:\daima\9\example5”，下面开始讲解本实例的具体实现流程。

9.5.1 Geocoder 基础

什么是 Geocoder，简单说就是根据某个 key（键值）寻找地理位置的坐标，这个 key 可以是地址。近年来 Google Map API 做了很多改进，已经自动整合了中文地址的定位功能，不再需要以前的 mashup 了。以前只有 6 个国家提供了街道级别的定位，Google 从不让人失望。虽然存在每日只能 50000 次查询的限制，但是一般也足够了，何况还有客户端的 built-in cache 改善用户体验。

这个 key 还可以是 ip 地址，例如做一个 library 可以读取本地数据库转换 ip 为坐标，这个坐标只是城市中心的坐标，不过一般也够用了。同时 library 也将提供基于 http 请求的公共服务，地址定位和 ip 定位现在都很容易实现了。

9.5.2 具体实现

本实例的主文件是 example5.java，功能是在地图中定位输入的地址。首先通过方法 getGeoByAddress() 获取输入地址的字符串，然后传入这个地址，并且使用 Geocoder.getFromLocationName()方法获取从 Google 服务器找到的结果。文件 example5.java 的具体实现代码如下。

```
public class example5 extends MapActivity
{
  private MapController mMapController01;
  private MapView mMapView01;
  private Button mButton01,mButton02,mButton03;
  private EditText mEditText01;
  private int intZoomLevel=15;
  private String TAG = "HIPPO_GEO_DEBUG";

  @Override
  protected void onCreate(Bundle icicle)
  {
    super.onCreate(icicle);
    setContentView(R.layout.main);

    mEditText01 = (EditText)findViewById(R.id.myEditText1);
    mEditText01.setText
    (
```

```
        getResources().getText(R.string.str_default_address).toString()
    );

    /* 创建 MapView 对象 */
    mMapView01 = (MapView)findViewById(R.id.myMapView1);
    mMapController01 = mMapView01.getController();

/* 设置 MapView 的显示选项（卫星、街道）*/
    mMapView01.setSatellite(true);

    mMapView01.setStreetView(true);

    mButton01 = (Button)findViewById(R.id.myButton1);
    mButton01.setOnClickListener(new Button.OnClickListener()
    {
      @Override
      public void onClick(View v)
      {
        if(mEditText01.getText().toString()!="")
        {
          refreshMapViewByGeoPoint
          (
            getGeoByAddress
            (
              mEditText01.getText().toString()
            ),mMapView01,intZoomLevel,true
          );
        }
      }
    });

    /* 放大 */
    mButton02 = (Button)findViewById(R.id.myButton2);
    mButton02.setOnClickListener(new Button.OnClickListener()
    {
      @Override
      public void onClick(View v)
      {
        intZoomLevel++;
        if(intZoomLevel>mMapView01.getMaxZoomLevel())
        {
          intZoomLevel = mMapView01.getMaxZoomLevel();
        }
        mMapController01.setZoom(intZoomLevel);
      }
    });
```

```
            /* 缩小 */
            mButton03 = (Button)findViewById(R.id.myButton3);
            mButton03.setOnClickListener(new Button.OnClickListener()
            {
              @Override
              public void onClick(View v)
              {

                // TODO Auto-generated method stub
                intZoomLevel--;
                if(intZoomLevel<1)
                {
                  intZoomLevel = 1;
                }
                mMapController01.setZoom(intZoomLevel);
              }
            });

            /* 初次查询地点 */
            refreshMapViewByGeoPoint
            (
              getGeoByAddress
              (
                  getResources().getText(R.string.str_default_address).toString()
              ),mMapView01,intZoomLevel,true
            );
          }

          private GeoPoint getGeoByAddress(String strSearchAddress)
          {
            GeoPoint gp = null;
            try
            {
              if(strSearchAddress!="")
              {
                Geocoder mGeocoder01 = new Geocoder(example5.this, Locale.getDefault());
                List<Address> lstAddress = mGeocoder01.getFromLocationName(strSearchAddress, 1);
                if (!lstAddress.isEmpty())
                {
                  // Address[addressLines=[0:"U.S PIZZA",1:"15th Main Rd, Phase II, J P
Nagar",2:"Bengaluru, Karnataka",3:"India"],feature=U.S PIZZA,admin=Karnataka,sub-admin=Bengaluru,locality=
Bengaluru,thoroughfare=15th Main Rd,postalCode=null,countryCode=IN,countryName= India,hasLatitude= true,
latitude=18.508933,hasLongitude=true,longitude=73.8042,phone=null,url=null,extras=null]
                  /*
                  for (int i = 0; i < lstAddress.size(); ++i)
                  {
                    Address adsLocation = lstAddress.get(i);
```

```
                Log.i(TAG, "Address found = " + adsLocation.toString());
                lat = adsLocation.getLatitude();
                lon = adsLocation.getLongitude();
              }
              */
              Address adsLocation = lstAddress.get(0);
              double geoLatitude = adsLocation.getLatitude()*1E6;

              double geoLongitude = adsLocation.getLongitude()*1E6;
              gp = new GeoPoint((int) geoLatitude, (int) geoLongitude);
            }
            else
            {
              Log.i(TAG, "Address GeoPoint NOT Found.");
            }
          }
        }
        catch (Exception e)
        {
          e.printStackTrace();
        }
        return gp;
      }

      public static void refreshMapViewByGeoPoint(GeoPoint gp, MapView mv, int zoomLevel, boolean
bIfSatellite)
      {
        try
        {
          mv.displayZoomControls(true);
          /* 取得 MapView 的 MapController */
          MapController mc = mv.getController();
          /* 移至该地理坐标地址 */
          mc.animateTo(gp);

          /* 放大地图层级 */
          mc.setZoom(zoomLevel);

          /* 设置 MapView 的显示选项（卫星、街道）*/
          if(bIfSatellite)
          {
            mv.setSatellite(true);
            mv.setStreetView(true);
          }
          else
          {
            mv.setSatellite(false);
```

```
            }
        }
        catch(Exception e)
        {
            e.printStackTrace();
        }
    }

    @Override
    protected boolean isRouteDisplayed()
    {
        // TODO Auto-generated method stub
        return false;
    }
}
```

执行后能够实现地址反查处理，如图 9-12 所示。

图 9-12　执行效果

9.6　Directions Route 实现路径导航

在 Android SDK 中，可以调用手机内置地图程序来传递导航坐标来规划路径。在本节的内容中，将通过一个具体实例的实现，介绍在 Android 中通过 Directions Route 实现路径导航的流程。本实例的源代码保存在“光盘:\daima\9\example6”，下面开始讲解本实例的具体实现流程。

9.6.1　实现原理

在具体实现上，先调用 getLocationProvider()方法来获取当前 Location（位置），以取得当前所在位置的地理坐标。并通过提供的 EditText 组件来让用户输入地址，通过地址取得目的地的地理坐标。通过 2 个 GeoPoint 对象，并通过 Intent 方式来调用内置的地图程序。在具体实现上，要把握如下 2 点。

- 以 Url.parse()方法传入 Google Map 的路径规划参数方法。
- 根据手机的移动状态，更改并更新当前 GeoPoint 的方法。

只要设置好了起点 GeoPoint 和终点 GeoPoint，通过重组 Google Map 的 GET 传输参数，便可以传入 Google Map 显示。实际上，map.google.com 能够接受的经纬度需要以“精度，维度”的字符串格式来传递，所以编写了一个 GeoPointToString()来重组 GeoPoint 的经纬度值。

9.6.2 具体实现

本实例的主文件是 example6.java，下面开始讲解其具体实现流程。

1）创建 MapView 对象，并创建 LocationManager 对象取得系统定位服务，然后设置 MapView 对象的显示选项（卫星、街道）。具体实现代码如下。

```
@Override
protected void onCreate(Bundle icicle)
{
  // TODO Auto-generated method stub
  super.onCreate(icicle);
  setContentView(R.layout.main);

  mTextView01 = (TextView)findViewById(R.id.myTextView1);

  mEditText01 = (EditText)findViewById(R.id.myEditText1);
  mEditText01.setText
  (
    getResources().getText
    (R.string.str_default_address).toString()
  );

  /* 创建 MapView 对象 */
  mMapView01 = (MapView)findViewById(R.id.myMapView1);
  mMapController01 = mMapView01.getController();

  /* 设置 MapView 的显示选项（卫星、街道）*/
  mMapView01.setSatellite(true);
  mMapView01.setStreetView(true);

  /* 放大的层级 */
  intZoomLevel = 15;
  mMapController01.setZoom(intZoomLevel);

  /* 创建 LocationManager 对象取得系统 LOCATION 服务 */
  mLocationManager01 =
```

```
(LocationManager)getSystemService(Context.LOCATION_SERVICE);

/*
 * 自定义方法，访问 Location Provider，
 * 并将之存储在 strLocationProvider 当中
 */
getLocationProvider();

/* 传入 Location 对象，显示于 MapView */
fromGeoPoint = getGeoByLocation(mLocation01);
refreshMapViewByGeoPoint(fromGeoPoint,
                         mMapView01, intZoomLevel);

/* 创建 LocationManager 对象，监听
 * Location 更改时事件，更新 MapView*/
mLocationManager01.requestLocationUpdates
(strLocationProvider, 2000, 10, mLocationListener01);
```

2）定义 toGeoPoint 对象获取用户要前往地址的 GeoPoint 对象，传入路径规划所需要的地标地址。具体实现代码如下。

```
mButton01 = (Button)findViewById(R.id.myButton1);
mButton01.setOnClickListener(new Button.OnClickListener()
{
  @Override
  public void onClick(View v)
  {
    if(mEditText01.getText().toString()!="")
    {
      /* 取得用户要前往地址的 GeoPoint 对象 */
      toGeoPoint =
      getGeoByAddress(mEditText01.getText().toString());

      /* 路径规划 Intent */
      Intent intent = new Intent();
      intent.setAction(android.content.Intent.ACTION_VIEW);

      /* 传入路径规划所需要的地标地址 */
      intent.setData
      (
        Uri.parse("http://maps.google.com/maps?f=d&saddr="+
        GeoPointToString(fromGeoPoint)+
        "&daddr="+GeoPointToString(toGeoPoint)+
        "&hl=cn" +
        "")
      );
      startActivity(intent);
```

```
            }
        }
    });
```

3）定义 mButton02 按钮处理事件，实现地图放大处理。定义 mButton03 按钮处理事件，实现地图缩小处理。具体代码如下。

```
    /* 放大地图 */
    mButton02 = (Button)findViewById(R.id.myButton2);
    mButton02.setOnClickListener(new Button.OnClickListener()
    {
      @Override
      public void onClick(View v)
      {
        intZoomLevel++;
        if(intZoomLevel>mMapView01.getMaxZoomLevel())
        {
          intZoomLevel = mMapView01.getMaxZoomLevel();
        }
        mMapController01.setZoom(intZoomLevel);
      }
    });

    /* 缩小地图 */
    mButton03 = (Button)findViewById(R.id.myButton3);
    mButton03.setOnClickListener(new Button.OnClickListener()
    {
      @Override
      public void onClick(View v)
      {
        intZoomLevel--;
        if(intZoomLevel<1)
        {
          intZoomLevel = 1;
        }
        mMapController01.setZoom(intZoomLevel);
      }
    });
  }
```

4）捕捉当手机 GPS 坐标更新时的事件，当手机收到位置更改时，将定位位置传入 getMyLocation 对象。具体代码如下。

```
    /* 捕捉当手机 GPS 坐标更新时的事件 */
    public final LocationListener mLocationListener01 =
    new LocationListener()
    {
```

```
        @Override
        public void onLocationChanged(Location location)
        {

          /* 当手机收到位置更改时，将 location 传入 getMyLocation */
          mLocation01 = location;
          fromGeoPoint = getGeoByLocation(location);
          refreshMapViewByGeoPoint(fromGeoPoint,
                  mMapView01, intZoomLevel);

        }

        @Override
        public void onProviderDisabled(String provider)
        {
          mLocation01 = null;
        }

        @Override
        public void onProviderEnabled(String provider)
        {

        }

        @Override
        public void onStatusChanged(String provider,
                    int status, Bundle extras)
        {

        }
      };
```

5）定义方法 getGeoByLocation(Location location)，当传入定位位置对象时取回其 GeoPoint 对象。具体代码如下。

```
      /* 传入 Location 对象，取回其 GeoPoint 对象 */
      private GeoPoint getGeoByLocation(Location location)
      {
        GeoPoint gp = null;
        try
        {
          /* 当 Location 存在 */
          if (location != null)
          {
            double geoLatitude = location.getLatitude()*1E6;
            double geoLongitude = location.getLongitude()*1E6;
            gp = new GeoPoint((int) geoLatitude, (int) geoLongitude);
```

```
        }
    }
    catch(Exception e)
    {
        e.printStackTrace();
    }

    return gp;
}
```

6）定义方法 getGeoByAddress(String strSearchAddress)，当输入地址时取得其 GeoPoint 对象。具体代码如下。

```
/* 输入地址，取得其 GeoPoint 对象 */
private GeoPoint getGeoByAddress(String strSearchAddress)
{
    GeoPoint gp = null;
    try
    {
        if(strSearchAddress!="")
        {
            Geocoder mGeocoder01 = new Geocoder
            (example6.this, Locale.getDefault());

            List<Address> lstAddress = mGeocoder01.getFromLocationName
                                (strSearchAddress, 1);
            if (!lstAddress.isEmpty())
            {
                Address adsLocation = lstAddress.get(0);
                double geoLatitude = adsLocation.getLatitude()*1E6;
                double geoLongitude = adsLocation.getLongitude()*1E6;
                gp = new GeoPoint((int) geoLatitude, (int) geoLongitude);
            }
        }
    }
    catch (Exception e)
    {
        e.printStackTrace();
    }
    return gp;
}
```

7）定义方法 refreshMapViewByGeoPoint()和方法 refreshMapViewByCode()，传入 geoPoint 对象更新 MapView 里的 Google Map 和传入经纬度更新 MapView 里的 Google Map。具体代码如下。

```
/* 传入 geoPoint 更新 MapView 里的 Google Map */
public static void refreshMapViewByGeoPoint
```

```
(GeoPoint gp, MapView mapview, int zoomLevel)
{
  try
  {
    mapview.displayZoomControls(true);
    MapController myMC = mapview.getController();

    myMC.animateTo(gp);
    myMC.setZoom(zoomLevel);
    mapview.setSatellite(false);
  }
  catch(Exception e)
  {
    e.printStackTrace();
  }
}

/* 传入经纬度更新 MapView 里的 Google Map*/
public static void refreshMapViewByCode
(double latitude, double longitude,
    MapView mapview, int zoomLevel)
{
  try
  {
    GeoPoint p = new GeoPoint((int) latitude, (int) longitude);
    mapview.displayZoomControls(true);
    MapController myMC = mapview.getController();
    myMC.animateTo(p);
    myMC.setZoom(zoomLevel);
    mapview.setSatellite(false);
  }
  catch(Exception e)
  {
    e.printStackTrace();
  }
}
```

8）定义方法 GeoPointToString(GeoPoint gp)，将 GeoPoint 里的经纬度以“String,String”形式返回。具体代码如下。

```
/* 将 GeoPoint 里的经纬度以 String，String 返回 */
private String GeoPointToString(GeoPoint gp)
{
  String strReturn="";
  try
  {
    /* 当 Location 存在 */
```

```
            if (gp != null)
            {
                double geoLatitude = (int)gp.getLatitudeE6()/1E6;
                double geoLongitude = (int)gp.getLongitudeE6()/1E6;
                strReturn = String.valueOf(geoLatitude)+","+
                    String.valueOf(geoLongitude);

            }
        }
        catch(Exception e)
        {
            e.printStackTrace();
        }
        return strReturn;
    }
```

9）定义方法 getLocationProvider()，用于获取定位者。具体代码如下。

```
    /* 取得 LocationProvider */
    public void getLocationProvider()
    {
        try
        {
            Criteria mCriteria01 = new Criteria();
            mCriteria01.setAccuracy(Criteria.ACCURACY_FINE);
            mCriteria01.setAltitudeRequired(false);
            mCriteria01.setBearingRequired(false);
            mCriteria01.setCostAllowed(true);
            mCriteria01.setPowerRequirement(Criteria.POWER_LOW);
            strLocationProvider =
            mLocationManager01.getBestProvider(mCriteria01, true);

            mLocation01 = mLocationManager01.getLastKnownLocation
            (strLocationProvider);
        }
        catch(Exception e)
        {
            mTextView01.setText(e.toString());
            e.printStackTrace();
        }
    }

    @Override
    protected boolean isRouteDisplayed()
    {
        return false;
    }
}
```

执行后的效果如图 9-13 所示；单击“开始规划路径”按钮后弹出选择对话框，如图 9-14 所示；在此选择“Maps”后弹出规划界面，如图 9-15 所示；在第一个文本框中设置出发位置，例如“beijing”，在第二个文本框中设置目的地位置，例如“tianjin”，选择汽车前往标志按钮 ，如图 9-16 所示；单击按钮“Go”，系统将实现由北京出发，目的地到天津的线路规划，产生线路规划图，最终的执行界面如图 9-17 所示。

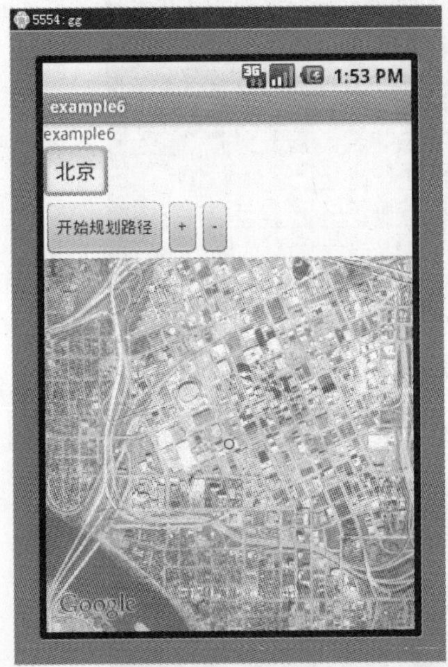

图 9-13　执行效果

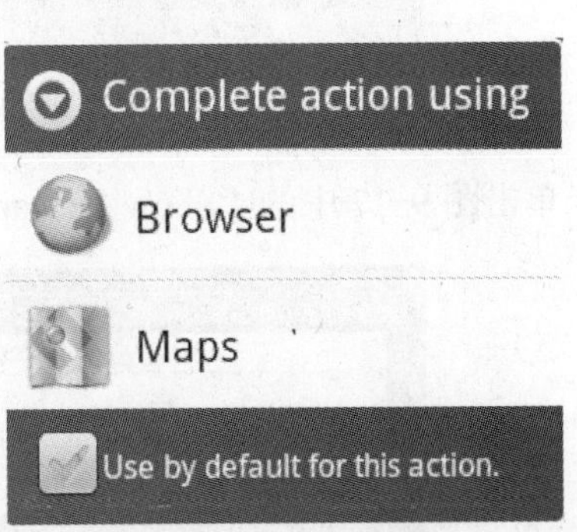

图 9-14　选择对话框

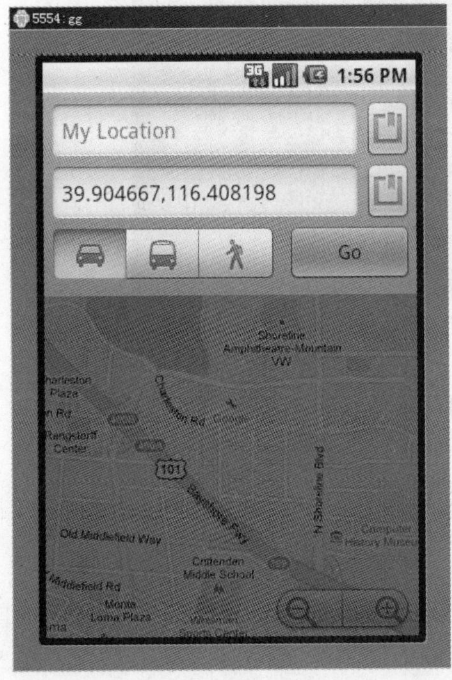

图 9-15　规划界面

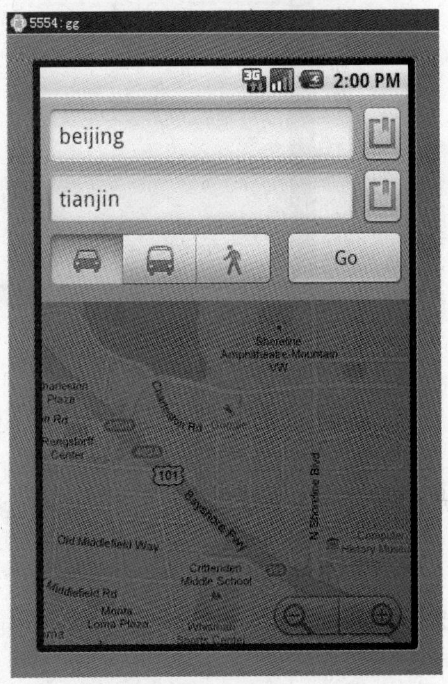

图 9-16　设置出发地和目的地

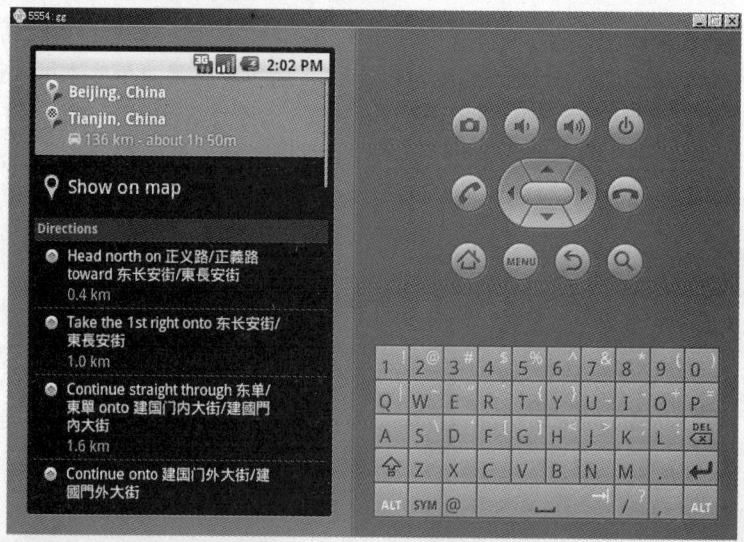

图 9-17　生成的线路规划

当单击图 9-17 中的“Show on map”后，将会在地图中显示行走线路，如图 9-18 所示。

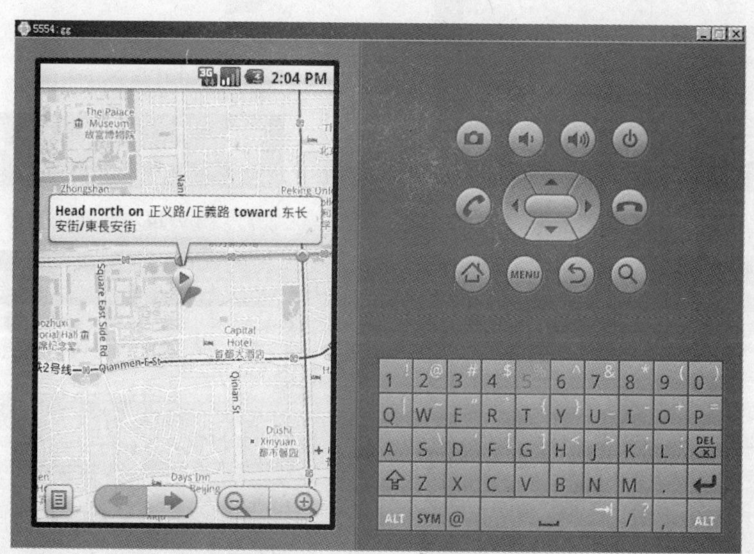

图 9-18　地图中的线路规划

注意：在此事发地和目的地不能属于两个不同的国家，否则将会产生错误提示。

9.7　LocationListener 和 MapView 实时更新

在现实应用中，GPS 的使用越来越广泛。但是每一个位置和路况都不是固定不变的，这就要求系统能够根据各种变化而变化，不能误导了人们。在本节的内容中，将通过一个具体实例的实现，介绍在 Android 中实现 GPS 实时更新的处理流程。本实例的源代码保存在“光盘:\daima\9\example7”，下面开始讲解本实例的具体实现流程。

9.7.1　实现原理

在 Android SDK 中，支持手机 GPS 定位事件处理。在 Android 手机中内置了 Google Map，但是不能随着手机的移动而更新。在本实例中，将插入一个 TextView 和 MapView，当手机移动时，会触发内置的 GPS 定位坐标的改变事件。只要程序发现地理位置的变化，便实时更新 MapView 里的 Google Map，并反查地理坐标系统的信息。

9.7.2　具体实现

本实例的主文件是 example7.java，下面开始介绍其实现流程。

1）先创建 MapView 对象 mMapView01，然后创建 LocationManager 对象 mLocationManager01 来获取系统 LOCATION 服务。具体代码如下。

```
protected void onCreate(Bundle icicle)
{
  super.onCreate(icicle);
  setContentView(R.layout.main);

  mTextView01 = (TextView)findViewById(R.id.myTextView1);
  /* 创建 MapView 对象 */
  mMapView01 = (MapView)findViewById(R.id.myMapView1);

  /* 创建 LocationManager 对象取得系统 LOCATION 服务 */
  mLocationManager01 =
  (LocationManager)getSystemService(Context.LOCATION_SERVICE);
```

2）通过 getLocationProvider 取得当前 Location，然后创建 LocationManager 对象用于监听 Location 更改事件，并更新 MapView。具体代码如下。

```
/* 第一次运行从 Location Provider 取得 Location */
mLocation01 = getLocationProvider(mLocationManager01);

if(mLocation01!=null)
{
  processLocationUpdated(mLocation01);
  }
  else
  {
    mTextView01.setText
    (
      getResources().getText(R.string.str_err_location).toString()
    );
  }
  /* 创建 LocationManager 对象，监听 Location 更改事件，更新 MapView */
  mLocationManager01.requestLocationUpdates
```

```
    (strLocationProvider, 2000, 10, mLocationListener01);
}
```

3）通过 LocationListener()监听定位信息，当手机收到位置更改时，将位置传入取得地理坐标。具体代码如下。

```
public final LocationListener
mLocationListener01 = new LocationListener()
{
  @Override
  public void onLocationChanged(Location location)
  {
    // TODO Auto-generated method stub

    /* 当手机收到位置更改时，将位置传入取得地理坐标 */
    processLocationUpdated(location);
  }

  @Override
  public void onProviderDisabled(String provider)
  {
    /* 当 Provider 已离开服务范围时 */
  }

  @Override
  public void onProviderEnabled(String provider)
  {
  }

  @Override
  public void onStatusChanged
  (String provider, int status, Bundle extras)
  {

  }
};
```

4）定义方法 getAddressbyGeoPoint(GeoPoint gp)，用于获取定位地址的信息。先创建 Geocoder 对象并取出地理坐标经纬度，然后判断地址是否为多行，最后将获取的地址组合后放在 StringBuilder 对象中输出。具体代码如下。

```
public String getAddressbyGeoPoint(GeoPoint gp)
{
  String strReturn = "";
  try
```

```
{
  /* 当 GeoPoint 不等于空时 */
  if (gp != null)
  {
    /* 创建 Geocoder 对象 */
    Geocoder gc = new Geocoder
    (example7.this, Locale.getDefault());

    /* 取出地理坐标经纬度 */
    double geoLatitude = (int)gp.getLatitudeE6()/1E6;
    double geoLongitude = (int)gp.getLongitudeE6()/1E6;


    /* 从经纬度取得地址（可能有多行地址） */
    List<Address> lstAddress =
    gc.getFromLocation(geoLatitude, geoLongitude, 1);

    StringBuilder sb = new StringBuilder();

    /* 判断地址是否为多行 */
    if (lstAddress.size() > 0)
    {
      Address adsLocation = lstAddress.get(0);

      for(int i=0;i<adsLocation.getMaxAddressLineIndex();i++)
      {
        sb.append(adsLocation.getAddressLine(i)).append("\n");
      }
      sb.append(adsLocation.getLocality()).append("\n");
      sb.append(adsLocation.getPostalCode()).append("\n");
      sb.append(adsLocation.getCountryName());
    }

    /*
    * 将获取的地址
    * 组合后放在 StringBuilder 对象中输出用
    */
    strReturn = sb.toString();
  }
}
catch(Exception e)
{
  e.printStackTrace();
}
```

```
    return strReturn;
  }

  public Location getLocationProvider(LocationManager lm)
  {
    Location retLocation = null;
    try
    {
      Criteria mCriteria01 = new Criteria();
      mCriteria01.setAccuracy(Criteria.ACCURACY_FINE);
      mCriteria01.setAltitudeRequired(false);
      mCriteria01.setBearingRequired(false);
      mCriteria01.setCostAllowed(true);
      mCriteria01.setPowerRequirement(Criteria.POWER_LOW);
      strLocationProvider = lm.getBestProvider(mCriteria01, true);


      retLocation = lm.getLastKnownLocation(strLocationProvider);
    }
    catch(Exception e)
    {
      mTextView01.setText(e.toString());
      e.printStackTrace();
    }
    return retLocation;
  }

  private GeoPoint getGeoByLocation(Location location)
  {
    GeoPoint gp = null;
    try
    {
      /* 当位置存在 */
      if (location != null)
      {
        double geoLatitude = location.getLatitude()*1E6;
        double geoLongitude = location.getLongitude()*1E6;
        gp = new GeoPoint((int) geoLatitude, (int) geoLongitude);
      }
    }
    catch(Exception e)
    {
      e.printStackTrace();
    }
    return gp;
```

```
}

public static void refreshMapViewByGeoPoint
(GeoPoint gp, MapView mv, int zoomLevel, boolean bIfSatellite)
{
  try
  {
    mv.displayZoomControls(true);
    /* 取得 MapView 的 MapController */
    MapController mc = mv.getController();
    /* 移至该地理坐标地址 */
    mc.animateTo(gp);

    /* 放大地图层级 */
    mc.setZoom(zoomLevel);

    /* 设置 MapView 的显示选项（卫星、街道）*/
    if(bIfSatellite)

    {
      mv.setSatellite(true);
      mv.setStreetView(true);
    }
    else
    {
      mv.setSatellite(false);
    }
  }
  catch(Exception e)
  {
    e.printStackTrace();
  }
}
```

5）定义方法 processLocationUpdated(Location location)，当手机收到位置更改时，将位置传入 GeoPoint 及 MapView。具体代码如下。

```
/* 当手机收到位置更改，将位置传入 GeoPoint 及 MapView */
private void processLocationUpdated(Location location)
{
  /* 传入 Location 对象，取得 GeoPoint 地理坐标 */
  currentGeoPoint = getGeoByLocation(location);

  /* 更新 MapView 显示 Google Map */
  refreshMapViewByGeoPoint
  (currentGeoPoint, mMapView01, intZoomLevel, true);
```

```
        mTextView01.setText
        (
          getResources().getText(R.string.str_my_location).toString()+
          "\n"+getAddressbyGeoPoint(currentGeoPoint)
        );
      }

      @Override
      protected boolean isRouteDisplayed()
      {
        return false;
      }
    }
```

执行后将显示当前位置的定位信息。如图 9-19 所示，并能实现及时更新功能。

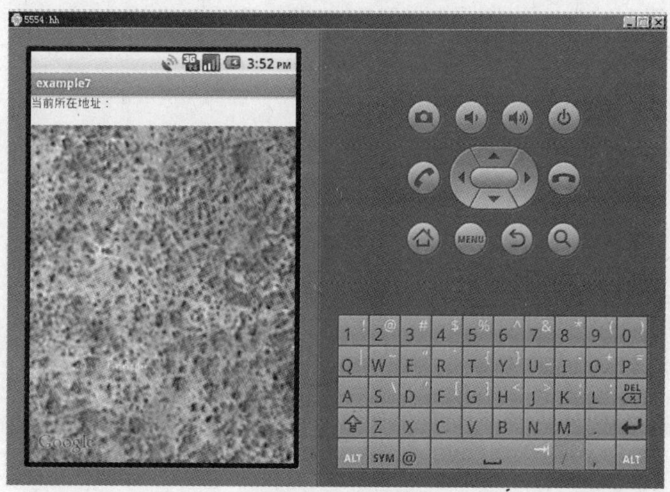

图 9-19 执行效果

9.8 Google Translate API 翻译

在 Android 系统中，是没有手机翻译功能的。但是 Android 官方为其提供了 API 支持，通过调用 Google Translate API 即可实现翻译功能。在本节的内容中，将通过一个具体实例的实现，介绍在 Android 中通过 Google Translate API 实现翻译处理的过程。本实例的源代码保存在“光盘:\daima\9\example8”，下面开始讲解本实例的具体实现流程。

9.8.1 Google Translate API 介绍

Google 的在线翻译功能十分强大，现在 Google 已经开放了其 Ajax 的 API。使用 Ajax 技术的 API，可以仅使用 JavaScript 来翻译和检测网页中文本块的语言。此外，可以在网页的任何文本字段或文本区域启用音译。例如，如果您已音译为北印度语，则该 API 会允许用户使

用英语按照发音拼出北印度语单词，并在北印度语脚本中显示出来。

语言 API 旨在简单容易地用来在脱机翻译不可用时翻译和检测使用的语言。Google 公司计划以后向 Ajax 语言 API 添加更多令人激动的的功能，敬请期待。

google-api-translate-java 是 Java 语言对 Google 翻译引擎的封装类库，具体使用方法如下。

```
import com.google.api.translate.Language;
import com.google.api.translate.Translate;

public class Main {
  public static void main(String[] args) {
    try {
      String translatedText =
          Translate.translate("Salut le monde", Language.FRENCH, Language.ENGLISH);
      System.out.println(translatedText);
    } catch (Exception ex) {
      ex.printStackTrace();

    }
  }
}
```

9.8.2　具体实现

本实例的主文件是 example8.java，其实现流程如下。

第 1 步：在 Activity 中设置一个 EditText 组件，用于接收用户欲翻译的字符串。

第 2 步：编写和 Google Translate API 通信的 JavaScript，并以 HTML 格式存储在“asssts”文件夹中。

第 3 步：在 HTML 网页中编写一个<a href>的链接。

第 4 步：当用户在 EditText 输入英文后，单击“中文”链接后开始翻译处理，并将翻译结果显示在 WebView 中。

下面开始讲解文件 example8.java 的具体实现流程。

1）定义 WebSettings 对象 webSettings，用于获取 WebSettings。具体代码如下。

```
myEditText1 = (EditText) findViewById(R.id.myEditText1);
myWebView1 = (WebView) findViewById(R.id.myWebView1);

/* 取得 WebSettings */
WebSettings webSettings = myWebView1.getSettings();
/* 设置可运行 JavaScript */
webSettings.setJavaScriptEnabled(true);
webSettings.setSaveFormData(false);
webSettings.setSavePassword(false);
webSettings.setSupportZoom(false);
myWebView1.setWebChromeClient(new MyWebChromeClient());
/* 设置给 Html 调用的对象及名称 */
```

```
        myWebView1.addJavascriptInterface(new runJavaScript(), "irdc");
        /* 将 assets/google_translate.html 载入 */
        String url = "file:///android_asset/google_translate.html";
        myWebView1.loadUrl(url);
    }
```

2）定义 runJavaScript()方法，用于调用 google_translate.html 里的 javaScript 以显示结果。具体代码如下。

```
final class runJavaScript
{
    public void runOnAndroidJavaScript()
    {
        mHandler01.post(new Runnable()
        {
            public void run()
            {
                if (myEditText1.getText().toString() != "")
                {
                    /* 调用 google_translate.html 里的 javaScript */
                    myWebView1.loadUrl("javascript:translate('"
                            + myEditText1.getText().toString() + "')");
                }
            }
        });
    }
}
```

3）定义 onJsAlert()方法，用于捕捉网页里的 JavaScript 提示语句作为.js 调试之用，并输出至 LogCat 对象。具体代码如下。

```
/**
 * 捕捉网页里的 alert javascript 作为.js 调试之用，并输出至 LogCat
 */
final class MyWebChromeClient extends WebChromeClient
{
    @Override
    public boolean onJsAlert(WebView view, String url,
            String message, JsResult result)
    {
        // TODO Auto-generated method stub
        Log.d(LOG_TAG, message);
        // result.confirm();
        return super.onJsAlert(view, url, message, result);
    }
}
}
```

执行后的效果如图 9-20 所示；当输入英文字符并单击“中文”链接后，将分别弹出 2 个对话框，分别如图 9-21 和图 9-22 所示；依次单击“OK”按钮，即可实现翻译处理，如图 9-23 所示显示的翻译结果。

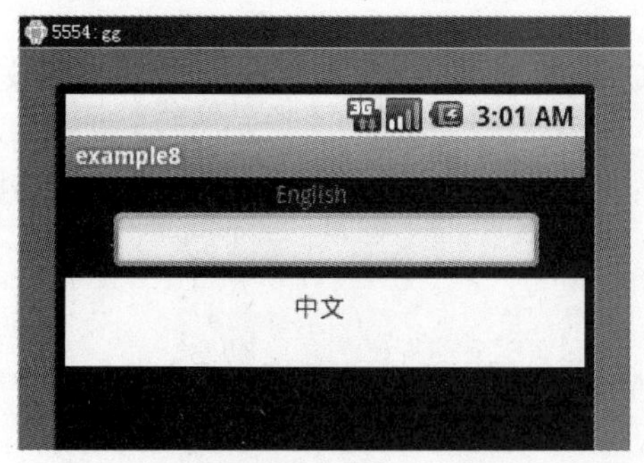

图 9-20　初始效果

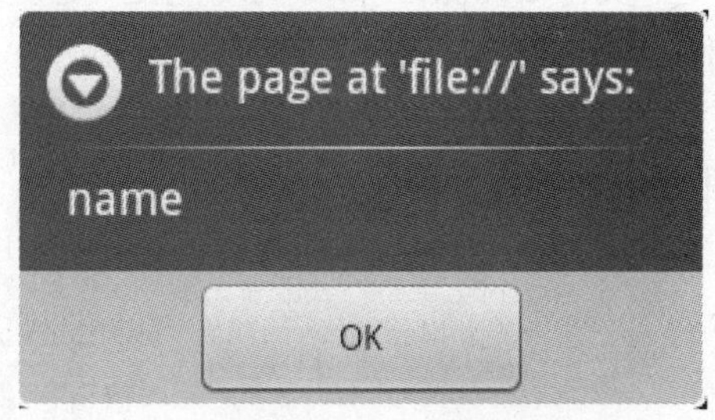

图 9-21　单击“OK”按钮

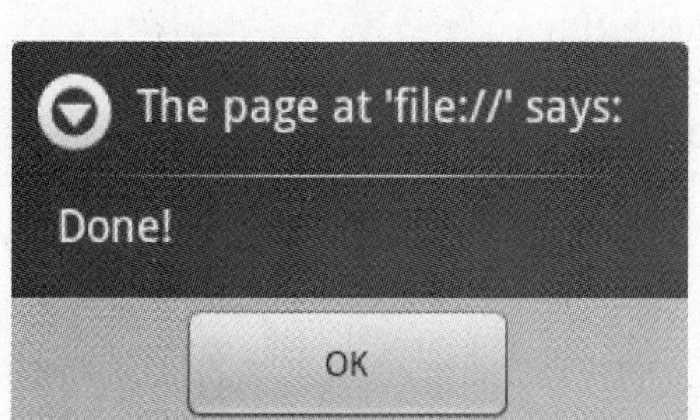

图 9-22　单击“OK”按钮

图 9-23　翻译结果

9.9　画图并计算距离

通过前面实例的学习，大家知道 Android 实现 GPS 导航十分容易。实际上除了导航之外，还可以在地图上绘制线路，计算距离，实现完整的导航功能。在本节的内容中，将通过一个具体实例的实现，介绍在地图上画图并计算距离的具体流程。本实例的源代码保存在“光盘:\daima\9\example9”，下面开始讲解本实例的具体实现流程。

9.9.1　实现原理

Google Web Toolkit（GWT）引入了 JavaScript overlay 类型以简化将整个 JavaScript 对象家族集成到 GWT 项目的过程。该技术有很多优势，如利用 Java IDE 的代码完成和重构能力，甚至当你在编写无类型的 JavaScript 对象时也可以充分利用这一优势。

通过 overlay 类可以在地图上绘制图形或添加图片，在使用前需要引用，格式如下。

```
import com.google.android.maps.Overlay;
```

在本实例中，设置了一个继承了 com.google.android.maps.Overlay 的类 MyOverLay，并对方法 onDraw()进行了重写，这样实现了在 MapView 添加轨迹的效果。

9.9.2 具体实现

本实例的主文件是 example9.java 和 OverLay.java，下面分别介绍其实现流程。

1．文件 example8.java

文件 example1.java 的功能是，通过自定义的 MyOveryLay 类在 MapView 中画上标记。文件 example8.java 的具体实现流程如下。

1）创建 MapView 对象，分别对象初始化 mTextView、mButton01、mButton02、mButton03 和 mButton04。具体代码如下。

```
/* 创建 MapView 对象 */
mMapView = (MapView)findViewById(R.id.myMapView1);
mMapController = mMapView.getController();
/* 对象初始化 */
mTextView = (TextView)findViewById(R.id.myText1);
mButton01 = (Button)findViewById(R.id.myButton1);
mButton02 = (Button)findViewById(R.id.myButton2);
mButton03 = (Button)findViewById(R.id.myButton3);
mButton04 = (Button)findViewById(R.id.myButton4);
```

2）设置默认的放大层级为 17 级，对 Provider 初始化处理并分别获取提供者与位置，取得当前位置。具体代码如下。

```
/* 设置默认的放大层级 */
zoomLevel = 17;
mMapController.setZoom(zoomLevel);

/* Provider 初始化 */
mLocationManager = (LocationManager)
                    getSystemService(Context.LOCATION_SERVICE);
/* 取得 Provider 与 Location */
getLocationPrivider();
if(mLocation!=null)

{
  /* 取得目前的经纬度 */
  gp1=getGeoByLocation(mLocation);
  gp2=gp1;
  /* 将 MapView 的中点移至目前位置 */
  refreshMapView();
  /* 设置事件的 Listener */
  mLocationManager.requestLocationUpdates(mLocationPrivider,
      2000, 10, mLocationListener);
}
else
{
  new AlertDialog.Builder(example9.this).setTitle("系统信息")
```

```
.setMessage(getResources().getString(R.string.str_message))
.setNegativeButton("确定",new DialogInterface.OnClickListener()
 {
   public void onClick(DialogInterface dialog, int which)
   {
     example9.this.finish();
   }
 })
 .show();
}
```

3）定义方法 mButton01.setOnClickListener()，用于响应单击“开始记录”按钮后的处理事件。具体代码如下。

```
/* 开始记录的 Button */
mButton01.setOnClickListener(new Button.OnClickListener()
{
  @Override
  public void onClick(View v)
  {
    gp1=gp2;
    /* 清除 Overlay */
    resetOverlay();
    /* 画起点 */
    setStartPoint();
    /* 更新 MapView */
    refreshMapView();
    /* 重设移动距离为 0，并更新 TextView */
    distance=0;
    mTextView.setText("移动距离：0M");
    /* 启动画路线的机制 */
    _run=true;
  }
});
```

4）定义方法 mButton02.setOnClickListener()，用于响应单击“结束记录”按钮后的处理事件。具体代码如下。

```
/* 结束记录的 Button */
mButton02.setOnClickListener(new Button.OnClickListener()
{
  @Override
  public void onClick(View v)
  {
    /* 画终点 */
    setEndPoint();
    /* 更新 MapView */
```

```
            refreshMapView();
            /* 终止画路线的机制 */
            _run=false;
        }
    });
```

5）定义方法 mButton03.setOnClickListener()，用于响应单击“缩小地图”按钮后的处理事件。具体代码如下。

```
    /* 缩小地图的 Button */
    mButton03.setOnClickListener(new Button.OnClickListener()
    {
      @Override
      public void onClick(View v)
      {
        zoomLevel--;
        if(zoomLevel<1)
        {
          zoomLevel = 1;
        }
        mMapController.setZoom(zoomLevel);
      }
    });
```

6）定义方法 mButton04.setOnClickListener()，用于响应单击“放大地图”按钮后的处理事件。具体代码如下。

```
    /* 放大地图的 Button */
    mButton04.setOnClickListener(new Button.OnClickListener()
    {
      @Override
      public void onClick(View v)
      {
        zoomLevel++;
        if(zoomLevel>mMapView.getMaxZoomLevel())


        {
          zoomLevel = mMapView.getMaxZoomLevel();
        }
        mMapController.setZoom(zoomLevel);
      }
    });
  }
```

7）定义方法 onLocationChanged(Location location)，用于监听当前位置的变化，如果变化则记下轨迹线路。具体代码如下。

```
/* MapView 的 Listener */
public final LocationListener mLocationListener =
  new LocationListener()
{
  @Override
  public void onLocationChanged(Location location)
  {
    /* 如果记录进行中，就画路线并更新移动距离 */
    if(_run)
    {
      /* 记下移动后的位置 */
      gp2=getGeoByLocation(location);
      /* 画路线 */
      setRoute();
      /* 更新 MapView */
      refreshMapView();
      /* 取得移动距离 */
      distance+=GetDistance(gp1,gp2);
      mTextView.setText("移动距离："+format(distance)+"M");

      gp1=gp2;
    }
  }

  @Override
  public void onProviderDisabled(String provider)
  {
  }
  @Override
  public void onProviderEnabled(String provider)
  {
  }
  @Override
  public void onStatusChanged(String provider,int status,
                                         Bundle extras)

  {
  }
};
```

8）定义方法 getGeoByLocation(Location location)，用于取得 GeoPoint 的方法。具体代码如下。

```
/* 取得 GeoPoint 的方法 */
private GeoPoint getGeoByLocation(Location location)
{
  GeoPoint gp = null;
```

```
    try
    {
        if (location != null)
        {
            double geoLatitude = location.getLatitude()*1E6;
            double geoLongitude = location.getLongitude()*1E6;
            gp = new GeoPoint((int) geoLatitude, (int) geoLongitude);
        }
    }
    catch(Exception e)
    {
        e.printStackTrace();
    }
    return gp;
}
```

9）定义方法 getLocationPrivider()，用于获取 LocationProvider。具体代码如下。

```
/* 取得 LocationProvider */
public void getLocationPrivider()
{
    Criteria mCriteria01 = new Criteria();
    mCriteria01.setAccuracy(Criteria.ACCURACY_FINE);
    mCriteria01.setAltitudeRequired(false);
    mCriteria01.setBearingRequired(false);
    mCriteria01.setCostAllowed(true);
    mCriteria01.setPowerRequirement(Criteria.POWER_LOW);

    mLocationPrivider = mLocationManager
                        .getBestProvider(mCriteria01, true);
    mLocation = mLocationManager
                .getLastKnownLocation(mLocationPrivider);
}
```

10）分别设置起点方法 setStartPoint()、路线的方法 setRoute()、终点的方法 setEndPoint()、重设 Overlay 的方法 resetOverlay()和更新 MapView 的方法 setEndPoint()。具体代码如下。

```
/* 设置起点的方法 */
private void setStartPoint()
{
    int mode=1;
    OverLay mOverlay = new OverLay(gp1,gp2,mode);
    List<Overlay> overlays = mMapView.getOverlays();
    overlays.add(mOverlay);
}
/* 设置路线的方法 */
private void setRoute()
```

```
{
  int mode=2;
  OverLay mOverlay = new OverLay(gp1,gp2,mode);
  List<Overlay> overlays = mMapView.getOverlays();
  overlays.add(mOverlay);
}
/* 设置终点的方法 */
private void setEndPoint()
{
  int mode=3;
  OverLay mOverlay = new OverLay(gp1,gp2,mode);
  List<Overlay> overlays = mMapView.getOverlays();
  overlays.add(mOverlay);
}
/* 重设 Overlay 的方法 */
private void resetOverlay()
{
  List<Overlay> overlays = mMapView.getOverlays();
  overlays.clear();
}
/* 更新 MapView 的方法 */
public void refreshMapView()
{
  mMapView.displayZoomControls(true);
  MapController myMC = mMapView.getController();
  myMC.animateTo(gp2);
  myMC.setZoom(zoomLevel);
  mMapView.setSatellite(false);
}
```

11）定义方法 GetDistance(GeoPoint gp1,GeoPoint gp2)，用于获取两点间的距离，并通过方法 format(double num)处理移动的距离。具体代码如下。

```
/* 取得两点间的距离的方法 */
public double GetDistance(GeoPoint gp1,GeoPoint gp2)
{

  double Lat1r = ConvertDegreeToRadians(gp1.getLatitudeE6()/1E6);
  double Lat2r = ConvertDegreeToRadians(gp2.getLatitudeE6()/1E6);
  double Long1r= ConvertDegreeToRadians(gp1.getLongitudeE6()/1E6);
  double Long2r= ConvertDegreeToRadians(gp2.getLongitudeE6()/1E6);
  /* 地球半径(KM) */
  double R = 6371;
  double d = Math.acos(Math.sin(Lat1r)*Math.sin(Lat2r)+
            Math.cos(Lat1r)*Math.cos(Lat2r)*
            Math.cos(Long2r-Long1r))*R;
  return d*1000;
```

```
    }

    private double ConvertDegreeToRadians(double degrees)
    {
      return (Math.PI/180)*degrees;
    }

    /* format 移动距离的方法 */
    public String format(double num)
    {
      NumberFormat formatter = new DecimalFormat("###");
      String s=formatter.format(num);
      return s;
    }

    @Override
    protected boolean isRouteDisplayed()
    {
      return false;
    }
}
```

本实例的主文件是 example9.java 和 OverLay.java，下面分别介绍其实现流程。

2．文件 OverLay.java

文件 OverLay.java 的功能是，定一个继承自 Overlay 的子类 OverLay，并在 MapView 上绘制图形，并以 getProjection()方法获取 Projection 对象，再以 projection.toPixels(gp1, point)方法将 getProjection()方法转换成 Point（点），再利用 Point 对象的对应位置来绘制图形。文件 OverLay.java 的具体代码如下所示。

```
public class OverLay extends Overlay
{
  private GeoPoint gp1;
  private GeoPoint gp2;
  private int mRadius=6;
  private int mode=0;

  /* 构造器，传入起点与终点的 GeoPoint 与 mode */
  public OverLay(GeoPoint gp1,GeoPoint gp2,int mode)
  {
    this.gp1 = gp1;
    this.gp2 = gp2;
    this.mode = mode;
  }

  @Override
  public boolean draw
  (Canvas canvas, MapView mapView, boolean shadow, long when)
```

```
{
  Projection projection = mapView.getProjection();
  if (shadow == false)
  {
    /* 设置笔刷 */
    Paint paint = new Paint();
    paint.setAntiAlias(true);
    paint.setColor(Color.BLUE);

    Point point = new Point();
    projection.toPixels(gp1, point);
    /* mode=1：创建起点 */
    if(mode==1)
    {
      /* 定义 RectF 对象 */
      RectF oval=new RectF(point.x - mRadius, point.y - mRadius,
                           point.x + mRadius, point.y + mRadius);
      /* 绘制起点的圆形 */
      canvas.drawOval(oval, paint);
    }
    /* mode=2：画路线 */
    else if(mode==2)
    {
      Point point2 = new Point();
      projection.toPixels(gp2, point2);

      paint.setColor(Color.BLACK);
      paint.setStrokeWidth(5);
      paint.setAlpha(120);
      /* 画线 */
      canvas.drawLine(point.x, point.y, point2.x,point2.y, paint);
    }
    /* mode=3：创建终点 */
    else if(mode==3)
    {
      /* 避免误差，先画最后一段的路线 */
      Point point2 = new Point();
      projection.toPixels(gp2, point2);
      paint.setStrokeWidth(5);
      paint.setAlpha(120);
      canvas.drawLine(point.x, point.y, point2.x,point2.y, paint);

      /* 定义 RectF 对象 */
      RectF oval=new RectF(point2.x - mRadius,point2.y - mRadius,
                           point2.x + mRadius,point2.y + mRadius);
      /* 绘制终点的圆形 */
      paint.setAlpha(255);
```

```
                canvas.drawOval(oval, paint);
            }
        }
        return super.draw(canvas, mapView, shadow, when);
    }
}
```

执行后依次单击“开始记录”和“结束记录”按钮，能实现 GPS 轨迹记录。效果如图 9-24 所示。

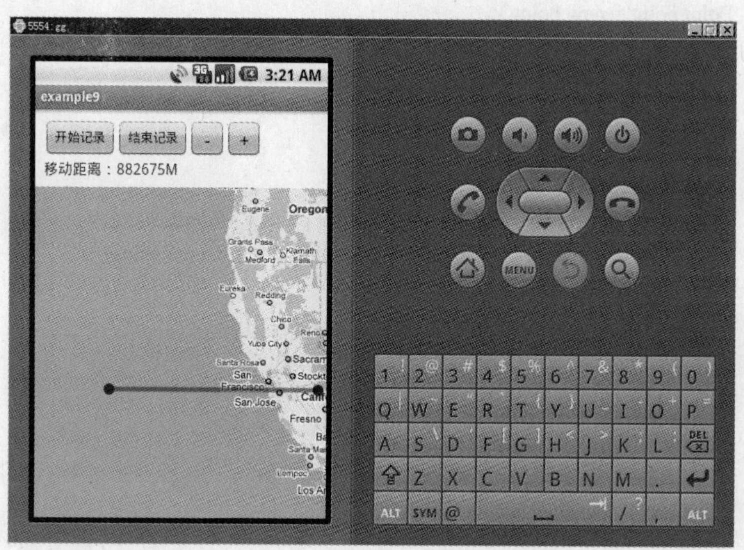

图 9-24　执行效果

9.10　生成二维条码

在现实应用中，二维条码的应用比较常见。在本节的内容中，将通过一个具体实例的实现，介绍在 Android 中通过 swetake 实现二维条码的具体流程。本实例的源代码保存在“光盘:\daima\9\example10”，下面开始讲解本实例的具体实现流程。

9.10.1　实现原理

在本书前面的实例中，作者介绍了通过 Google Chat API 实现二维条码的流程。但是我们不能保证手机都处于联网状态，也就不能确保使用 Google Chat API 了。为了解决这个问题，可以使用开放的 Library，例如最常见的 swetake，读者可以去 http://www.swetake.com/网站下载获取。下载后将其引入到 Android 工程中，并将文件名称改为 SwetakeQRCode.jar。

9.10.2　具体实现

本实例的主文件是 example1.java，下面分别介绍其实现流程。

1）设置应用程序全屏幕运行，而不使用标题栏。具体代码如下。

```
/* 应用程序全屏幕运行，不使用标题栏*/
requestWindowFeature(Window.FEATURE_NO_TITLE);
setContentView(R.layout.main);
```

2）取得屏幕解析像素，并以 SurfaceView 作为相机预览之用，绑定 SurfaceView，取得 SurfaceHolder 对象，并产生二维条码的按钮事件处理。具体代码如下。

```
/* 取得屏幕解析像素 */
DisplayMetrics dm = new DisplayMetrics();
getWindowManager().getDefaultDisplay().getMetrics(dm);

mTextView01 = (TextView) findViewById(R.id.myTextView1);
mTextView01.setText(R.string.str_qr_gen);

/* 以 SurfaceView 作为相机 Preview 之用 */
mSurfaceView01 = (SurfaceView) findViewById(R.id.mSurfaceView1);

/* 绑定 SurfaceView，取得 SurfaceHolder 对象 */
mSurfaceHolder01 = mSurfaceView01.getHolder();

/* Activity 必须实现 SurfaceHolder.Callback */
mSurfaceHolder01.addCallback(example1.this);

/* 产生 QRCode 的按钮事件处理 */
mButton01 = (Button)findViewById(R.id.myButton1);
mButton01.setOnClickListener(new Button.OnClickListener()
{
  @Override
  public void onClick(View arg0)
  {
    // TODO Auto-generated method stub
    if(mEditText01.getText().toString()!="")
    {
      /* 传入 setQrcodeVersion 为 4，仅能接受 62 个字符 */
      AndroidQREncode(mEditText01.getText().toString(), 4);
    }
  }
});

mEditText01 = (EditText)findViewById(R.id.myEditText1);
mEditText01.setText("DavidLanz");
mEditText01.setOnKeyListener(new EditText.OnKeyListener()
{
  @Override
  public boolean onKey(View v, int keyCode, KeyEvent event)
  {
    // TODO Auto-generated method stub
```

```
            return false;
        }
    });
}
```

3）自定义产生 QRCode 的方法 AndroidQREncode()，实现二维编码处理。具体代码如下。

```
/* 自定义产生 QRCode 的方法 */
public void AndroidQREncode(String strEncoding, int qrcodeVersion)
{
  try
  {
    /* 建构 QRCode 编码对象 */
    com.swetake.util.Qrcode testQrcode =
    new com.swetake.util.Qrcode();

    /* 'L','M','Q','H' */
    testQrcode.setQrcodeErrorCorrect('M');
    /* "N","A" 或者其他*/
    testQrcode.setQrcodeEncodeMode('B');
    /* 0-20 */
    testQrcode.setQrcodeVersion(qrcodeVersion);

    byte[] bytesEncoding = strEncoding.getBytes("utf-8");

    if (bytesEncoding.length>0 && bytesEncoding.length <120)
    {
      /* 将字符串通过 calQrcode 方法转换成 boolean 数组 */
      boolean[][] bEncoding = testQrcode.calQrcode(bytesEncoding);

      /* 依据编码后的 boolean 数组，绘图 */
      drawQRCode
      (bEncoding, getResources().getColor(R.drawable.black));
    }
  }

  catch (Exception e)
  {
    e.printStackTrace();
  }
}
```

4）定义方法 drawQRCode()，用于在 SurfaceView 上绘制 QRCode 条形码，解锁 SurfaceHolder 并绘图。具体代码如下。

```
/* 在 SurfaceView 上绘制 QRCode 条形码 */
private void drawQRCode(boolean[][] bRect, int colorFill)
```

```
{
  /* test Canvas*/
  int intPadding = 20;

  /* 欲在 SurfaceView 上绘图，需先锁定 SurfaceHolder */
  Canvas mCanvas01 = mSurfaceHolder01.lockCanvas();

  /* 设置画布绘制颜色 */
  mCanvas01.drawColor(getResources().getColor(R.drawable.white));

  /* 创建画笔 */
  Paint mPaint01 = new Paint();

  /* 设置画笔颜色及模式 */
  mPaint01.setStyle(Paint.Style.FILL);
  mPaint01.setColor(colorFill);
  mPaint01.setStrokeWidth(1.0F);

  /* 逐一加载 2 维 boolean 数组 */
  for (int i=0;i<bRect.length;i++)
  {
    for (int j=0;j<bRect.length;j++)
    {
      if (bRect[j][i])
      {
        /* 依据数组值，绘出条形码方块 */
        mCanvas01.drawRect
        (
          new Rect
          (
            intPadding+j*3+2,
            intPadding+i*3+2,
            intPadding+j*3+2+3,
            intPadding+i*3+2+3
          ), mPaint01


        );
      }
    }
  }
  /* 解锁 SurfaceHolder，并绘图 */
  mSurfaceHolder01.unlockCanvasAndPost(mCanvas01);
}

public void mMakeTextToast(String str, boolean isLong)
{
  if(isLong==true)
  {
```

```
            Toast.makeText(example1.this, str, Toast.LENGTH_LONG).show();
        }
        else
        {
            Toast.makeText(example1.this, str, Toast.LENGTH_SHORT).show();
        }
    }

    @Override
    public void surfaceChanged
    (SurfaceHolder surfaceholder, int format, int w, int h)
    {
        Log.i(TAG, "Surface Changed");
    }

    @Override
    public void surfaceCreated(SurfaceHolder surfaceholder)
    {
        Log.i(TAG, "Surface Changed");
    }

    @Override
    public void surfaceDestroyed(SurfaceHolder surfaceholder)
    {
        Log.i(TAG, "Surface Destroyed");
    }
}
```

执行后可以对输入的文本转换为二维条形码。如图 9-25 所示。

图 9-25　执行效果

9.11　动态二维条码扫描仪

在手机中可以开发一个二维码的扫描程序，这样可以随时随地解码二维条码了。在本节的内容中，将通过一个具体实例的实现，介绍在 Android 中编写动态二维条码扫描程序的具体流程。本实例的源代码保存在“光盘:\daima\9\example11”，下面开始讲解本实例的具体实现流程。

9.11.1　实现原理

本实例同样需要使用第三方开放的 Library，在此需要引用 QRCode 项目，在下载.jar 文件之后，将文件名修改为 SourceForgeQRCode.jar，并导入到 Android 工程中去。当前的二维码标准是 QRCode，QRCode 码是由日本 Denso 公司于 1994 年 9 月研制的一种矩阵二维码符号，它具有一维条码及其他二维条码所具有的信息容量大、可靠性高、可表示汉字及图像多种文字信息、保密防伪性强等优点。

9.11.2　具体实现

本实例的主文件是 example2.java，下面分别介绍其实现流程。

1）分别创建私有 Camera 对象 mCamera01，mButton01，mButton02 和 mButton03。具体代码如下。

```
    /* 创建私有 Camera 对象 */
    private Camera mCamera01;
    private Button mButton01, mButton02, mButton03;
```

2）分别创建变量 mImageView01、TAG、mSurfaceView01 和 mSurfaceHolder01 作为 review 照下来的照片使用，并设置默认相机预览模式为 false。具体代码如下。

```
    /* 作为 review 照下来的相片之用 */
    private ImageView mImageView01;
    private String TAG = "HIPPO";
    private SurfaceView mSurfaceView01;

    private SurfaceHolder mSurfaceHolder01;

    /* 默认相机预览模式为 false */
    private boolean bIfPreview = false;

    /** Called when the activity is first created. */
```

3）设置应用程序全屏幕运行，并添加红色正方形方框 View 供用户对准条形码，然后将创建的红色方框添加至此 Activity 中。具体代码如下。

```
  @Override
  public void onCreate(Bundle savedInstanceState)
```

```
{
  super.onCreate(savedInstanceState);

  /* 应用程序全屏幕运行，不使用标题栏*/
  requestWindowFeature(Window.FEATURE_NO_TITLE);

  setContentView(R.layout.main);

  /* 添加红色正方形方框 View，供用户对准条形码 */
  DrawCaptureRect mDraw = new DrawCaptureRect
  (
    example2.this,
    110, 10, 100, 100,
    getResources().getColor(R.drawable.lightred)
  );

  /* 将创建的红色方框添加至此 Activity 中 */
  addContentView
  (
    mDraw,
    new LayoutParams
    (
      LayoutParams.WRAP_CONTENT, LayoutParams.WRAP_CONTENT
    )
  );
```

4）分别取得屏幕解析像素，绑定 SurfaceView 对象，设置预览大小。具体代码如下。

```
/* 取得屏幕解析像素 */
DisplayMetrics dm = new DisplayMetrics();
getWindowManager().getDefaultDisplay().getMetrics(dm);

mImageView01 = (ImageView) findViewById(R.id.myImageView1);

/* 以 SurfaceView 作为相机预览之用 */

mSurfaceView01 = (SurfaceView) findViewById(R.id.mSurfaceView1);

/* 绑定 SurfaceView，取得 SurfaceHolder 对象 */
mSurfaceHolder01 = mSurfaceView01.getHolder();

/* Activity 必须实现 SurfaceHolder.Callback */
mSurfaceHolder01.addCallback(example2.this);

/* 额外的设置预览大小设置，在此不使用 */
mSurfaceHolder01.setFixedSize(320, 240);
```

```
/*
 * 以 SURFACE_TYPE_PUSH_BUFFERS(3)
 * 作为 SurfaceHolder 显示类型
 * */
mSurfaceHolder01.setType
(SurfaceHolder.SURFACE_TYPE_PUSH_BUFFERS);

mButton01 = (Button)findViewById(R.id.myButton1);
mButton02 = (Button)findViewById(R.id.myButton2);
mButton03 = (Button)findViewById(R.id.myButton3);
```

5）定义方法 mButton01.setOnClickListener，用于打开相机及预览二维条形码。具体代码如下。

```
/* 打开相机及预览二维条形码 */
mButton01.setOnClickListener(new Button.OnClickListener()
{
  @Override
  public void onClick(View arg0)
  {

    /* 自定义初始化打开相机方法 */
    initCamera();
  }
});
```

6）定义方法 mButton02.setOnClickListener，用于停止预览。具体代码如下。

```
/* 停止预览 */
mButton02.setOnClickListener(new Button.OnClickListener()
{
  @Override
  public void onClick(View arg0)
  {

    /* 自定义重置相机，并关闭相机预览方法 */
    resetCamera();
  }
});
```

7）定义方法 mButton03.setOnClickListener，用于拍照 QRCode 二维条形码。具体代码如下。

```
/* 拍照 QRCode 二维条形码 */
mButton03.setOnClickListener(new Button.OnClickListener()
{
  @Override
```

```
        public void onClick(View arg0)
        {
          /* 自定义拍照方法 */
          takePicture();
        }
      });
    }
```

8）定义方法 initCamera()，用于自定义初始相机方法。具体代码如下。

```
  /* 自定义初始相机方法 */
  private void initCamera()
  {
    if(!bIfPreview)
    {
      /* 若相机不在预览模式，则打开相机 */
      mCamera01 = Camera.open();
    }

    if (mCamera01 != null && !bIfPreview)
    {
      Log.i(TAG, "inside the camera");

      /* 创建 Camera.Parameters 对象 */
      Camera.Parameters parameters = mCamera01.getParameters();

      /* 设置相片格式为 JPEG */
      parameters.setPictureFormat(PixelFormat.JPEG);

      /* 指定 preview 的屏幕大小 */
      parameters.setPreviewSize(160, 120);

      /* 设置图片分辨率大小 */

      parameters.setPictureSize(160, 120);

      /* 将 Camera.Parameters（相机参数）设置为 Camera */
      mCamera01.setParameters(parameters);

      /* setPreviewDisplay 唯一的参数为 SurfaceHolder */
      mCamera01.setPreviewDisplay(mSurfaceHolder01);

      /* 立即运行 Preview */
      mCamera01.startPreview();
      bIfPreview = true;
    }
  }
```

9）定义方法 takePicture()，用于拍照处理并获取图像。具体代码如下。

```
/* 拍照获取图像 */
private void takePicture()
{
  if (mCamera01 != null && bIfPreview)
  {
    /* 调用 takePicture()方法拍照 */
    mCamera01.takePicture
    (shutterCallback, rawCallback, jpegCallback);
  }
}
```

10）定义方法 resetCamera()，实现相机重置，并需要释放 Camera 对象。具体代码如下。

```
/* 相机重置 */
private void resetCamera()
{
  if (mCamera01 != null && bIfPreview)
  {
    mCamera01.stopPreview();
    /*释放 Camera 对象 */
    //mCamera01.release();
    mCamera01 = null;
    bIfPreview = false;
  }
}

private ShutterCallback shutterCallback = new ShutterCallback()
{
  public void onShutter()
  {

  }
};

private PictureCallback rawCallback = new PictureCallback()
{
  public void onPictureTaken(byte[] _data, Camera _camera)
  {
  }
};
```

11）定义方法 onPictureTaken()，对传入的图片进行处理。具体流程如下。

第 1 步：设置 onPictureTaken 传入的第一个参数即为照片的 byte。

第 2 步：使用 Matrix.postScale 方法缩小 Bitmap Size。

第 3 步：创建新的 Bitmap 对象。

第 4 步：获取 4:3 图片的居中红色框部分 100×100 像素。

第 5 步：将拍照的图文件以 ImageView 显示出来。

第 6 步：将传入的图文件译码成字符串。

第 7 步：定义方法 mMakeTextToast 输出提示。

具体代码如下。

```
private PictureCallback jpegCallback = new PictureCallback()
{
  public void onPictureTaken(byte[] _data, Camera _camera)
  {
    try
    {
      /* onPictureTaken 传入的第一个参数即为相片的 byte */
      Bitmap bm =
      BitmapFactory.decodeByteArray(_data, 0, _data.length);

      int resizeWidth = 160;
      int resizeHeight = 120;
      float scaleWidth = ((float) resizeWidth) / bm.getWidth();
      float scaleHeight = ((float) resizeHeight) / bm.getHeight();

      Matrix matrix = new Matrix();
      /* 使用 Matrix.postScale 方法缩小 Bitmap Size*/
      matrix.postScale(scaleWidth, scaleHeight);

      /* 创建新的 Bitmap 对象 */
      Bitmap resizedBitmap = Bitmap.createBitmap
      (bm, 0, 0, bm.getWidth(), bm.getHeight(), matrix, true);

      /* 获取 4:3 图片的居中红色框部分 100×100 像素 */
      Bitmap resizedBitmapSquare = Bitmap.createBitmap
      (resizedBitmap, 30, 10, 100, 100);

      /* 将拍照的图文件以 ImageView 显示出来 */
      mImageView01.setImageBitmap(resizedBitmapSquare);

      /* 将传入的图文件译码成字符串 */
      String strQR2 = decodeQRImage(resizedBitmapSquare);
      if(strQR2!="")
      {
        if (URLUtil.isNetworkUrl(strQR2))
        {
          /* OMIA 规范，网址条形码，打开浏览器上网 */
          mMakeTextToast(strQR2, true);
          Uri mUri = Uri.parse(strQR2);
          Intent intent = new Intent(Intent.ACTION_VIEW, mUri);
```

```
            startActivity(intent);
          }
          else if(eregi("wtai://",strQR2))
          {
            /* OMIA 规范，手机拨打电话格式 */
            String[] aryTemp01 = strQR2.split("wtai://");
            Intent myIntentDial = new Intent
            (
              "android.intent.action.CALL",
              Uri.parse("tel:"+aryTemp01[1])
            );
            startActivity(myIntentDial);
          }
          else if(eregi("TEL:",strQR2))
          {
            /* OMIA 规范，手机拨打电话格式 */
            String[] aryTemp01 = strQR2.split("TEL:");
            Intent myIntentDial = new Intent
            (
              "android.intent.action.CALL",
              Uri.parse("tel:"+aryTemp01[1])
            );
            startActivity(myIntentDial);
          }
          else
          {
            /* 若仅是文字，则以 Toast 显示出来 */
            mMakeTextToast(strQR2, true);
          }

        }

        /* 显示图文件后，立即重置相机，并关闭预览 */
        resetCamera();

        /* 再重新启动相机继续预览 */
        initCamera();
      }
      catch (Exception e)
      {
        Log.e(TAG, e.getMessage());
      }
    }
  };

  public void mMakeTextToast(String str, boolean isLong)
  {
```

```
    if(isLong==true)
    {
      Toast.makeText(example2.this, str, Toast.LENGTH_LONG).show();
    }
    else
    {
      Toast.makeText(example2.this, str, Toast.LENGTH_SHORT).show();
    }
  }
```

12）定义方法 checkSDCard()，判断记忆卡是否存在。具体代码如下。

```
private boolean checkSDCard()
{
  /* 判断记忆卡是否存在 */
  if(android.os.Environment.getExternalStorageState().equals
  (android.os.Environment.MEDIA_MOUNTED))
  {
    return true;
  }
  else
  {
    return false;
  }
}
```

13）定义方法 decodeQRImage(Bitmap myBmp)，用于解码传入的 Bitmap 图片。具体代码如下。

```
/* 解码传入的 Bitmap 图片 */

public String decodeQRImage(Bitmap myBmp)
{
  String strDecodedData = "";
  try
  {
    QRCodeDecoder decoder = new QRCodeDecoder();
    strDecodedData    = new String
    (decoder.decode(new AndroidQRCodeImage(myBmp)));
  }
  catch(Exception e)
  {
    e.printStackTrace();
  }
  return strDecodedData;
}
```

14）自定义实现二维条码图像类 AndroidQRCodeImage，具体代码如下。

```
/* 自定义实现 QRCodeImage 类 */
class AndroidQRCodeImage implements QRCodeImage
{
  Bitmap image;

  public AndroidQRCodeImage(Bitmap image)
  {
    this.image = image;
  }

  public int getWidth()
  {
    return image.getWidth();
  }

  public int getHeight()
  {
    return image.getHeight();
  }

  public int getPixel(int x, int y)
  {
    return image.getPixel(x, y);
  }
}
```

15）定义类 DrawCaptureRect，用于绘制相机预览画面里的正方形方框。具体代码如下。

```
/* 绘制相机预览画面里的正方形方框 */
class DrawCaptureRect extends View
{
  private int colorFill;
  private int intLeft,intTop,intWidth,intHeight;

  public DrawCaptureRect
  (
    Context context, int intX, int intY, int intWidth,
    int intHeight, int colorFill
  )
  {
    super(context);
    this.colorFill = colorFill;
    this.intLeft = intX;
    this.intTop = intY;
    this.intWidth = intWidth;
```

```
        this.intHeight = intHeight;
    }

    @Override
    protected void onDraw(Canvas canvas)
    {
        Paint mPaint01 = new Paint();
        mPaint01.setStyle(Paint.Style.FILL);
        mPaint01.setColor(colorFill);
        mPaint01.setStrokeWidth(1.0F);
        /* 在画布上绘制红色的四条方边框作为瞄准器 */
        canvas.drawLine
        (
            this.intLeft, this.intTop,
            this.intLeft+intWidth, this.intTop, mPaint01
        );
        canvas.drawLine
        (
            this.intLeft, this.intTop,
            this.intLeft, this.intTop+intHeight, mPaint01
        );
        canvas.drawLine
        (
            this.intLeft+intWidth, this.intTop,
            this.intLeft+intWidth, this.intTop+intHeight, mPaint01
        );
        canvas.drawLine
        (
            this.intLeft, this.intTop+intHeight,

            this.intLeft+intWidth, this.intTop+intHeight, mPaint01
        );
        super.onDraw(canvas);
    }
}
```

16）定义方法 eregi(String strPat, String strUnknow)，实现自定义比较字符串处理。具体代码如下。

```
/* 自定义比较字符串方法 */
public static boolean eregi(String strPat, String strUnknow)
{
    String strPattern = "(?i)"+strPat;
    Pattern p = Pattern.compile(strPattern);
    Matcher m = p.matcher(strUnknow);
    return m.find();
}
```

```
        @Override
        public void surfaceChanged
        (SurfaceHolder surfaceholder, int format, int w, int h)
        {
          Log.i(TAG, "Surface Changed");
        }

        @Override
        public void surfaceCreated(SurfaceHolder surfaceholder)
        {
          Log.i(TAG, "Surface Changed");
        }

        @Override
        public void surfaceDestroyed(SurfaceHolder surfaceholder)
        {
          Log.i(TAG, "Surface Destroyed");
        }

        @Override
        protected void onPause()
        {
          super.onPause();
        }
      }
```

执行后能够通过手机拍照实现二维码解析。如图如图 9-26 所示。

图 9-26　执行效果

9.12　设置手机屏幕颜色

在现实应用中，可以设计屏幕的显示颜色。在本节的内容中，将通过一个具体实例的实现，介绍在 Android 中设置手机屏幕颜色的具体实现流程。本实例的源代码保存在“光盘:\daima\9\example12”，下面开始讲解本实例的具体实现流程。

9.12.1 实现原理

在 Android 中，可以通过 Android.os.PowerManage 可以控制手机的 WakeLock（手机的锁机制），这样可以让手机的屏幕保持在恒亮状态，再通过程序将手机亮度调到最高 255。

9.12.2 具体实现

本实例的主文件是 example12.java 和 MyAdapter，下面分别介绍其实现流程。

1．文件 example12.java

在文件 example12.java 中，先将屏幕设置为全屏显示，然后以 PowerManager.newWakeLock()方法来获取 WakeLock 对象，并记下 Activity 启动前的屏幕亮度。当启动 Activity 时调用 onResume()方法，并运行 wakeLock()方法，设置屏幕亮度为 255；当暂停或停止 Activity 时调用 onPause()，并运行 wakeUnlock 方法，设置屏幕亮度为程序启动时的亮度。文件 example3.java 的具体代码如下。

```
public class example12 extends Activity
{
  private boolean ifLocked = false;
  private PowerManager.WakeLock mWakeLock;
  private PowerManager mPowerManager;
  private LinearLayout mLinearLayout;
  /* 定义 menu 选项标识，用于识别每个选项的事件 */
  static final private int M_CHOOSE = Menu.FIRST;
  static final private int M_EXIT = Menu.FIRST+1;
  /* 颜色与文字数组*/
  private int[] color={R.drawable.white,R.drawable.blue,
                      R.drawable.pink,R.drawable.green,
                      R.drawable.orange,R.drawable.yellow};
  private int[] text={R.string.str_white,R.string.str_blue,
                     R.string.str_pink,R.string.str_green,
                     R.string.str_orange,R.string.str_yellow};

  @Override
  public void onCreate(Bundle savedInstanceState)
  {
    super.onCreate(savedInstanceState);

    /* 必须在 setContentView 之前实现回屏幕显示 */
    requestWindowFeature(Window.FEATURE_NO_TITLE);

    this.getWindow().setFlags
    (
      WindowManager.LayoutParams.FLAG_FULLSCREEN,
      WindowManager.LayoutParams.FLAG_FULLSCREEN
    );
```

```
  setContentView(R.layout.main);

  /* 在 Activity 启动时将屏幕亮度调整为最亮
   */
  WindowManager.LayoutParams lp = getWindow().getAttributes();
  lp.screenBrightness = 1.0f;
  getWindow().setAttributes(lp);



  /* 初始化 mLinearLayout */
  mLinearLayout=(LinearLayout)findViewById(R.id.myLinearLayout1);

  /* 取得 PowerManager */
  mPowerManager = (PowerManager)
                  getSystemService(Context.POWER_SERVICE);
  /* 取得 WakeLock */
  mWakeLock = mPowerManager.newWakeLock
  (
    PowerManager.SCREEN_BRIGHT_WAKE_LOCK, "BackLight"
  );
}

@Override
public boolean onCreateOptionsMenu(Menu menu)
{
  /* menu 群组 ID */
  int idGroup1 = 0;
  /* menuItemID */
  int orderMenuItem1 = Menu.NONE;
  int orderMenuItem2 = Menu.NONE+1;
  /* 建立 menu */
  menu.add(idGroup1,M_CHOOSE,orderMenuItem1,R.string.str_title);
  menu.add(idGroup1,M_EXIT,orderMenuItem2,R.string.str_exit);
  menu.setGroupCheckable(idGroup1, true, true);

  return super.onCreateOptionsMenu(menu);
}

@Override
public boolean onOptionsItemSelected(MenuItem item)
{
  switch(item.getItemId())
  {
    case (M_CHOOSE):
      /* 弹出选择背后颜色的 AlertDialog */
      new AlertDialog.Builder(example12.this)
        .setTitle(getResources().getString(R.string.str_title))
```

```
                .setAdapter(new MyAdapter(this,color,text),listener1)
                .setPositiveButton("取消",
                    new DialogInterface.OnClickListener()
                {
                  public void onClick(DialogInterface dialog, int which)
                  {
                  }


                })
                .show();
            break;
          case (M_EXIT):
            /* 离开程序 */
            this.finish();
            break;
        }
        return super.onOptionsItemSelected(item);
      }

      /* 选择背后颜色的 AlertDialog 的 OnClickListener */
      OnClickListener listener1=new DialogInterface.OnClickListener()
      {
        public void onClick(DialogInterface dialog,int which)
        {
          /* 更改背景颜色 */
          mLinearLayout.setBackgroundResource(color[which]);
          /* 通过 Toast 提示显示设定的颜色 */
          Toast.makeText(example3.this,
                          getResources().getString(text[which]),
                          Toast.LENGTH_LONG).show();
        }
      };

      @Override
      protected void onResume()
      {
        /* onResume()重启时运行 wakeLock()方法 */
        wakeLock();
        super.onResume();
      }

      @Override
      protected void onPause()
      {
        /* onPause()暂停时运行 wakeUnlock()方法 */
        wakeUnlock();
        super.onPause();
      }
```

```
/* 唤起 WakeLock 的方法*/
private void wakeLock()
{
  if (!ifLocked)
  {


    ifLocked = true;
    mWakeLock.acquire();
  }
}

/* 释放 WakeLock 的方法 */
private void wakeUnlock()
{
  if (ifLocked)
  {
    mWakeLock.release();
    ifLocked = false;
  }
}
}
```

2．文件 MyAdapter.java

在文件 MyAdapter.java 中，设置了背景颜色菜单 Adapter 继承来自 android.widget.BaseAdapter，使用 change_color.xml 作为 Layout，具体代码如下。

```
package irdc.example12;

/* 引入相关类 */
import irdc.example12.R;
import android.content.Context;
import android.view.LayoutInflater;
import android.view.View;
import android.view.ViewGroup;
import android.widget.BaseAdapter;
import android.widget.TextView;

/* 自定义的 Adapter，继承 android.widget.BaseAdapter */
public class MyAdapter extends BaseAdapter
{
  private LayoutInflater mInflater;
  private int[] color;
  private int[] text;

  public MyAdapter(Context context,int[] _color,int[] _text)
  {
```

```
        mInflater = LayoutInflater.from(context);
        color = _color;
        text = _text;
    }

    /* 继承 BaseAdapter，重写覆盖方法 */


    @Override
    public int getCount()
    {
        return text.length;
    }

    @Override
    public Object getItem(int position)
    {
        return text[position];
    }

    @Override
    public long getItemId(int position)
    {
        return position;
    }

    @Override
    public View getView(int position,View convertView,ViewGroup par)
    {
        ViewHolder holder;

        if(convertView == null)
        {
            /* 使用自定义的 change_color 实现布局*/
            convertView = mInflater.inflate(R.layout.change_color, null);
            /* 初始化 holder 的文字 */
            holder = new ViewHolder();
            holder.mText=(TextView)convertView.findViewById(R.id.myText);
            convertView.setTag(holder);
        }
        else
        {
            holder = (ViewHolder) convertView.getTag();
        }
        holder.mText.setText(text[position]);
        holder.mText.setBackgroundResource(color[position]);

        return convertView;
    }
```

```
        /* class ViewHolder */
        private class ViewHolder
        {
            TextView mText;

        }
    }
```

执行后将按照默认样式显示屏幕颜色，如图 9-27 所示；单击“MENU”后弹出 2 个选项卡，如图 9-28 所示；单击“选择背光颜色”选项后弹出设置对话框，在此可以设置要显示颜色，如图 9-29 所示。

图 9-27　默认效果

图 9-28　选项卡

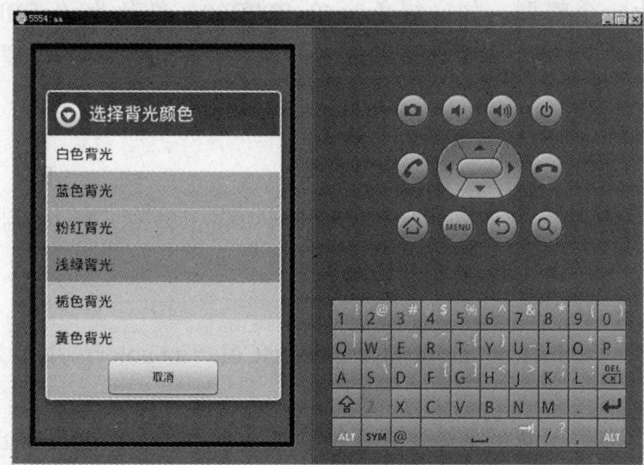

图 9-29　设置对话框

第 10 章　典型手机游戏应用

除了拨打电话和发送短信外，智能手机一般还具备音频/视频播放、移动上网、蓝牙、收音机、软件下载及游戏等功能。特别是游戏功能的强大与否，直接影响了手机设备的销量。在本章的内容中，将详细讲解 Android 手机游戏开发的基本流程，并通过几个典型实例的实现，详细介绍几个典型 Android 手机游戏应用的具体开发流程。

10.1　Graphics 绘图处理

在手机上开发娱乐游戏，首先要实现绘图功能。在 Android 系统中，绘图功能是通过 Graphics 类实现的。Graphics 类能够很方便地绘制 2D 图像，并填充颜色。在本书的前几章实例中，已经涉及了此类的使用知识，在本节的内容中，将进一步剖析 Graphics 类的基本知识。

10.1.1　Color 类

Color 类即 Android.Graphics.Color，在 Android 平台上表示颜色的方法有很多种，Color 提供了常规主要颜色的定义，比如 Color.BLACK 和 Color.GREEN 等，平时创建时主要使用以下静态方法。

1）static int argb(int alpha, int red, int green, int blue)：构造一个包含透明对象的颜色。

2）static int rgb(int red, int green, int blue)：构造一个标准的颜色对象。

3）static int parseColor(String colorString)：解析一种颜色字符串的值，比如传入 Color.BLACK。

本类返回的均为一个整型，如，绿色为 0xff00ff00，红色为 0xffff0000。可以将这个 DWORD 型看做 AARRGGBB，AA 代表 Aphla 透明色，后面的就不难理解，每个部分的取值范围是 0～255。

10.1.2　Paint 类

Paint 类即 Android.Graphics.Paint，我们可以理解 Paint 类为画笔、画刷的属性定义，本类中常用的方法如下。

1）void reset()：重置。

2）void setARGB(int a, int r, int g, int b)或 void setColor(int color)：用于为设置 Paint 对象的颜色。

3）void setAntiAlias(boolean aa)：用于抗锯齿。需要配合 void setFlags (Paint.ANTI_ALIAS_FLAG) 来帮助消除锯齿，使其边缘更平滑。

4）Shader setShader(Shader shader)：用于设置阴影。Shader 类是一个矩阵对象，如果为 NULL，将清除阴影。

5）void setStyle(Paint.Style style)：用于设置样式。一般为 FILL（填充）或者 STROKE（凹陷）效果。

6）void setTextSize(float textSize)：设置字体大小。

7）void setTextAlign(Paint.Align align)：文本对齐方式。

8）Typeface setTypeface(Typeface typeface)：设置字体，通过 Typeface 可以加载 Android 内部的字体，对于中文来说一般为宋体，还可以自己添加字体，例如雅黑等。

9）void setUnderlineText(boolean underlineText)：用于设置下画线，需要结合 void setFlags (Paint.UNDERLINE_TEXT_FLAG)方法来实现。

下面将通过一个具体实例的实现，介绍在 Android 中 Color 类和 Paint 类实现绘图处理的流程。本实例的功能是绘制一个矩形，源代码保存在“光盘:\daima\10\example1”，下面开始讲解本实例的具体实现流程。

第一步：编写主文件 main.xml。

第二步：编写文件 Activity.java，通过“mGameView = new GameView(this)”，用 Activity 类的 setContentView 方法来设置要显示的具体 View 类。文件 Activity.java 的具体实现代码如下。

```
package com.example1;

import Android.app.Activity;
import Android.os.Bundle;

public class Activity01 extends Activity
{

    private GameView mGameView;
    /** Called when the activity is first created. */
    @Override
    public void onCreate(Bundle savedInstanceState)
    {

        super.onCreate(savedInstanceState);

        mGameView = new GameView(this);

        setContentView(mGameView);
    }
}
```

第三步：编写主文件 draw.java，其功能是绘制出指定的图形。具体实现流程如下。

1）声明 Paint 对象 mPaint，定义 draw 分别用于构建对象和开启线程。具体代码如下。

```
    /* 声明 Paint 对象 */
    private Paint mPaint    = null;
```

```
public draw(Context context)
{
        super(context);
        /* 构建对象 */
        mPaint = new Paint();

        /* 开启线程 */
        new Thread(this).start();
}
```

2）定义方法 onDraw，先设置 Paint 格式和颜色，并根据提取的颜色、尺寸、风格、字体和属性实现绘制处理。具体代码如下。

```
public void onDraw(Canvas canvas)
{

        super.onDraw(canvas);

        /* 设置 Paint 为无锯齿 */
        mPaint.setAntiAlias(true);

        /* 设置 Paint 的颜色 */
        mPaint.setColor(Color.WHITE);
        mPaint.setColor(Color.BLUE);
        mPaint.setColor(Color.YELLOW);
        mPaint.setColor(Color.GREEN);
        /* 同样是设置颜色 */
        mPaint.setColor(Color.rgb(255, 0, 0));

        /* 提取颜色 */
        Color.red(0xcccccc);
        Color.green(0xcccccc);

        /* 设置 paint 的颜色和 Alpha 值(a,r,g,b) */
        mPaint.setARGB(255, 255, 0, 0);

        /* 设置 paint 的 Alpha 值 */
        mPaint.setAlpha(220);

        /* 这里可以设置为另外一个 paint 对象 */
        mPaint.set(new Paint());
        /* 设置字体的尺寸 */
        mPaint.setTextSize(14);
```

```
/* 设置 paint 的风格为“空心” */.
/* 当然也可以设置为“实心” (Paint.Style.FILL)*/
mPaint.setStyle(Paint.Style.STROKE);

/* 设置“空心”的外框的宽度*/
mPaint.setStrokeWidth(5);

/* 得到 Paint 的一些属性 */
Log.i(TAG, "paint 的颜色：" + mPaint.getColor());

Log.i(TAG, "paint 的 Alpha：" + mPaint.getAlpha());

Log.i(TAG, "paint 的外框的宽度：" + mPaint.getStrokeWidth());

Log.i(TAG, "paint 的字体尺寸：" + mPaint.getTextSize());

/* 绘制一个矩形 */
/* 肯定是一个空心的矩形*/

canvas.drawRect((320 - 80) / 2, 20, (320 - 80) / 2 + 80, 20 + 40, mPaint);

/* 设置风格为实心 */
mPaint.setStyle(Paint.Style.FILL);

mPaint.setColor(Color.GREEN);

/* 绘制绿色实心矩形 */
canvas.drawRect(0, 20, 40, 20 + 40, mPaint);
}
```

3）分别定义触屏事件 onTouchEvent，按键按下事件 onKeyDown，按键弹起事件 onKeyUp。具体代码如下。

```
/* 触屏事件*/
public boolean onTouchEvent(MotionEvent event)
{
    return true;
}

/* 按键按下事件*/
public boolean onKeyDown(int keyCode, KeyEvent event)
{
    return true;
}

/* 按键弹起事件*/
public boolean onKeyUp(int keyCode, KeyEvent event)
```

```
        {
            return false;
        }

        public boolean onKeyMultiple(int keyCode, int repeatCount, KeyEvent event)
        {
            return true;
        }
        public void run()
        {
            while (!Thread.currentThread().isInterrupted())
            {
                try
                {
                    Thread.sleep(100);
                }
                catch (InterruptedException e)
                {

                    Thread.currentThread().interrupt();
                }
                /* 使用 postInvalidate 可以直接在线程中更新界面*/
                postInvalidate();
            }
        }
    }
```

执行后的效果如图 10-1 所示。

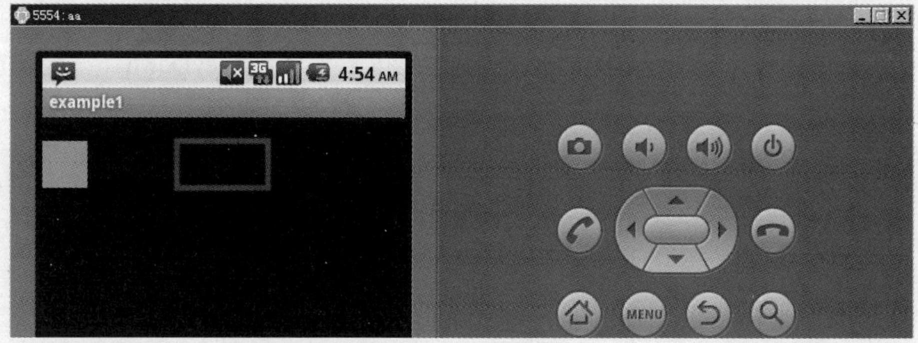

图 10-1 执行效果

10.1.3 Canvas

Canvas 类 Canvas 名为画布，可以看作是一种处理过程，使用各种方法来管理 Bitmap、GL 或者 Path 路径，同时它可以配合 Matrix 矩阵类给图像做旋转、缩放等操作，同时 Canvas 类还提供了裁剪、选取等操作。

下面将通过一个具体实例的实现，介绍在 Android 中使用 Canvas 类的流程。本实例的源代码保存在“光盘:\daima\10\example2”，其主文件是 example2.java，具体代码如下。

```
public class example2 extends View implements Runnable
{
    /* 声明 Paint 对象 */
    private Paint mPaint    = null;



    public example2(Context context)
    {
        super(context);
        /* 构建对象 */
        mPaint = new Paint();

        /* 开启线程 */
        new Thread(this).start();
    }

    public void onDraw(Canvas canvas)
    {
        super.onDraw(canvas);

        /* 设置画布的颜色 */
        canvas.drawColor(Color.BLACK);

        /* 设置取消锯齿效果 */
        mPaint.setAntiAlias(true);

        /* 设置裁剪区域 */
        canvas.clipRect(10, 10, 280, 260);

        /* 先锁定画布 */
        canvas.save();
        /* 旋转画布 */
        canvas.rotate(45.0f);

        /* 设置颜色及绘制矩形 */
        mPaint.setColor(Color.RED);
        canvas.drawRect(new Rect(15,15,140,70), mPaint);

        /* 解除画布的锁定 */
        canvas.restore();

        /* 设置颜色及绘制另一个矩形 */
        mPaint.setColor(Color.GREEN);
        canvas.drawRect(new Rect(150,75,260,120), mPaint);
```

```
    }

    /* 触屏事件*/
    public boolean onTouchEvent(MotionEvent event)
    {
        return true;
    }

    /* 按键按下事件*/
    public boolean onKeyDown(int keyCode, KeyEvent event)


    {
        return true;
    }


    /* 按键弹起事件*/
    public boolean onKeyUp(int keyCode, KeyEvent event)
    {
        return false;
    }


    public boolean onKeyMultiple(int keyCode, int repeatCount, KeyEvent event)
    {
        return true;
    }


    public void run()
    {
        while (!Thread.currentThread().isInterrupted())
        {
            try
            {
                Thread.sleep(100);
            }
            catch (InterruptedException e)
            {
                Thread.currentThread().interrupt();
            }
            /* 使用 postInvalidate 可以直接在线程中更新界面*/
            postInvalidate();
        }
    }
}
```

执行后的效果如图 10-2 所示。

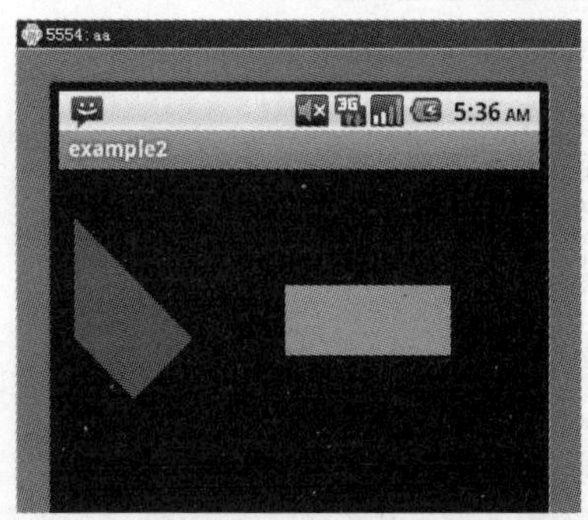

图 10-2　执行效果

10.1.4　Rect 类

Rect 类即 Android.Graphics.Rect，可以理解为矩形区域，Rect 类除了表示一个矩形区域位置描述外，Android 提示主要可以帮助用户计算图形之间是否有碰撞（包含）关系，对于 Android 游戏开发比较有用，其主要的成员 contains 有如下 3 种重载方法来判断包含关系。

```
boolean contains(int left, int top, int right, int bottom)
boolean contains(int x, int y)
boolean contains(Rect r)
```

下面将通过一个具体实例的实现，介绍在 Android 中使用 Canvas 类的流程。本实例的源代码保存在“光盘:\daima\10\example3”，其主文件是 example.java，具体代码如下。

```
/* 声明 Paint 对象 */
private Paint mPaint = null;

private example3_1 mGameView2 = null;
public example(Context context)
{
	super(context);
	/* 构建对象 */
	mPaint = new Paint();

	mGameView2 = new example3_1(context);

	/* 开启线程 */
	new Thread(this).start();
}

public void onDraw(Canvas canvas)

{
	super.onDraw(canvas);

	/* 设置画布为黑色背景 */
```

```
canvas.drawColor(Color.BLACK);
/* 取消锯齿 */
mPaint.setAntiAlias(true);

mPaint.setStyle(Paint.Style.STROKE);

{
    /* 定义矩形对象 */
    Rect rect1 = new Rect();
    /* 设置矩形大小 */
    rect1.left = 5;
    rect1.top = 5;
    rect1.bottom = 25;
    rect1.right = 45;

    mPaint.setColor(Color.BLUE);
    /* 绘制矩形 */
    canvas.drawRect(rect1, mPaint);

    mPaint.setColor(Color.RED);
    /* 绘制矩形 */
    canvas.drawRect(50, 5, 90, 25, mPaint);

    mPaint.setColor(Color.YELLOW);
    /* 绘制圆形（圆心 x,圆心 y,半径 r,p） */
    canvas.drawCircle(40, 70, 30, mPaint);

    /* 定义椭圆对象 */
    RectF rectf1 = new RectF();
    /* 设置椭圆大小 */
    rectf1.left = 80;
    rectf1.top = 30;
    rectf1.right = 120;
    rectf1.bottom = 70;

    mPaint.setColor(Color.LTGRAY);
    /* 绘制椭圆 */
    canvas.drawOval(rectf1, mPaint);

    /* 绘制多边形 */
    Path path1 = new Path();


    /*设置多边形的点*/
    path1.moveTo(150+5, 80-50);
    path1.lineTo(150+45, 80-50);
    path1.lineTo(150+30, 120-50);
```

```
        path1.lineTo(150+20, 120-50);
        /* 使这些点构成封闭的多边形 */
        path1.close();

        mPaint.setColor(Color.GRAY);
        /* 绘制这个多边形 */
        canvas.drawPath(path1, mPaint);

        mPaint.setColor(Color.RED);
        mPaint.setStrokeWidth(3);
        /* 绘制直线 */
        canvas.drawLine(5, 110, 315, 110, mPaint);
    }
    /*
    下面绘制实心几何体
    */
    mPaint.setStyle(Paint.Style.FILL);
    {
        /* 定义矩形对象 */
        Rect rect1 = new Rect();
        /* 设置矩形大小 */
        rect1.left = 5;
        rect1.top = 130+5;
        rect1.bottom = 130+25;
        rect1.right = 45;

        mPaint.setColor(Color.BLUE);
        /* 绘制矩形 */
        canvas.drawRect(rect1, mPaint);

        mPaint.setColor(Color.RED);
        /* 绘制矩形 */
        canvas.drawRect(50, 130+5, 90, 130+25, mPaint);

        mPaint.setColor(Color.YELLOW);
        /* 绘制圆形(圆心 x,圆心 y,半径 r,p) */
        canvas.drawCircle(40, 130+70, 30, mPaint);

        /* 定义椭圆对象 */
        RectF rectf1 = new RectF();
        /* 设置椭圆大小 */
        rectf1.left = 80;


        rectf1.top = 130+30;
```

```
                rectf1.right = 120;
                rectf1.bottom = 130+70;

                mPaint.setColor(Color.LTGRAY);
                /* 绘制椭圆 */
                canvas.drawOval(rectf1, mPaint);

                /* 绘制多边形 */
                Path path1 = new Path();

                /*设置多边形的点*/
                path1.moveTo(150+5, 130+80-50);
                path1.lineTo(150+45, 130+80-50);
                path1.lineTo(150+30, 130+120-50);
                path1.lineTo(150+20, 130+120-50);
                /* 使这些点构成封闭的多边形 */
                path1.close();

                mPaint.setColor(Color.GRAY);
                /* 绘制这个多边形 */
                canvas.drawPath(path1, mPaint);

                mPaint.setColor(Color.RED);
                mPaint.setStrokeWidth(3);
                /* 绘制直线 */
                canvas.drawLine(5, 130+110, 315, 130+110, mPaint);
            }

            /* 通过 ShapeDrawable 来绘制几何图形 */
            mGameView2.DrawShape(canvas);
        }

        /* 触屏事件*/
        public boolean onTouchEvent(MotionEvent event)
        {
            return true;
        }

        /* 按键按下事件*/
        public boolean onKeyDown(int keyCode, KeyEvent event)
        {
            return true;
        }

        /* 按键弹起事件*/
```

```
        public boolean onKeyUp(int keyCode, KeyEvent event)
        {
            return false;
        }

        public boolean onKeyMultiple(int keyCode, int repeatCount, KeyEvent event)
        {
            return true;
        }


        public void run()
        {
            while (!Thread.currentThread().isInterrupted())
            {
                try
                {
                    Thread.sleep(100);
                }
                catch (InterruptedException e)
                {
                    Thread.currentThread().interrupt();
                }
                /*使用 postInvalidate 可以直接在线程中更新界面*/
                postInvalidate();
            }
        }
    }
```

执行后的效果如图 10-3 所示。

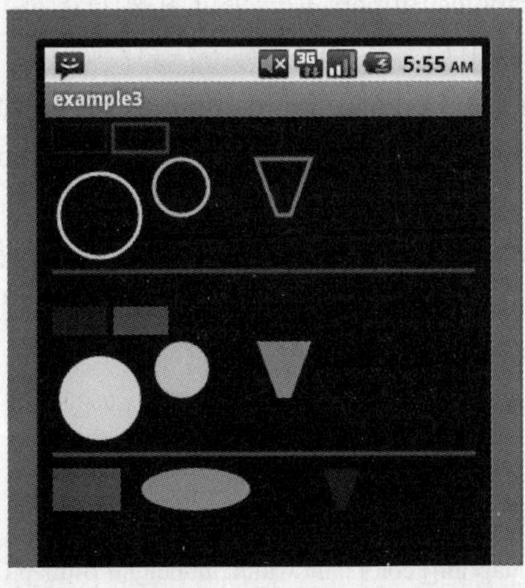

图 10-3　执行效果

10.1.5　NinePatch 类

NinePatch 类即 Android.Graphics.NinePatch，NinePatch 是 Android 平台特有的一种非矢量图形自然拉伸处理方法，能够实现在拉伸常规的图形时不会自动缩放的效果，实例中 Android 开发网提示大家对于 Toast 的显示就是该原理，同时 SDK 中提供了一个工具名为 Draw 9-Patch，有关该工具的使用可以参考已经发布的 Draw 9-Patch 使用方法。由于该类提供了高质量支持透明的缩放方式，所以图形格式为 PNG，文件命名方式为.9.png 的后缀比如 Android123.9.png。

10.1.6　Matrix 类

Matrix 类即 Android.Graphics.Matrix，能够实现图形的变换操作，例如常见的缩放和旋转处理。Matrix 中有关图形的变换、缩放等相关操作常用的方法有如下几种。

1）void reset()：重置一个 matrix 对象。

2）void set(Matrix src)：复制一个源矩阵，和本类的构造方法 Matrix(Matrix src)一样。

3）boolean isIdentity()：返回这个矩阵是否定义（已经有意义）。

4）void setRotate(float degrees)：指定一个角度以（0,0）为坐标进行旋转。

5）void setRotate(float degrees, float px, float py)：指定一个角度以（px,py）为坐标进行旋转。

6）void setScale(float sx, float sy)：缩放处理。

7）void setScale(float sx, float sy, float px, float py)：以坐标（px,py）进行缩放。

8）void setTranslate(float dx, float dy)：平移。

9）void setSkew (float kx, float ky, float px, float py)：以坐标（px，py）进行倾斜。

10）void setSkew (float kx, float ky)：倾斜处理。

有关 Matrix 类实现图片缩放处理的操作实例请读者参阅本书的 4.17 内容。

10.1.7　Bitmap 类

Bitmap 类即 Android.Graphics.Bitmap，是一个位图操作类，实现对位图的基本操作。Bitmap 中提供了很多实用的方法，其中最为常用的几种方法总结如下。

1）boolean compress(Bitmap.CompressFormat format, int quality, OutputStream stream)：压缩一个 Bitmap 对象，根据相关的编码、画质保存到一个 OutputStream 中。其中第一个参数格式目前有 JPG 和 PNG。

2）void copyPixelsFromBuffer(Buffer src)：从一个 Buffer 缓冲区复制位图像素。

3）void copyPixelsToBuffer(Buffer dst)：将当前位图像素内容复制到一个 Buffer 缓冲区。

目前在 Android 2.1 SDK 中，创建位图对象包含了 6 种方法。当然使用的 API Level 均为 1，所以说从 Android 1.0 SDK 开始就支持了，大家可以放心使用。

4）下列方法用于创建一个可以缩放的位图对象。

```
static Bitmap createBitmap(Bitmap src)
static Bitmap createBitmap(int[] colors, int width, int height, Bitmap.Config config)
static Bitmap createBitmap(int[] colors, int offset, int stride, int width, int height, Bitmap.Config config)
```

```
static Bitmap createBitmap(Bitmap source, int x, int y, int width, int height, Matrix m, boolean filter)
static Bitmap createBitmap(int width, int height, Bitmap.Config config)
static Bitmap createBitmap(Bitmap source, int x, int y, int width, int height)
static Bitmap createScaledBitmap(Bitmap src, int dstWidth, int dstHeight, boolean filter)
```

5）final int getHeight()：获取高度。

6）final int getWidth()：获取宽度。

7）final boolean hasAlpha()：是否有透明通道。

8）void setPixel(int x, int y, int color)：设置某像素的颜色。

9）int getPixel(int x, int y)：获取某像素的颜色，此处返回的 int 型参数是指 color 的定义。

在本书 4.18 内中，已经通过具体实例介绍了 Bitmap 类实现图片旋转处理的具体流程。

下面将通过一个具体实例的实现，介绍在 Android 中使用 Bitmap 类实现模拟水纹效果的流程。本实例的源代码保存在“光盘:\daima\10\example4”，其主文件是 example4.java，具体代码如下。

```
public class example4 extends View implements Runnable
{
    int BACKWIDTH;

    int BACKHEIGHT;

    short[] buf2;

    short[] buf1;

    int[] Bitmap2;

    int[] Bitmap1;

     public example4(Context context)
     {
          super(context);

          /* 加载图片 */
          Bitmap      image = BitmapFactory.decodeResource(this.getResources(),R.drawable.qq);
          BACKWIDTH = image.getWidth();
          BACKHEIGHT = image.getHeight();

          buf2 = new short[BACKWIDTH * BACKHEIGHT];
          buf1 = new short[BACKWIDTH * BACKHEIGHT];
          Bitmap2 = new int[BACKWIDTH * BACKHEIGHT];
          Bitmap1 = new int[BACKWIDTH * BACKHEIGHT];

          /* 加载图片的像素到数组中 */
```

```
            image.getPixels(Bitmap1, 0, BACKWIDTH, 0, 0, BACKWIDTH, BACKHEIGHT);

            new Thread(this).start();
    }


    void DropStone(int x,// x 坐标
                    int y,// y 坐标
                    int stonesize,// 波源半径
                    int stoneweight)// 波源能量
    {
        for (int posx = x - stonesize; posx < x + stonesize; posx++)
            for (int posy = y - stonesize; posy < y + stonesize; posy++)
                if ((posx - x) * (posx - x) + (posy - y) * (posy - y) < stonesize * stonesize)
                    buf1[BACKWIDTH * posy + posx] = (short) -stoneweight;
    }


    void RippleSpread()
    {
        for (int i = BACKWIDTH; i < BACKWIDTH * BACKHEIGHT - BACKWIDTH; i++)
        {
            // 波能扩散
            buf2[i] = (short) (((buf1[i - 1] + buf1[i + 1] + buf1[i - BACKWIDTH] + buf1[i +
BACKWIDTH]) >> 1) - buf2[i]);
            // 波能衰减
            buf2[i] -= buf2[i] >> 5;
        }

        // 交换波能数据缓冲区
        short[] ptmp = buf1;
        buf1 = buf2;
        buf2 = ptmp;
    }
    /* 渲染水纹效果 */
    void render()
    {
        int xoff, yoff;
        int k = BACKWIDTH;
        for (int i = 1; i < BACKHEIGHT - 1; i++)
        {
            for (int j = 0; j < BACKWIDTH; j++)
            {
                /* 计算偏移量*/
                xoff = buf1[k - 1] - buf1[k + 1];
                yoff = buf1[k - BACKWIDTH] - buf1[k + BACKWIDTH];
```

```
                    /* 判断坐标是否在窗口范围内*/
                    if ((i + yoff) < 0)
                    {
                        k++;
                        continue;
                    }
                    if ((i + yoff) > BACKHEIGHT)
                    {
                        k++;
                        continue;
                    }
                    if ((j + xoff) < 0)
                    {
                        k++;
                        continue;
                    }
                    if ((j + xoff) > BACKWIDTH)
                    {
                        k++;
                        continue;
                    }

                    /* 计算出偏移像素和原始像素的内存地址偏移量*/
                    int pos1, pos2;
                    pos1 = BACKWIDTH * (i + yoff) + (j + xoff);
                    pos2 = BACKWIDTH * i + j;
                    Bitmap2[pos2++] = Bitmap1[pos1++];
                    k++;
                }
            }
        }

        public void onDraw(Canvas canvas)


        {
            super.onDraw(canvas);

            /* 绘制经过处理的图片效果 */
            canvas.drawBitmap(Bitmap2, 0, BACKWIDTH, 0, 0, BACKWIDTH, BACKHEIGHT, false,
null);
        }

        /* 触屏事件*/
        public boolean onTouchEvent(MotionEvent event)
        {

            return true;
```

```
    }


    /* 按键按下事件*/
    public boolean onKeyDown(int keyCode, KeyEvent event)
    {
        return true;
    }


    /* 按键弹起事件*/
    public boolean onKeyUp(int keyCode, KeyEvent event)
    {
        DropStone(BACKWIDTH/2, BACKHEIGHT/2, 10, 30);
        return false;
    }


    public boolean onKeyMultiple(int keyCode, int repeatCount, KeyEvent event)
    {
        return true;
    }

    /**
     * 线程处理
     */
    public void run()
    {
        while (!Thread.currentThread().isInterrupted())
        {
            try
            {
                Thread.sleep(50);


            }
            catch (InterruptedException e)
            {
                Thread.currentThread().interrupt();
            }
           RippleSpread();
           render();
             /*使用 postInvalidate 可以直接在线程中更新界面*/
             postInvalidate();
        }
    }
}
```

执行后将通过对图像像素的操作数来模拟水纹效果，效果如图 10-4 所示。

图 10-4　执行效果

10.1.8　BitmapFactory 类

BitmapFactory 类即 Android.Graphics.BitmapFactory，作为 Bitmap 对象的 I/O 类，BitmapFactory 类提供了丰富的构造 Bitmap 对象的方法，比如从一个字节数组、文件系统、资源 ID 以及输入流中创建一个 Bitmap 对象，下面是本类的全部成员，除了 decodeFileDescriptor 外，其他的重载方法都很常用。

（1）从字节数组创建

```
static Bitmap decodeByteArray(byte[] data, int offset, int length)
static Bitmap decodeByteArray(byte[] data, int offset, int length, BitmapFactory.Options opts)
```

（2）从文件创建，路径要写全

```
static Bitmap decodeFile(String pathName, BitmapFactory.Options opts)
static Bitmap decodeFile(String pathName)
```

（3）从输入流句柄创建

```
static Bitmap decodeFileDescriptor(FileDescriptor fd, Rect outPadding, BitmapFactory.Options opts)
static Bitmap decodeFileDescriptor(FileDescriptor fd)
```

（4）从 Android 的 APK 文件资源中创建

Android 提示是从"/res/"的 drawable 中创建：

```
static Bitmap decodeResource(Resources res, int id)
static Bitmap decodeResource(Resources res, int id, BitmapFactory.Options opts)
static Bitmap decodeResourceStream(Resources res, TypedValue value, InputStream is, Rect pad, BitmapFactory.Options opts)
```

（5）从一个输入流中创建

```
static Bitmap decodeStream(InputStream is)
static Bitmap decodeStream(InputStream is, Rect outPadding, BitmapFactory.Options opts)
```

10.1.9 Region 类

Region 类即 Android.Graphics.Region，Region 在 Android 平台中表示一个区域，和 Rect 不同的是，它表示的是一个不规则的图形，可以是椭圆、多边形等，而 Rect 仅仅是矩形。同样 Region 的 boolean contains(int x, int y)成员可以判断一个点是否在该区域内。

10.1.10 Typeface 类

Typeface 类即 Android.Graphics.Typeface，Typeface 类帮助描述一个字体对象，在 TextView 中通过使用 setTypeface()方法来指定一种输出文本的字体，其调用成员 create 方法可以直接指定一种字体名称和样式，例如，

```
static Typeface create(Typeface family, int style)
static Typeface create(String familyName, int style)
```

同时使用 isBold 和 isItalic 方法可以判断出是否包含粗体或斜体的字形。

```
final boolean   isBold()
final boolean   isItalic()
```

除了上述创建方法外，该类还可以从 APK 的资源或从一个具体的文件路径获取字体，其具体方法如下。

```
static Typeface   createFromAsset(AssetManager mgr, String path)
static Typeface   createFromFile(File path)
static Typeface   createFromFile(String path)
```

10.1.11 Shader 类

Shader 类用于渲染图像和一些几何图形。下面将通过一个具体实例的实现，介绍在 Android 中使用 Bitmap 类实现模拟水纹效果的流程。本实例的源代码保存在“光盘:\daima\10\example5”，其主文件是 example5.java，具体代码如下。

```
public class example5 extends View implements Runnable
{
    /* 声明 Bitmap 对象 */
    Bitmap      mBitQQ      = null;
    int         BitQQwidth      = 0;
    int         BitQQheight     = 0;

    Paint    mPaint = null;

    /* Bitmap 渲染 */
    Shader mBitmapShader = null;

    /* 线性渐变渲染 */
```

```
        Shader mLinearGradient = null;

        /* 混合渲染 */
        Shader mComposeShader = null;

        /* 放射性渐变渲染 */
        Shader mRadialGradient = null;

        /* 梯度渲染 */
        Shader mSweepGradient = null;


        ShapeDrawable mShapeDrawableQQ = null;

        public example5(Context context)
        {
            super(context);

            /* 加载资源 */
            mBitQQ = ((BitmapDrawable) getResources().getDrawable(R.drawable.qq)).getBitmap();

            /* 得到图片的宽度和高度 */
            BitQQwidth = mBitQQ.getWidth();
            BitQQheight = mBitQQ.getHeight();

            /* 创建 BitmapShader 对象 */
            mBitmapShader = new BitmapShader(mBitQQ,Shader.TileMode. REPEAT,Shader.
TileMode.MIRROR);

            /* 创建线性渐变并设置渐变的颜色数组 */
            mLinearGradient = new LinearGradient(0,0,100,100,
            new int[]{Color.RED,Color.GREEN, Color.BLUE,Color.WHITE},
                                              null,Shader.TileMode.REPEAT);
            /* 这里笔者理解为“混合渲染”--大家可以有自己的理解，能明白这个意思就好*/
            mComposeShader = new ComposeShader(mBitmapShader, mLinearGradient,PorterDuff.Mode.
DARKEN);

            /* 构建放射性渐变对象，设置半径的属性 */
            /* 这里使用了 BitmapShader 和 LinearGradient 进行混合*/
            /* 当然也可以使用其他的组合*/
            /* 混合渲染的模式很多，可以根据自己需要来选择*/
            mRadialGradient = new RadialGradient(50,200,50,
            new  int[]{Color.GREEN,Color.RED,  Color.BLUE,Color.WHITE},null,Shader.TileMode.
REPEAT);
            /* 构建梯度渐变对象 */
```

```
            mSweepGradient = new SweepGradient(30,30,new int[]{Color.GREEN,Color.RED,Color.BLUE,
Color. WHITE},null);

            mPaint = new Paint();

            /* 开启线程 */
            new Thread(this).start();
        }

        public void onDraw(Canvas canvas)
        {

            super.onDraw(canvas);

            //将图片裁剪为椭圆形
            /* 构建 ShapeDrawable 对象并定义形状为椭圆 */
            mShapeDrawableQQ = new ShapeDrawable(new OvalShape());

            /* 设置要绘制的椭圆形的东西为 ShapeDrawable 图片 */
            mShapeDrawableQQ.getPaint().setShader(mBitmapShader);

            /* 设置显示区域 */
            mShapeDrawableQQ.setBounds(0,0, BitQQwidth, BitQQheight);

            /* 绘制 ShapeDrawableQQ */
            mShapeDrawableQQ.draw(canvas);

            /* 绘制渐变的矩形*/
            mPaint.setShader(mLinearGradient);
            canvas.drawRect(BitQQwidth, 0, 320, 156, mPaint);

            /* 显示混合渲染效果*/
            mPaint.setShader(mComposeShader);
            canvas.drawRect(0, 300, BitQQwidth, 300+BitQQheight, mPaint);

            /* 绘制环形渐变*/
            mPaint.setShader(mRadialGradient);
            canvas.drawCircle(50, 200, 50, mPaint);

            /* 绘制梯度渐变*/
            mPaint.setShader(mSweepGradient);
            canvas.drawRect(150, 160, 300, 300, mPaint);

        }

        /* 触屏事件*/
```

```
    public boolean onTouchEvent(MotionEvent event)
    {
        return true;
    }


    /* 按键按下事件*/
    public boolean onKeyDown(int keyCode, KeyEvent event)
    {
        return true;
    }

    /* 按键弹起事件*/
    public boolean onKeyUp(int keyCode, KeyEvent event)


    {
        return false;
    }


    public boolean onKeyMultiple(int keyCode, int repeatCount, KeyEvent event)
    {
        return true;
    }


    /**
     * 线程处理
     */
    public void run()
    {
        while (!Thread.currentThread().isInterrupted())
        {
            try
            {
                Thread.sleep(100);
            }
            catch (InterruptedException e)
            {
                Thread.currentThread().interrupt();
            }
            /* 使用 postInvalidate 可以直接在线程中更新界面*/
            postInvalidate();
        }
    }
}
```

执行后的效果如图 10-5 所示。

图 10-5　执行效果

10.2　游戏框架

大多数知名企业的游戏框架都是基于 Apache Struts、Spring 和 Hibernate 等开发框架的。这些框架都是基于 MVC 设计模式，业务和逻辑被分开，这已经成为当前游戏开发的主流模式。在本节的内容中，作者将简单分析 Android 平台提供的 View 和 Surfaceview 类来作为 MVC 中视图基类的开发框架。

10.2.1　View 类

在下面的内容中，将通过一个具体实例的实现，介绍在 Android 中使用 View 类实现屏幕更新显示的流程。本实例的源代码保存在“光盘:\daima\10\example7”，其主文件是 example7.java，具体代码如下。

```
public class example7 extends View
{
    int   miCount = 0;
    int   y = 0;
    public example7(Context context)
    {
        super(context);
    }
    /* 绘图处理*/
    public void onDraw(Canvas canvas)
    {
        if (miCount < 100)
        {
            miCount++;
```

```
            }
            else
            {
                miCount = 0;
            }
            /* 绘图*/
            Paint mPaint = new Paint();
            switch (miCount%4)
            {
            case 0:
                mPaint.setColor(Color.BLUE);
                break;
            case 1:
                mPaint.setColor(Color.GREEN);
                break;
            case 2:
                mPaint.setColor(Color.RED);
                break;
            case 3:
                mPaint.setColor(Color.YELLOW);
                break;
            default:
                mPaint.setColor(Color.WHITE);
                break;
            }
          /* 绘制矩形--后面将详细讲解*/
          canvas.drawRect((320-80)/2, y, (320-80)/2+80, y+40, mPaint);
        }
    }
```

执行后将在屏幕内绘制一个矩形，并随着线程的变化矩形的填充颜色也随之变化，从而实现闪烁效果，如图 10-6 所示；可以通过键盘上的上、下方向键来移动矩形，如图 10-7 所示。

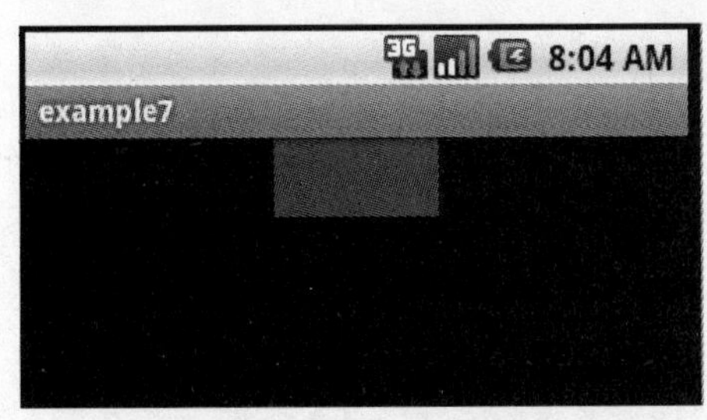

图 10-6　闪烁效果

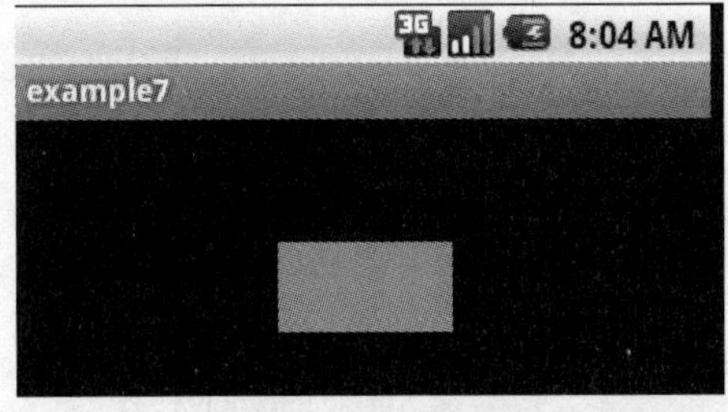

图 10-7　上下移动

10.2.2　SurfaceView 类

SurfaceView 类在游戏开发中有着举足轻重的地位，它对于画面的控制有着更大的自由度。SurfaceView 类是一个双缓冲机制，在开发游戏时会经常用到。

1. SurfaceView 类基础

一般来说，在 Android 中开发，复杂一点的游戏，是必须用 SurfaceView 的。SurfaceView 提供了直接访问一个可画图的界面，可以控制界面顶部的子视图层。

在 Android 中开发游戏，一般来说，或想写一个复杂一点的游戏，是必须用到 SurfaceView 来开发的。SurfaceView 提供直接访问一个可画图的界面，可以控制在界面顶部的子视图层。SurfaceView 是视图（View）的继承类，这个视图里内嵌了一个专门用于绘制的 Surface。你可以控制这个 Surface 的格式和尺寸。Surfaceview 控制这个 Surface 的绘制位置。Android 图形系统中一个重要的概念和线索是 surface。View 及其子类（如 TextView, Button）。

每个 surface 创建一个 Canvas 对象（属性时常改变），用来管理 view 在 surface 上的绘图操作，如画点、画线。还要注意的是，使用它的时候，一般都是出现在最顶层的。

使用的 SurfaceView 的时候，一般情况下还要对其进行创建、销毁及改变时的情况进行监视，格式如下。

```
//在 surface 的大小发生改变时激发
public void surfaceChanged(SurfaceHolder holder,int format,int width,int height){}
//在创建时激发，一般在这里调用画图的线程
public void surfaceCreated(SurfaceHolder holder){}
//销毁时激发，一般在这里将画图的线程停止、释放
public void surfaceDestroyed(SurfaceHolder holder) {}
  surfaceCreated 会先被调用，然后是 surfaceChanged，当程序结束时会调用 surfaceDestroyed
```

由于 SurfaceHolder 是一个共享资源，因此在对其操作时都应该实行“互斥操作”，即需要使用 synchronized 进行“封锁”机制。

渲染文字的工作实际上是主线程（也就是 LunarView 类）的父类 View 的工作，而并不属于工作线程 LunarThread，因此在工作线程中是无法控制的。所以我们改为向主线程发送一个消息来代替，让主线程通过 Handler 对接收到的消息进行处理，从而更新界面文字信息。

下面将通过一个具体实例的实现，介绍在 Android 中使用 SurfaceView 类实现屏幕更新显示的流程。本实例的源代码保存在“光盘:\daima\10\example8”，其主文件是 example8.java，具体代码如下。

```
public class example8 extends SurfaceView
                                  implements SurfaceHolder.Callback,Runnable
{
    //控制循环
    boolean mbLoop = false;

    //定义 SurfaceHolder 对象
    SurfaceHolder    mSurfaceHolder  = null;
    Int miCount = 0;
    Int y = 50;
```

```
public example8(Context context)
{
	super(context);

	// 实例化 SurfaceHolder
	mSurfaceHolder = this.getHolder();

	// 添加回调
	mSurfaceHolder.addCallback(this);
	this.setFocusable(true);

	mbLoop = true;
}


// 在 surface 的大小发生改变时激发
public void surfaceChanged(SurfaceHolder holder, int format, int width, int height)
{

}

// 在 surface 创建时激发
public void surfaceCreated(SurfaceHolder holder)
{
	//开启绘图线程
	new Thread(this).start();
}

// 在 surface 销毁时激发
public void surfaceDestroyed(SurfaceHolder holder)
{
	// 停止循环
	mbLoop = false;
}

// 绘图循环
public void run()
{
	while (mbLoop)
	{
		try
		{
			Thread.sleep(200);
		}
		catch (Exception e)
		{
```

```
            }

            synchronized( mSurfaceHolder )
            {
                Draw();
            }

        }
    }

    // 绘图方法
    public void Draw()
    {
        //锁定画布，得到 canvas
        Canvas canvas= mSurfaceHolder.lockCanvas();

        if (mSurfaceHolder==null || canvas == null )
        {
            return;
        }

        if (miCount < 100)
        {
            miCount++;
        }
        else
        {
            miCount = 0;
        }
        // 绘图
        Paint mPaint = new Paint();
        mPaint.setAntiAlias(true);
        mPaint.setColor(Color.BLACK);
        //绘制矩形--清屏作用
        canvas.drawRect(0, 0, 320, 480, mPaint);
        switch (miCount % 4)
        {
        case 0:
            mPaint.setColor(Color.BLUE);
            break;
        case 1:
            mPaint.setColor(Color.GREEN);
            break;
        case 2:
            mPaint.setColor(Color.RED);
            break;
```

```
            case 3:
                mPaint.setColor(Color.YELLOW);

                break;
            default:
                mPaint.setColor(Color.WHITE);
                break;
            }
            // 绘制矩形--后面将详细讲解
            canvas.drawCircle((320 - 25) / 2, y, 50, mPaint);
            // 绘制后解锁，绘制后必须解锁才能显示
            mSurfaceHolder.unlockCanvasAndPost(canvas);
        }
    }
```

执行后将在屏幕内绘制一个圆形，并随着线程的变化圆形的填充颜色也随之变化，从而实现闪烁效果，如图 10-8 所示；可以通过键盘上的上、下方向键来移动圆形，如图 10-9 所示。

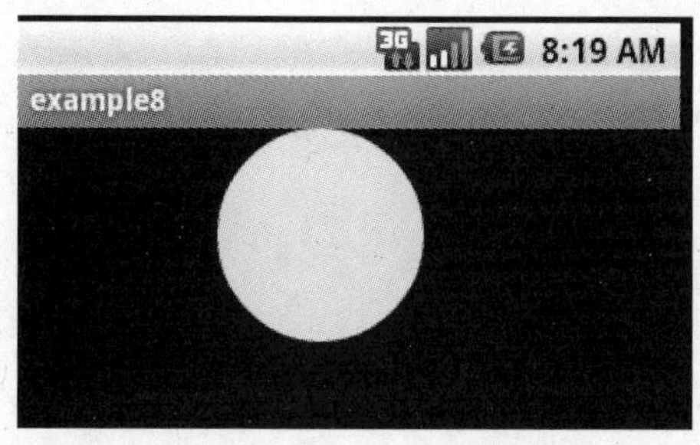

图 10-8　闪烁效果

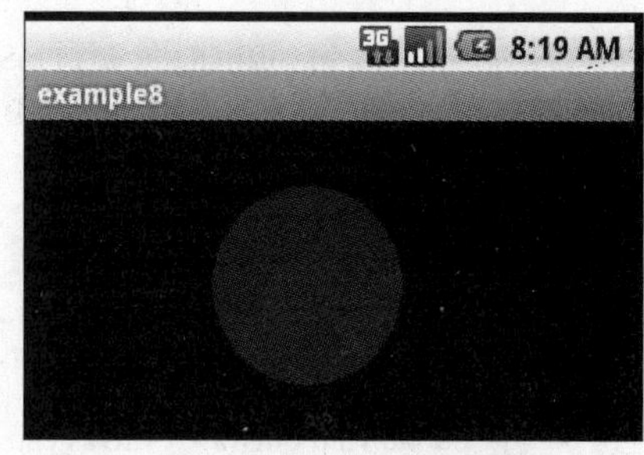

图 10-9　上下移动

2．双缓冲

双缓冲的核心是先通过 setBitmap 方法将要绘制的所有图形绘制到一个 Bitmap 上，然后调用 drawBitmap 方法绘制出这个 Bitmap，并在屏幕上显示出来。

下面将通过一个具体实例的实现，介绍在 Android 中使用 SurfaceView 类实现双缓冲的流程。本实例的源代码保存在“光盘:\daima\10\example6”，其主文件是 example6.java，具体代码如下。

```
    /* 声明 Bitmap 对象 */
    Bitmap    mBitQQ    = null;

    Paint    mPaint = null;

    /* 创建一个缓冲区 */
    Bitmap    mSCBitmap = null;

    /* 创建 Canvas 对象 */
    Canvas mCanvas = null;
```

```
public example6(Context context)
{
    super(context);

    /* 加载资源 */
    mBitQQ = ((BitmapDrawable) getResources().getDrawable(R.drawable.qq)).getBitmap();

    /* 创建屏幕大小的缓冲区 */
    mSCBitmap=Bitmap.createBitmap(320, 480, Config.ARGB_8888);

    /* 创建 Canvas */
    mCanvas = new Canvas();

    /* 设置将内容绘制在 mSCBitmap 上 */
    mCanvas.setBitmap(mSCBitmap);

    mPaint = new Paint();

    /* 将 mBitQQ 绘制到 mSCBitmap 上 */
    mCanvas.drawBitmap(mBitQQ, 0, 0, mPaint);

    /* 开启线程 */
    new Thread(this).start();
}

public void onDraw(Canvas canvas)
{
    super.onDraw(canvas);

    /* 将 mSCBitmap 显示到屏幕上 */
    canvas.drawBitmap(mSCBitmap, 0, 0, mPaint);
}

/* 触屏事件*/
public boolean onTouchEvent(MotionEvent event)
{
    return true;
}


/* 按键按下事件*/
public boolean onKeyDown(int keyCode, KeyEvent event)
{
    return true;
}
```

```
        /* 按键弹起事件*/
        public boolean onKeyUp(int keyCode, KeyEvent event)
        {
            return false;
        }


        public boolean onKeyMultiple(int keyCode, int repeatCount, KeyEvent event)
        {
            return true;
        }


        /**
         * 线程处理
         */
        public void run()
        {
            while (!Thread.currentThread().isInterrupted())
            {
                try
                {
                    Thread.sleep(100);
                }
                catch (InterruptedException e)
                {
                    Thread.currentThread().interrupt();
                }
                //使用 postInvalidate 可以直接在线程中更新界面
                postInvalidate();
            }
        }
    }
```

执行后的效果如图 10-10 所示。

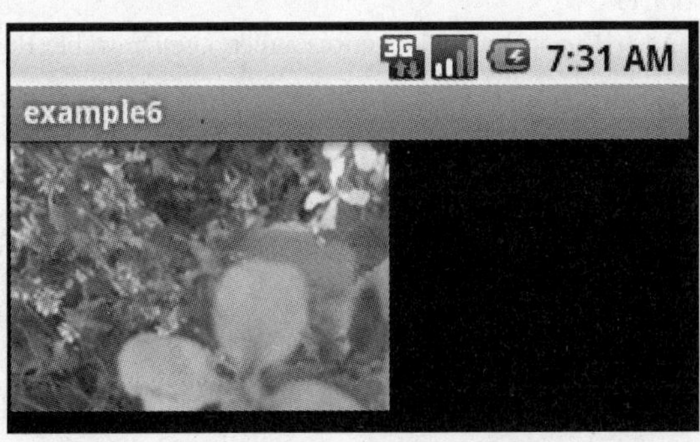

图 10-10　执行效果

10.3 动画处理

在 Android 平台中提供了 2 类动画，分别是 Tween 动画和 Frame 动画。Tween 动画用于对场景里的图像进行变换来产生动画效果；Frame 动画用于顺序播放事先做好的图像。在本节的内容中，将通过具体实例的实现，来详细讲解上述 2 种动画的实现流程。

10.3.1 Tween 动画

下面将通过一个具体实例的实现过程，介绍在 Android 中使用 Tween 动画的流程。本实例的源代码保存在“光盘:\daima\10\example9”，其主文件是 example9.java，具体代码如下。

```
public class example9 extends View
{
    /* 定义 Alpha 动画 */
    private Animation       mAnimationAlpha         = null;

    /* 定义 Scale 动画 */
    private Animation       mAnimationScale         = null;

    /* 定义 Translate 动画 */
    private Animation       mAnimationTranslate  = null;

    /* 定义 Rotate 动画 */
    private Animation       mAnimationRotate     = null;

    /* 定义 Bitmap 对象 */
    Bitmap                  mBitQQ                  = null;

    public example9(Context context)
    {
        super(context);

        /* 加载资源 */
        mBitQQ = ((BitmapDrawable) getResources().getDrawable(R.drawable.qq)).getBitmap();
    }

    public void onDraw(Canvas canvas)
    {
        super.onDraw(canvas);

        /* 绘制图片 */
        canvas.drawBitmap(mBitQQ, 0, 0, null);
    }
```

```
    public boolean onKeyUp(int keyCode, KeyEvent event)
    {
        switch ( keyCode )
        {
        case KeyEvent.KEYCODE_DPAD_UP:
            /* 创建 Alpha 动画 */
            mAnimationAlpha = new AlphaAnimation(0.1f, 1.0f);
            /* 设置动画的时间 */
            mAnimationAlpha.setDuration(3000);
            /* 开始播放动画 */
            this.startAnimation(mAnimationAlpha);
            break;
        case KeyEvent.KEYCODE_DPAD_DOWN:
            /* 创建 Scale 动画 */
            mAnimationScale =new ScaleAnimation(0.0f, 1.0f, 0.0f, 1.0f,
                                    Animation.RELATIVE_TO_SELF, 0.5f,
                                    Animation.RELATIVE_TO_SELF, 0.5f);

            /* 设置动画的时间 */
            mAnimationScale.setDuration(500);
            /* 开始播放动画 */
            this.startAnimation(mAnimationScale);
            break;
        case KeyEvent.KEYCODE_DPAD_LEFT:
            /* 创建 Translate 动画 */
            mAnimationTranslate = new TranslateAnimation(10, 100,10, 100);
            /* 设置动画的时间 */
            mAnimationTranslate.setDuration(1000);
            /* 开始播放动画 */
            this.startAnimation(mAnimationTranslate);
            break;
        case KeyEvent.KEYCODE_DPAD_RIGHT:
            /* 创建 Rotate 动画 */
            mAnimationRotate=new RotateAnimation(0.0f, +360.0f,
                                    Animation.RELATIVE_TO_SELF,0.5f,
                                    Animation.RELATIVE_TO_SELF, 0.5f);
            /* 设置动画的时间 */
            mAnimationRotate.setDuration(1000);
            /* 开始播放动画 */
            this.startAnimation(mAnimationRotate);
            break;
        }
        return true;
    }
}
```

程序执行后，将会把指定的目标图片模拟为动画显示，如图 10-11 所示。

图 10-11　执行效果

10.3.2　Frame 动画

下面将通过一个具体实例的实现，介绍在 Android 中使用 Frame 动画的流程。本实例的源代码保存在“光盘:\daima\10\example10”，其主文件是 example10.java，具体代码如下。

```
/* 定义 AnimationDrawable 动画 */
private AnimationDrawable   frameAnimation   = null;
Context mContext = null;

/* 定义一个 Drawable 对象 */
Drawable mBitAnimation = null;
public example10(Context context)
{
        super(context);

        mContext = context;

        /* 实例化 AnimationDrawable 对象 */
        frameAnimation = new AnimationDrawable();

        /* 装载资源 */
        /* 这里用一个循环了装载所有名字类似的资源*/
        /* 如“a1.......15.png”的图片*/
        /* 这个方法用处非常大*/
        for (int i = 1; i <= 15; i++)
        {
                int id = getResources().getIdentifier("a" + i, "drawable", mContext.getPackageName());
                mBitAnimation = getResources().getDrawable(id);
                /* 为动画添加一帧 */
                /* 参数 mBitAnimation 是该帧的图片*/
                /* 参数 500 是该帧显示的时间，按毫秒计算*/
                frameAnimation.addFrame(mBitAnimation, 500);
        }

        /* 设置播放模式是否循环，false 表示循环，true 表示不循环 */
```

```
            frameAnimation.setOneShot( false );

            /* 设置本类将要显示这个动画 */
            this.setBackgroundDrawable(frameAnimation);
        }

        public void onDraw(Canvas canvas)
        {
            super.onDraw(canvas);

        }

        public boolean onKeyUp(int keyCode, KeyEvent event)
        {
            switch ( keyCode )
            {
            case KeyEvent.KEYCODE_DPAD_UP:
                /* 开始播放动画 */
                frameAnimation.start();
                break;
            }
            return true;
        }
    }
```

实例执行后，通过按下键盘的上、下方向键，能够实现动画效果。如图 10-12 所示。

图 10-12　执行效果

10.4 手机游戏——魔塔游戏

手机游戏是指运行于手机上的游戏软件。目前用来编写手机游戏，使用最多的语言是 Java 语言。其次是 C 语言。随着科技的发展，现在手机的功能也越来越多，越来越强大。手机游戏也远远不是我们印象中的“俄罗斯方块”、“贪吃蛇”这类画面简陋，规则简单的游戏，而是发展到了可以和掌上游戏机媲美，具有很强的娱乐性和交互性的复杂形态了。于是，抛弃你的随身听和 Gameboy，买一个好手机吧，你会发现，一个手机已经可以满足你大部分娱乐需要。

在本节的内容中，将通过一个具体实例的实现，介绍在 Android 平台中开发“魔塔”游戏的实现流程。本实例的源代码保存在“光盘:\daima\10\example11”。

10.4.1 Java 游戏开发流程

一款 J2ME 典型游戏的开发流程如图 10-13 所示。

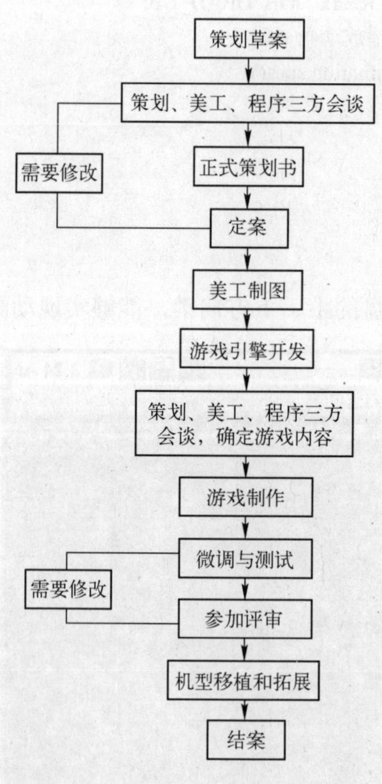

图 10-13 典型 Java 游戏开发流程

1. 立项

在制作游戏之前，策划首先要确定一点：到底想要制作一个什么样的游戏？而制作一个游戏并不是闭门造车，一个策划说了就算数的简单事情。制作一款游戏受到多方面的限制。

1）市场：即将开发的游戏是不是具备市场潜力？在市场上推出以后会不会被大家所接

受？是否能够取得良好的市场回报？

2）技术：即将开发的游戏从程序和美术上是不是完全能够实现？如果不能实现，是不是能够有折中的办法？

3）规模：以现有的资源是否能很好的协调并完成即将要开发的游戏？是否需要另外增加人员或设备？

4）周期：游戏的开发周期长短是否合适？能否在开发结束时正好赶上游戏的销售旺季？

5）产品：即将做的游戏在其同类产品中是否有新颖的设计？是否能有吸引玩家的地方？如果在游戏设计上达不到创新，是否能够在美术及程序方面加以补足？如果同类型的游戏市场上已经有了很多，那么即将做的游戏的卖点在哪里？

以上各个问题都是要经过开发组全体成员反复进行讨论才能够确定下来的，集思广益，共同探讨一个可行的方案。如果对上述全部问题都能够有肯定的答案的话，那么这个项目基本是可行的。但是即便项目获得了通过，在进行过程中也可能会有种种不可预知的因素导致意外情况的发生，所以项目能够成立，只是游戏制作的开始。

在项目确立了以后，下一步要进行的就是进行游戏的大纲策划工作。

2．大纲策划的进行

游戏大纲关系到游戏的整体面貌，当大纲策划案定稿以后，没有特殊的情况，是不允许进行更改的。程序和美术工作人员将按照策划所构思的游戏形式来架构整个游戏，因此，在制定策划案时一定要做到慎重，尽量考虑成熟。

3．游戏的正式制作

当游戏大纲策划案完成并讨论通过后，游戏就由三方面同时开始进行制作了。在这一阶段，策划的主要任务是在大纲的基础上对游戏的所有细节进行完善，将游戏大纲逐步填充为完整的游戏策划案。根据不同的游戏种类，所要进行细化的部分也不尽相同。

在正式制作的过程中，策划、程序、美工人员进行及时和经常性的交流，了解工作进展以及是否有难以克服的困难，并且根据现实情况有目的的变更工作计划或设计思想。三方面的配合在游戏正式制作过程中是最重要的。

4．配音、配乐

在程序和美工进行得差不多要结束的时候，就要进行配音和配乐的工作了。虽然音乐和音效是游戏的重要组成部分，能够起到很好的烘托游戏气氛的作用，但是限于 J2ME 游戏的开发成本和设置的处理能力，这部份已经被弱化到可有可无的地步了。应选择跟游戏风格能很好配合的音乐当作游戏背景音乐，这个工作交给策划比较合适。

5．检测、调试

游戏刚制作完成，在程序上肯定会有很多的错误，错误严重时会导致游戏完全没有办法进行下去。同样，策划的设计也会有不完善的地方，主要在游戏的参数部分。参数部分的不合理，会影响游戏的可玩性。此时测试人员需检测程序上的漏洞和通过试玩，调整游戏的各个部分参数使之基本平衡。

10.4.2　设计游戏框架

转回正文，开始讲解“魔塔”游戏的具体设计。因为所有游戏是基于框架的，所以设计一个合适的框架尤为重要。为了正确设计框架，我们先看市面上魔塔游戏的界面，如图 10-14 所示。

由游戏界面可知，游戏中包含了地图、角色、屏幕界面和道具等元素，上述元素构成了一个视图，例如屏幕视图、道具视图和角色视图等。

图 10-14　游戏界面

1．界面视图

在 Android 中，视图是通过继承 View 类实现的，在 View 类中包含了各种绘制图形的方法和事件，这些知识在本章的 10.1 和 10.2 中已经进行了讲解。这样构建一个游戏界面类将变得轻而易举。界面类 GameView 类的具体代码如下。

```
package com.example11;
import android.content.Context;
import android.graphics.Canvas;
import android.view.View;
public abstract class GameView extends View
{
	public GameView(Context context)
	{
		super(context);
	}
	/**
	 * 绘图
	 *
	 * @param	N/A
	 *
	 * @return	null
	 */
```

```
        protected abstract void onDraw(Canvas canvas);
        /**
         * 按键按下
         *
         * @param      N/A
         *
         * @return     null
         */
        public abstract boolean onKeyDown(int keyCode);
        /**
         * 按键弹起
         *
         * @param      N/A
         *
         * @return     null
         */
        public abstract boolean onKeyUp(int keyCode);
        /**
         * 回收资源
         *
         */
        protected abstract void reCycle();

        /**
         * 刷新
         *
         */
        protected abstract void refurbish();
}
```

2. 屏幕处理

屏幕会伴随玩家的操作而变化，所以游戏整体屏幕框架设计完毕后，还需要设计屏幕的控制类。在此类中，可以根据不同的游戏状态来设置屏幕的具体显示内容。在此编写 MainGame 类，具体代码如下。

```
package com.example11;

import android.app.Activity;
import android.content.Context;

public class MainGame
{
    private static GameView     m_GameView        = null;          // 当前需要显示的对象
    private Context                  m_Context         = null;
    private MagicTower               m_MagicTower  = null;
```

```
private int                    m_status        = -1;        //游戏状态
public   CMIDIPlayer mCMIDIPlayer;
public byte mbMusic = 0;
public MainGame(Context context)
{

        m_Context = context;
        m_MagicTower = (MagicTower)context;
        m_status = -1;

        initGame();
}

/* 初始化游戏*/
public void initGame()
{
        controlView(yarin.GAME_SPLASH);
        mCMIDIPlayer = new CMIDIPlayer(m_MagicTower);
}
/* 得到游戏状态*/
public int getStatus()
{
        return m_status;
}
/* 设置游戏状态*/
public void setStatus(int status)
{
        m_status = status;
}
/* 得到主类对象*/
public Activity getMagicTower()
{
        return m_MagicTower;
}

/* 得到当前需要显示的对象*/
public static GameView getMainView()
{
        return m_GameView;
}

/* 控制显示什么界面*/
public void controlView(int status)
{
        if(m_status != status)
        {
                if(m_GameView != null)
```

```
                {
                    m_GameView.reCycle();
                    System.gc();
                }
            }

            freeGameView(m_GameView);
            switch (status)
            {
            case yarin.GAME_SPLASH:
                m_GameView = new SplashScreen(m_Context,this);
                break;
            case yarin.GAME_MENU:
                m_GameView = new MainMenu(m_Context,this);
                break;
            case yarin.GAME_HELP:
                m_GameView = new HelpScreen(m_Context,this);
                break;
            case yarin.GAME_ABOUT:
                m_GameView = new AboutScreen(m_Context,this);
                break;
            case yarin.GAME_RUN:
                m_GameView = new GameScreen(m_Context,m_MagicTower,this,true);
                break;
            case yarin.GAME_CONTINUE:
                m_GameView = new GameScreen(m_Context,m_MagicTower,this,false);
                break;
            }
            setStatus(status);
        }

        /* 释放界面对象*/
        public void freeGameView(GameView gameView)
        {
            if(gameView != null)
            {
                gameView = null;
                System.gc();
            }
        }
    }
```

3. 线程更新

屏幕界面设计告一段落，但是要实现界面的真正更新，则需要线程来实现。在此可以为游戏开启一个主线程，并通过 getMainView()方法来获取当前显示的界面，并根据不同的界面来进行游戏更新。此更新能在文件 ThreadCanvas.java 中实现，具体代码如下。

```
package com.example11;

import android.content.Context;
import android.graphics.Canvas;

import android.util.Log;
import android.view.View;

public class ThreadCanvas extends View implements Runnable
{
	private String	m_Tag	= "ThreadCanvas_Tag";


	public ThreadCanvas(Context context)
	{
		super(context);
	}


	/**
	 * 绘图
	 *
	 * @param N
	 *			/A
	 *
	 * @return null
	 */
	protected void onDraw(Canvas canvas)
	{
		if (MainGame.getMainView() != null)
		{
			MainGame.getMainView().onDraw(canvas);
		}
		else
		{
			Log.i(m_Tag, "null");
		}
	}


	/**
	 * 绘图显示
	 *
	 */
	public void start()
	{
		Thread t = new Thread(this);
```

```
        t.start();
    }


    // 刷新界面
    public void refurbish()
    {
        if (MainGame.getMainView() != null)
        {
            MainGame.getMainView().refurbish();
        }
    }


    /**
     * 游戏循环
     *
     * @param N
     *              /A
     *
     * @return null
     */
    public void run()
    {
        while (true)
        {
            try
            {
                Thread.sleep(yarin.GAME_LOOP);
            }
            catch (Exception e)
            {
                e.printStackTrace();
            }

            refurbish(); // 更新显示
            postInvalidate(); // 刷新屏幕
        }
    }


    /* 按键处理(按键按下)*/
    boolean onKeyDown(int keyCode)
    {
        if (MainGame.getMainView() != null)
        {
```

```
                MainGame.getMainView().onKeyDown(keyCode);
            }
            else
            {
                Log.i(m_Tag, "null");
            }
            return true;
        }


        /* 按键弹起*/
        boolean onKeyUp(int keyCode)
        {
            if (MainGame.getMainView() != null)
            {
                MainGame.getMainView().onKeyUp(keyCode);
            }
            else
            {
                Log.i(m_Tag, "null");
            }
            return true;
        }
    }
```

4. 具体显示

经过上述类的设计后，还需要进行最后一步，即用一个 Activity 来显示具体的界面。因为是在 ThreadCanvas 中控制界面的，所以只需用 setContentView 来显示一个 ThreadCanvas 对象即可。上述功能通过文件 MagicTower.java 实现，具体代码如下。

```
package com.example11;

import android.app.Activity;
import android.os.Bundle;
import android.view.KeyEvent;
import android.view.Window;
import android.view.WindowManager;

public class MagicTower extends Activity
{
    private ThreadCanvas   mThreadCanvas   = null;


    public void onCreate(Bundle savedInstanceState)
    {
```

```
            super.onCreate(savedInstanceState);
            setTheme(android.R.style.Theme_Black_NoTitleBar_Fullscreen);
            requestWindowFeature(Window.FEATURE_NO_TITLE);
            getWindow().setFlags(WindowManager.LayoutParams.FLAG_FULLSCREEN, WindowManager.
LayoutParams.FLAG_FULLSCREEN);

            new MainGame(this);
            mThreadCanvas = new ThreadCanvas(this);

            setContentView(mThreadCanvas);
        }


        /**
         * 暂停
         *
         * @param N
         *              /A
         *
         * @return null
         */
        protected void onPause()
        {
            super.onPause();
        }


        /**
         * 重绘
         *
         * @param N
         *              /A
         *
         * @return null
         */
        protected void onResume()
        {
            super.onResume();
            mThreadCanvas.requestFocus();
            mThreadCanvas.start();
        }


        /**
         * 按键按下
         *
         * @param N
```

```
         *                    /A
         *
         * @return null
         */
        public boolean onKeyDown(int keyCode, KeyEvent event)

        {
                mThreadCanvas.onKeyDown(keyCode);
                return false;
        }


        /**
         *  按键弹起
         *
         * @param N
         *                    /A
         *
         * @return null
         */
        public boolean onKeyUp(int keyCode, KeyEvent event)
        {
                mThreadCanvas.onKeyUp(keyCode);
                return false;
        }
}
```

通过上述处理，一个基本的游戏框架建设完毕。在后续的开发中，只需直接继承上面的类界面，即继承 GameView，然后在 MainView 中更改游戏状态即可。因为本书篇幅有限，后续代码读者可以参考本书配套的源代码。

参 考 文 献

[1] 朱桂英. Android 开发应用从入门到精通[M]. 北京：中国铁道出版社，2011.

[2] 张元亮. Android 开发应用实战详解[M]. 北京：中国铁道出版社，2011.

[3] 余志龙. Android SDK 开发范例大全[M]. 北京：人民邮电出版社，2011.

[4] 徐娜子. Android 江湖[M]. 北京：电子工业出版社，2011.

[5] 李佐彬. Android 开发入门与实战体验[M]. 北京：机械工业出版社，2011.